Geotechnologies and the Environment

Volume 13

The "Geotechnologies and the Environment" series is intended to provide specialists in the geotechnologies and academics who utilize these technologies, with an opportunity to share novel approaches, present interesting (sometimes counterintuitive) case studies, and most importantly to situate GIS, remote sensing, GPS, the internet, new technologies, and methodological advances in a real world context. In doing so, the books in the series will be inherently applied and reflect the rich variety of research performed by geographers and allied professionals.

Beyond the applied nature of many of the papers and individual contributions, the series interrogates the dynamic relationship between nature and society. For this reason, many contributors focus on human-environment interactions. The series are not limited to an interpretation of the environment as nature per se. Rather, the series "places" people and social forces in context and thus explore the many sociospatial environments humans construct for themselves as they settle the landscape. Consequently, contributions will use geotechnologies to examine both urban and rural landscapes.

More information about this series at http://www.springer.com/series/8088

Marco Helbich • Jamal Jokar Arsanjani
Michael Leitner
Editors

Computational Approaches for Urban Environments

Springer

Editors
Marco Helbich
Department of Human Geography
and Spatial Planning
Utrecht University
Utrecht, The Netherlands

Jamal Jokar Arsanjani
Institute of Geography
Heidelberg University
Heidelberg, Germany

Michael Leitner
Department of Geography and Anthropology
Louisiana State University
Baton Rouge, LA, USA

Geotechnologies and the Environment
ISBN 978-3-319-11468-2 ISBN 978-3-319-11469-9 (eBook)
DOI 10.1007/978-3-319-11469-9

Library of Congress Control Number: 2014958019

Springer Cham Heidelberg New York Dordrecht London

Printed on acid-free paper

Springer International Publishing AG Switzerland is part of Springer Science+Business Media (www.springer.com)

Preface

Although the last decade has witnessed a significant increase in the volume and complexity of geospatial data for exploring urban environments, the analysis of cities is becoming more challenging than ever. In this context, rapid developments in the fields of spatial statistics, spatial data mining, and geosimulation, among others, provide highly valuable and promising tools that enhance our understanding of how cities function and evolve in space-time. In light of these recent trends, the editors of this excellent volume bring a body of knowledge together into a volume that is an important resource for those who are interested in exploring cities through interdisciplinary perspectives. The editors have put together an outstanding collection of cutting-edge methods for urban modeling.

This book covers a wide range of topics, including methodological advances as well as specific urban-related themes including housing and real estate markets, urbanization, and transportation systems, among others. To deal with these topics properly, extensive expertise is required. The editors have commissioned leading authors from a wide variety of institutions for each particular topic across different geographical regions. The volume represents an excellent collection of the state-of-the-art research methods and knowledge, and it offers the readers a comprehensive overview on how contemporary cities can be investigated. Besides, the book shows a variety of complementing aspects about how we can utilize and integrate novel datasets, advanced geospatial information technologies, and computational methods to analyze essential aspects of urban environments. This volume strives to address challenging methodological and dataset issues in order to identify promising future research directions and challenges for more effective city planning and management.

The volume is no doubt an impressive collection of chapters and a significant contribution to an emerging and dynamic interdisciplinary field that lies at the intersection of urban research, geographic information science, and computational science. It is not only a well-balanced interdisciplinary collection of current research, it also introduces visionary concepts and outlines promising avenues for future research. It will be an inspiring and highly useful volume for both researchers and students in the field.

Champaign, IL, USA Mei-Po Kwan

Contents

About the Editors

Marco Helbich received his master and doctoral degrees at the Department of Geography and Regional Research, University of Vienna, Austria. He is currently an Assistant Professor at the Department of Human Geography and Spatial Planning, Utrecht University, The Netherlands, and previously a recipient of an Alexander von Humboldt Fellowship. His research interests are in cities, computational modeling, spatial analysis, real estate, crime mapping, and health geography. He has published 30+ refereed articles in top international journals in his discipline, such as *Urban Geography*, *International Journal of Geographical Information Science*, *Annals of the Association of American Geographers*, *International Journal of Health Geographics*, *Journal of Transport Geography* and others.

Jamal Jokar Arsanjani received his doctoral degree in Geographic Information Science (GISc) from the Department of Geography and Regional Research, University of Vienna, Austria. He is currently an Alexander von Humboldt Fellow at the Institute of Geography, Heidelberg University, Germany. His interdisciplinary research interests are in volunteered geographic information and crowdsourcing, geocomputation and spatial planning, remote sensing of the environment, and disaster management. He has published articles in leading international journals of his discipline, including *International Journal of Applied Earth Observation and Geoinformation*, *International Journal of Digital Earth*, *Transactions in GIS*, *Cities* and single-authored a book on *Dynamic Land Use/Cover Change Simulation: Geosimulation and Agent-based Modelling* with Springer publication.

Michael Leitner received a master degree at the Department of Geography and Regional Research, University of Vienna, Austria, and a second master and a doctoral degree in GISc at the Department of Geography, State University of New York at Buffalo, USA. He is currently a Professor of Geography in the Department of Geography and Anthropology, Louisiana State University, USA, and a faculty member in the Doctoral College "GIScience" at the University of Salzburg, Austria. He has previously received a Fulbright Scholarship and is the recipient of the 2007

Meredith F. Burrill Award from the Association of American Geographers. His research interests are in GISc and its application to spatial crime analysis, medical geography, and geospatial privacy. He has published two co-authored books, one single-edited book, and 40+ refereed articles and book chapters. He is the current editor of *Cartography and Geographic Information Science* (*CaGIS*).

Chapter 1
Computational Approaches for Urban Environments: An Editorial

Marco Helbich, Jamal Jokar Arsanjani, and Michael Leitner

Abstract Cities are under continuous pressure due to an increasing urbanization which will have far-reaching consequences for housing, transportation, retail, etc. To cope with these challenges, methodological advances in quantitative modeling coupled with growing amounts of spatial and spatiotemporal data can add significantly to our understanding of how cities function. Because the added value of data-driven approaches to analyze urban environments is promising but still in its infancy, the present volume aims to promote the application of advanced computational methodologies to achieve a better understanding of our cities and the underlying mechanisms.

Keywords Urban environments • Geographic information science • Spatial statistics • New science of cities

1.1 Prologue

Today's cities are highly complex, dynamic, and vibrant environments subject to frequent changes. The impacts of globalization, information economies, as well as social and demographic changes, among others, have given cities a distinct appearance (e.g., Hall 1993; Batty and Longley 1994; Anas et al. 1998; Batty 2013; Fujita and Thisse 2013). Urban environments are places where most of the people in our contemporary society live. In addition, the United Nations (2014)

M. Helbich (✉)
Department of Human Geography and Spatial Planning, Utrecht University, Heidelberglaan 2, 3584 CS Utrecht, The Netherlands
e-mail: m.helbich@uu.nl

J. Jokar Arsanjani
Institute of Geography, Heidelberg University, Berliner Straße 48, D-69120 Heidelberg, Germany
e-mail: jamaljokar@gmail.com

M. Leitner
Department of Geography and Anthropology, Louisiana State University, 227 Howe-Russell-Kniffen Geoscience Complex, Baton Rouge, LA 70803, USA
e-mail: mleitne@lsu.edu

M. Helbich et al. (eds.), *Computational Approaches for Urban Environments*, Geotechnologies and the Environment 13, DOI 10.1007/978-3-319-11469-9_1

predict that the worldwide proportion of people living in urban environments will double by 2030 compared to the 1950s. According to this trend, in Europe, for instance, the percentage of the population living in urban areas increased from 1950 to 2007 from 42 to 72 %. More important is that this trend of urban growth will continue within the next one or two decades. The United Nations forecast that the European urban population will increase to more than 78 % until 2030. Without doubt, these urbanization processes, due to their far-reaching impacts on the economy, climate change, and health, among others, will intensify the pressure on cities and urban societies even more. In order to be prepared for and to manage these future challenges, well-thought-out planning and policy strategies are required (Shafizadeh-Moghadam and Helbich 2015; Clarke 2014). In this regard, quantitative modeling can add significantly to our understanding of how cities function. Equally important is to undertake research linking cities with computational approaches and put such research on policy agendas throughout the world.

As discussed by Pacione (2009), urban environments are highly multifaceted areas consisting of several subcomponents, including housing markets (e.g., Helbich et al. 2013a, c), transportation systems (e.g., De Vos and Witlox 2013) etc., shaping our cities across different scales. Each of these components comprises of its own spatial and spatiotemporal patterns and processes. As indicated in Fig. 1.1, the complexity is further amplified through interrelationships between these individual components. For illustration purposes, let us examine the following example: We assume that the accessibility of an urban fringe is improved by means of a highway construction. Due to this new transportation infrastructure provision, the improved linkage between the suburbs and the core city reduces commuting time and thus makes these metropolitan outskirts more attractive for new residents to move in. This accompanied increase in the demand for new residential areas results in raising suburban land and housing prices, which makes residential land affordable for higher-income classes, only. This rather simple example illustrates that the components of urban environments are highly connected to each other and comprise of interrelated feedback loops.

Although cities have been widely studied during the last decades and meticulous research dealing with economical, infrastructural, environmental, political, social,

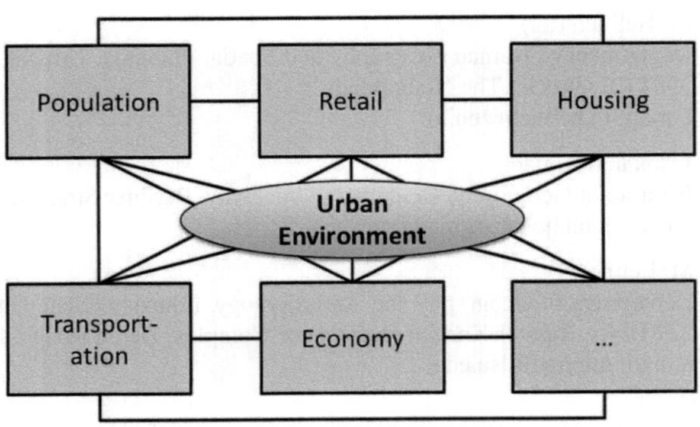

Fig. 1.1 Components of urban environments

and demographical aspects has been piled up promoting and supporting our understanding of the operation of cities (e.g., Batty 2008, 2013; Bettencourt 2013; Glaeser 2011; Clarke 2014), the underlying processes and the interactions between components are only partly understood thus far. Several academic disciplines are dedicated to urban research. Each of them has their own unique perspective on cities and utilizes their unique set of methods and tools. This results in fragmented knowledge and a lack of coherent insights (Solecki et al. 2013). While qualitative urban research has added significantly to the understanding of cities, past empirical quantitative research has benefited from advances in the field of geographic information systems (GIS; Goodchild 2010) technologies. At present, GIS-based analyses (e.g., Jokar Arsanjani et al. 2014) have reached some level of maturity and are an integral part of spatial sciences as well as of urban policy- and decision-making. Although the quantitative analysis of urban areas is not new and goes back to the quantitative revolution in geography (see Kwan and Schwanen 2009), the rapid methodological progress – including spatial statistics, remote sensing, data mining, and simulation-based modeling, among others – coupled with the recent accumulation of readily available spatial and spatiotemporal data on a detailed scale, i.e., volunteered geographic information (e.g., Jokar Arsanjani et al. 2013a), airborne laser scanning data (e.g., Xu et al. 2014), and cell phone data (e.g., Calabresea et al. 2013), among others, has stimulated and shifted the emphasis to a computationally oriented urban science (Batty 2013).

In the literature, this linkage between geography and computational science (Lazer et al. 2009) is referred to as geocomputation, coined by Openshaw and Abrahart (2000). While the prefix "geo" emphasizes that geocomputation deals with spatial theories, georeferenced data, and spatially explicit research problems, the latter term "computation" highlights how geographical science is conducted, namely, through a broad spectrum of computer-intensive methods, mathematical and spatial statistical models, simulations, etc. Thus, geocomputation aims to explore, extract, and generalize inherent urban patterns and processes, in data-driven fashion from spatial and spatiotemporal data to not only solve complex geographical urban problems but also to transform the implicit and hidden information in spatial databases into urban knowledge. As such, geocomputation is an umbrella term that includes, but is not at all limited to, agent-based modeling (e.g., Jokar Arsanjani et al. 2013b; Torrens 2012; Malleson et al. 2013), cellular automata (e.g., Vaz et al. 2012; Pijanowski et al. 2014), spatial (e.g., Helbich and Leitner 2012) and spatiotemporal cluster detection (e.g., Nakaya and Yano 2010; Hagenauer and Helbich 2013a, b), Bayesian models (e.g., Brunauer et al. 2013; Law and Quick 2013), fuzzy logic (e.g., Grekousis et al. 2013), local regression modelling (e.g., Leitner and Helbich 2011; Helbich et al. 2014), regionalization (e.g., Wang et al. 2012; Helbich et al. 2013c), and neurocomputing coupled with or without evolutionary algorithms (e.g., Arribas-Bel et al. 2011; Gu et al. 2011; Hagenauer et al. 2011; Helbich et al 2013b; Mimis et al. 2013). For a more detailed and thought-provoking theoretical discussion, the reader is referred to Couclelis (1998), Fischer (2006), as well as Birkin (2009).

However, nowadays the added value of the applications of cutting-edge computational approaches to analyze urban environments is promising but still in its infancy and far from being mainstream. Therefore, the editors of this volume believe that this book adds to the contemporary research agenda on cities from an interdisciplinary point of view and highlights the large potential of data-driven techniques to better understand how the individual urban subcomponents function and how cities as a whole operate. As such, it seems that the full potential of geocomputational approaches is still not entirely explored and more research is necessary. In this regard, the chapters in this volume make use of the ever-increasing and more precise geospatial urban data and linking them to up-to-date computational techniques (Miller and Goodchild 2014). This is consistent with the urgent call by Solecki et al. (2013) for a computationally integrated "urban science." As a consequence, this allows to formulate new and/or alternative hypotheses as well as to establish novel and universal "urban laws and theories." The authors hope that this edited volume can add to this recent trend that has been observed in the literature.

1.2 Objectives

The present volume entitled *Computational Approaches for Urban Environments* deals with the synergic usage of advanced computational methodologies in close relationship to geospatial information across cities of different scales. The main objective is to promote recent advances in the application of computational methods, beyond traditional urban analysis, to achieve a (hopefully) more appropriate understanding of the inherent complexity of our cities and underlying mechanisms. In doing so, the book seeks to offer a complementary perspective to the large body of literature dealing with the analysis of urban environments. To achieve more holistic insights into cities, their dynamics, shapes, morphologies, and residents, this collection of chapters subsumes research originating from disciplines such as geography, economics, computer science, statistics, geographic information science, remote sensing, and urban planning. It is anticipated that this book shades light on and contributes to contemporary problems cities are faced with and how it would be possible to tackle them to offer urban decision- and policy-making a sound and solid basis of understanding. The collection of chapters provides a selection of actual computational approaches useful for, but not limited to, audiences that include researchers, postgraduates, and professionals.

As indicated by the high response for the call for chapters, it appears that the intention of this book received wide approval. By the end of June 2013, a total of 32 chapter proposals were submitted. After an internal review by the editors, 30 authors of those originally 32 submitted proposals were invited to submit a chapter manuscript. After the final chapter submission deadline on November 1 2013, a total of 26 manuscripts were submitted. Subsequently, each of the 26 chapter manuscripts was evaluated through a *double-blind review process* by at least two international experts. For the review process, the standard Springer review guidelines were used. Besides the innovative aspect of the research, the scientific quality of the research weighted heavily on the decision whether or not a manuscript was accepted or rejected. In cases where major revisions were requested by the reviewers and to guarantee high scientific quality, a second round of review of the revised manuscript by one of the original reviewers or an alternative reviewer was conducted. If the reviews called for minor revisions, then a second round of reviews was not done. Instead, the editors made the decision whether or not the revised manuscript was fit for publication. In April 2014, 14 chapters were accepted and are now included in the present book.

1.3 Structure of the Book

The book integrates several areas of urban environments, each associated with a main theme of the book. The present volume has the following five sections: (1) spatial planning and decision-making, (2) housing and real estate, (3) urban transportation and mobility, (4) remote sensing, and (5) urban sensing, social networks, and social media. However, this structure should not be understood as fixed and definitive. Quite the contrary, the boundaries between these sections are partly fuzzy and overlap each other to some extent.

Part I on *Spatial Planning and Decision-Making* includes three chapters. Chapter 2 by Pierre Frankhauser reviews fractal geometry to explore the spatial organization of urban fabrics. He demonstrates how existing planning concepts can be enriched through fractal analysis. In Chap. 3, Martin Behnisch and Alfred Ultsch present an approach coupling machine learning and data mining techniques for discovering patterns in multidimensional building and land use data for urban districts in Germany. It is demonstrated how these techniques may serve as hypothesis generators for planning purposes. Closely related to this chapter is Chap. 4 by Julian Hagenauer. He proposes a method for clustering spatial data by integrating contextual neural gas and graph clustering. The efficiency of the method to derive meaningful and theoretically sound regions is demonstrated on synthetic data and a real-world case study dealing with demographic and socioeconomic data for Philadelphia (USA).

Part II deals with *Housing and Real Estate* and comprises of three chapters. In Chap. 5, Alexander Razen et al. review recent developments in structured additive regression models. Besides a multilevel structured additive regression for location scale and shape, a Bayesian version of the quantile regression is proposed. Investigating owner-occupied single family homes in Austrian urban areas, nonlinearities in the hedonic price function and spatial heterogeneity, among others, are observed. Next, in Chap. 6, Shipeng Sun and Steven Manson highlight the importance of the housing search in the context of intra-urban migration of domestic residents. In order to simulate the complex location decision-making, an agent-based model is formulated and validated against real-world housing vacancies for the Twin Cities of Minnesota (USA). The case study demonstrates how realistic intra-urban migration patterns emerge from rather simple behavioral rules of home searchers. Chapter 7 in Part II is written by Timothy Rosner and Kevin Curtin. The authors examine Jane

Jacobs' four generators of urban diversity – dwelling density, block length, mix of building age, and mix of uses – and develop a new composite urban livability index to measure social and economic characteristics of the built environment. The capability of the index is tested for the District of Columbia (USA).

Part III deals with another component of urban environments, namely, *Urban Transportation and Mobility*, and includes two chapters. In Chap. 8, Godwin Yeboah et al. propose a space-time analytical approach based on analyzing global positioning system data for cyclists. In addition, they advance policy strategies in Newcastle (UK) to improve cycling uptake as well as data processing methodologies through gaining a profound gender-based understanding of cycling behaviors. In contrast, Rashid Waraich et al. identify in Chap. 9 the challenges of utilizing agent-based traffic simulation frameworks. Focusing on the MATSim software environment, several methods to improve the simulation performance through a combination of reducing disk access, decoupling computational tasks, and making use of parallel computing are proposed. Additionally, an event-based model instead of a fixed time increment approach for the traffic simulation is propagated.

Remote Sensing advances and applications are the focus of Part IV, comprising of four chapters. In Chap. 10, Konstantinos Karantzalos gives a comprehensive review of the state of the art in the field of change detection to monitor the growth trajectories of urban areas. Essential change detection components, unsupervised and supervised classification methodologies, and object extraction, among others, are discussed in this chapter. On the contrary, Chap. 11 by Christian Berger et al. presents a data fusion technique in which airborne hyperspectral and light detection and ranging (LiDAR) data are combined in order to derive an urban surface material map required for a microclimate model. Two case studies underpin the potential of data fusion to derive key input parameters in this research domain. In Chap. 12, Junmei Tang monitors the spatiotemporal urban expansion of Houston's metropolitan area applying cellular automata models. It is concluded that the incorporation of socioeconomic data improves the predictive accuracy to simulate the growth of human-intervened landscapes.

Part V on *Urban Sensing, Social Network, and Social Media* contains three chapters. To begin with, in Chap. 13, Yaoli Wang et al. utilize mobile cell phone data from a Chinese city to investigate how social networks are embedded in the urban physical space. For instance, the authors find that higher degree users in the telephone contact network tend to congregate in the central business district and that the downtown area hosts many heterogeneous communities of social groups. In Chap. 14, Emily Schnebele et al. propose to integrate volunteered geographic information and social media data with authoritative sources to fill data gaps during environmental emergencies. Two applications are presented. While the first study applies an artificial neural network to transportation infrastructure flooding, the second study deals with the usage of mobile phone data during emergency evacuations. In Chap. 15, Sebastian Grauwin et al. look at the possibility of mapping space-time human activities in urban environments in the metropolises of New York, London, and Hong Kong based on the detection of mobile phone usage (i.e., number of calls, SMSs, and data transfers). The authors provide insights into both the universal structure of cities and the cultural, technological, and economical factors shaping human dynamics. Clustering identifies locations with similar patterns. Their findings confirm that while the economy becomes more global, common patterns emerge in business areas of different cities across the globe.

The final chapter by Paul Longley reflects about the potential of computational approaches discussed in this book, observes current development trends, and frames future research challenges. To sum up, the chapters introduced above tackle a variety of aspects of cities by means of innovative computational approaches. The collection of chapters indicates that cities are far too complex to deal with them from only one point of view and one set of argumentations. In addition, it is important to learn that different methodological and theoretical approaches enrich and complement each other. In this regard, it is desirable that the scientific outcome of this book will stimulate urban-related international and interdisciplinary research networks, paving the way to bring us closer to Michael Batty's (2013) vision of a "new science of cities."

Last but not least, the editors must express their gratefulness and gratitude to all reviewers for their support and their critical and constructive comments for each chapter. This has added significantly to the quality of the entire volume. We deeply appreciate the efforts of all authors, who submitted a full chapter manuscript and selected our book as a potential publication outlet for their research. Furthermore, we thank Mei-Po Kwan and Paul Longley for their comments. Marco Helbich and Jamal Jokar Arsanjani thank the Alexander von Humboldt foundation and Heidelberg University, Germany, for laying the foundation for this book. Finally, we acknowledge the Springer team as well as the series editors Jay D. Gatrell and Ryan R. Jensen for their great assistance throughout the whole publication process. Without all these helping hands, this volume would have never been published.

References

Anas A, Arnott R, Small K (1998) Urban spatial structure. J Econ Lit 36:1426–1464

Arribas-Bel D, Nijkamp P, Scholten H (2011) Multidimensional urban sprawl in Europe: A self-organizing map approach. Comput Environ Urban Syst 35:263–275

Batty M (2008) The size, scale, and shape of cities. Science 319:769–771

Batty M (2013) New science of cities. MIT Press, Cambridge

Batty M, Longley P (1994) Fractal cities: a geometry of form and function. Academic Press, London

Bettencourt L (2013) The origins of scaling in cities. Science 340:1438–1441

Birkin M (2009) Geocomputation. In: Kitchin R, Thrift N (eds) International encyclopedia of human geography. Elsevier, Oxford, pp 376–381

Brunauer W, Lang S, Umlauf N (2013) Modelling house prices using multilevel structured additive regression. Stat Model 13:95–123

Calabresea F, Diaob M, Lorenzo D, Ferreira J, Ratti C (2013) Understanding individual mobility patterns from urban sensing data: a mobile phone trace example. Trans Res Part C: Emerg Technol 26:301–313

Clarke K (2014) Why simulate cities? GeoJournal 79:129–136

Couclelis H (1998) Geocomputation in context. In: Longley P, Brooks S, McDonnell R, MacMillan B (eds) Geocomputation: a primer. Wiley, Chichester, pp 17–29

De Vos J, Witlox F (2013) Transportation policy as spatial planning tool; reducing urban sprawl by increasing travel costs and clustering infrastructure and public transportation. J Transp Geogr 33:117–125

Fischer M (2006) Spatial analysis and geocomputation. Springer, Berlin

Fujita M, Thisse J-F (2013) Economics of agglomeration: cities, industrial location, and globalization. Cambridge University Press, Cambridge

Glaeser E (2011) Cities, productivity, and quality of Life. Science 333:592–594

Goodchild M (2010) Twenty years of progress: GIScience in 2010. J Spat Inf Sci 1:3–20

Grekousis G, Manetos P, Photis YN (2013) Modeling urban evolution using neural networks, fuzzy logic and GIS: the case of the Athens metropolitan area. Cities 30:193–203

Gu J, Zhu M, Jiang L (2011) Housing price forecasting based on genetic algorithm and support vector machine. Expert Syst Appl 38:3383–3386

Hagenauer J, Helbich M (2013a) Hierarchical self-organizing maps for clustering spatiotemporal data. Int J Geogr Inf Sci 27:2026–2042

Hagenauer J, Helbich M (2013b) Contextual neural gas for spatial clustering and analysis. Int J Geogr Inf Sci 27:251–266

Hagenauer J, Helbich M, Leitner M (2011) Visualization of crime trajectories with self-organizing maps: a case study on evaluating the impact of hurricanes on spatio-temporal crime hotspots. In: 25th international cartographic conference, Paris, France

Hall P (1993) Forces shaping urban Europe. Urban Stud 30:883–898

Helbich M, Leitner M (2012) Evaluation of spatial cluster detection algorithms for crime locations. In: Gaul W, Geyer-Schulz A, Schmidt-Thieme L, Kunze J (ed) Challenges at the interface of data analysis, computer science, and optimization. Studies in classification, data analysis, and knowledge organization. Springer, Berlin, pp 193–201

Helbich M, Jochem A, Mücke W, Höfle B (2013a) Boosting the predictive accuracy of urban hedonic house price models through airborne laser scanning. Comput Environ Urban Syst 39:81–92

Helbich M, Hagenauer J, Leitner M, Edwards R (2013b) Exploration of unstructured narrative crime reports: an unsupervised neural network and point pattern analysis approach. Cartogr Geogr Inf Sci 40:326–336

Helbich M, Brunauer W, Hagenauer J, Leitner M (2013c) Data-driven regionalization of housing markets. Ann Assoc Am Geogr 103:871–889

Helbich M, Brunauer W, Vaz E, Nijkamp P (2014) Spatial heterogeneity in hedonic house price models: the case of Austria. Urban Stud 51:390–411

Jokar Arsanjani J, Helbich M, Bakillah M, Hagenauer J, Zipf A (2013a) Towards mapping land use patterns from volunteered geographic information. Int J Geogr Inf Sci 27:2264–2278

Jokar Arsanjani J, Helbich M, Vaz E (2013b) Spatiotemporal simulation of urban growth patterns using agent-based modeling: the case of Tehran. Cities 32:33–42

Jokar Arsanjani J, Helbich M, Mousivand A (2014) A morphological approach to predicting urban expansion. Trans GIS 18(2):219–233 (online first). doi: 10.111 1/tgis.12031

Kwan M-P, Schwanen T (2009) Quantitative revolution 2: the critical (re)turn. Prof Geogr 61: 283–291

Law J, Quick M (2013) Exploring links between juvenile offenders and social disorganization at a large map scale: a Bayesian spatial modeling approach. J Geogr Syst 15:89–113

Lazer D et al (2009) Computational social science. Sciences 323:721–723

Leitner M, Helbich M (2011) The impact of hurricanes on crime: a spatio-temporal analysis in the city of Houston, TX. Cartogr Geogr Inf Sci 38:214–222

Malleson N, Heppenstall A, See L (2013) Using an agent-based crime simulation to predict the effects of urban regeneration on individual household burglary risk. Environ Plan B 40:405–426

Miller H, Goodchild M (2014) Data-driven geography. GeoJournal (online first)

Mimis A, Rovolis A, Stamou M (2013) Property valuation with artificial neural network: the case of Athens. J Prop Res 30:128–143

Nakaya T, Yano K (2010) Visualising crime clusters in a space-time cube: an exploratory data-analysis approach using space-time kernel density estimation and scan statistics. Trans GIS 14:223–239

Openshaw S, Abrahart R (2000) GeoComputation. Taylor and Francis, London

Pacione M (2009) Urban geography: a global perspective. Routledge, New York

Pijanowski B, Tayyebi A, Doucette J, Pekin BK (2014) A big data urban growth simulation at a national scale: configuring the GIS and neural network based land transformation model to run in a high performance (HPC) environment. Environ Model Softw 51:250–268

Shafizadeh-Moghadam H, Helbich M (2015) Spatiotemporal variability of urban growth factors: a global and local perspective on the megacity of Mumbai. Int J Appl Earth Obs Geoinf 35(Part B):187–198

Solecki W, Seto K, Marcotullio P (2013) It's time for an urbanization science. Environment 55:12–16

Torrens P (2012) Moving agent-pedestrians through space and time. Ann Assoc Am Geogr 102:35–66

United Nations (2014) Department of economic and social affairs, population division, http://www.un.org/en/development/desa/population/

Vaz E, Caetano M, Nijkamp P, Painho M (2012) A multi-scenario prospection of urban change – a study on urban growth in the Algarve. Landsc Urban Plan 104:201–211

Wang F, Guo D, McLafferty S (2012) Constructing geographic areas for cancer data analysis: a case study on late-stage breast cancer risk in Illinois. Appl Geogr 35:1–11

Xu S, Vosselman G, Oude Elberink S (2014) Multiple-entity based classification of airborne laser scanning data in urban areas. ISPRS J Photogramm Remote Sens 88:1–15

Part I
Spatial Planning and Decision-Making

Chapter 2
From Fractal Urban Pattern Analysis to Fractal Urban Planning Concepts

Pierre Frankhauser

Abstract Fractal geometry can be used to develop a multiscale approach to investigate the spatial organization of urban fabrics. First, the concepts behind fractal reference models are introduced so as to provide a better understanding of the results obtained from empirical analyses of urban patterns. Then, different methods for conducting fractal analyses are presented and the results obtained for urban patterns are discussed. It turns out that, despite their irregular appearance, urban patterns are often organized by an inherent fractal order principle, at least across a certain range of scales. More detailed analysis of the findings reveals links between these fractal properties and the historical contexts in which cities or urban districts developed. The influence of specific urban planning concepts on fractal behavior may also be identified, whereas the national context has less of a hold.

Urban fabrics emerge from complex interactions among various types of decision makers and are, in most cases, the outcome of a self-organizing process. However, by considering particular features of such urban fabrics and by comparing them with social demand and against certain planning concepts, a new planning concept can be proposed based on fractal logic, but intended for the sustainable development of metropolitan areas without excluding periurbanization. Software tools are presented for developing and evaluating scenarios for further urbanization of metropolitan areas.

Keywords Fractal analysis of urban patterns • Fractal planning • Sustainable development • Urban modeling

2.1 Introduction

Since the industrial revolution in the nineteenth century, urban growth has never slowed what was made possible by improvements in transportation technologies. Indeed, before efficient transport systems were developed, the size of towns

P. Frankhauser (✉)
ThéMA Laboratory, Université de Franche-Comté/CNRS, 32, rue Mégevand,
F-25030 Besançon Cedex, France
e-mail: pierre.frankhauser@univ-fcomte.fr

M. Helbich et al. (eds.), *Computational Approaches for Urban Environments*,
Geotechnologies and the Environment 13, DOI 10.1007/978-3-319-11469-9_2

remained constrained (Bairoch 1985). It was the introduction of railways and later of tramways and suburban railway systems that made it possible to live in one place and work in another (Rodrigue 2013). The outcome was tentacular growth along public transportation network routes. Motorization changed accessibility dramatically. Road networks are being improved continuously covering space ever more uniformly. Increasing areas have come to be affected by that what is currently called "urban sprawl." So, if we speak here of "urban sprawl," we refer to the phenomenon of more or less uncontrolled urban growth in Western countries generating mainly low density zones consisting of individual housing, often localized in former rural areas which have developed since the use of private cars predominates. Town planners often deplore that growth like this cannot readily be controlled and that the patterns generated look rather "chaotic," more like ink splashes than compact shapes such as circles or squares, which are often thought of as the geometric paradigms of ancient cities.

Compact forms like circular or square-like patterns minimize boundary lengths and so are optimal for protecting cities against attack. Yet, even in the Middle Ages, poorer households often settled outside the city walls along the highways. Within the walls, land was often reserved with the result that the actual urban fabric looked less regular than might be claimed. But, of course, this remained a local phenomenon whereas nowadays urban fabrics form a patchwork of complex clusters connecting several ancient nuclei.

This contribution focuses first on the spatial distribution of built-up space in contemporary settlement patterns in Western Europe. We look how built-up space fills up surface across scales. Only in the last part we consider, in the context of planning strategies, the intensity of soil occupation, i.e., the degree of concentration of population in buildings. Hence, except in the last part, we restrict discussion to a two-dimensional approach.

In particular, we are interested in the extent to which these patterns are organized around a certain ordering principle. Indeed, despite their irregular shapes, some research since the late 1980s has shown that, at the macroscale, urban patterns obey rather precise distribution laws corresponding closely to fractal geometry. Much basic work on fractal investigations of urban patterns has been done since the 1980s, in particular by Batty and Longley (1986, 1994), by Goodchild and Mark (1987), Lam and de Cola (2002), as well as by White and Engelen (1994) and the present author Frankhauser (1994). Subsequent publications have deepened the methodological aspects and confirmed the value of this approach (e.g., Batty and Kim 1992; Batty and Xie 1996; Frankhauser 1998; Benguigui et al. 2000; Shen 2002; De Keersmaecker et al. 2003; Frankhauser 2004, 2008; Thomas et al. 2008a, b, 2010, 2012; Chen 2009, Chen and Feng 2010).

Thereafter, the focus shifts to particular features of fractal geometry liable to provide a better understanding of the spatial properties of such urban fabrics. This leads us to ask to what extent the emerging fractal shape of urban patterns cannot inspire planning concepts designed to manage peripheral urbanization intelligently without dismissing it a priori (Frankhauser 2008; Frankhauser et al. 2011). This will be done by introducing a planning concept inspired by fractal geometry.

The goal is to satisfy social demand for living in quiet surroundings, near open landscapes where natural resources can be safeguarded, while reducing car use at the same time. Beyond the theoretical aspect which makes use of the hierarchical principles peculiar to fractals, it is shown how the concept can be used practically for developing planning scenarios.

2.2 The Morphology of Sprawling Patterns: Some Preliminary Remarks

One of the most striking reasons why sprawling urban patterns look "amorphous" is that they are made up of elements belonging to a great variety of scales ranging from buildings to entire metropolitan areas. The metropolitan area of Stuttgart illustrates this. Even the rather coarse-grained map of the built-up area in Fig. 2.1a reveals details of very different sizes: along the boundary of any arbitrarily chosen settlement cluster, small bays alternate irregularly with larger ones. Moreover, metropolitan areas are made up of many clusters of very different sizes that are nonuniformly distributed across space. Ribbon-like built-up areas run along valleys or transportation network routes, but sparsely urbanized areas also occur. This shows the interaction between urbanization and natural conditions (Mohajeri et al. 2013). Similar characteristics appear at the microscale of towns as can be seen from the close-up of the northern urban fringe of the agglomeration of Stuttgart (Fig. 2.1b). Here buildings form clusters of different sizes and densities since industrial zones, historical town centers, and recent detached housing areas are mixed. But the size of the vacant areas between clusters also varies greatly.

Traditional density-based measures currently used in geography and planning are unsuitable for describing such features. Indeed, density measures *mean occupation*

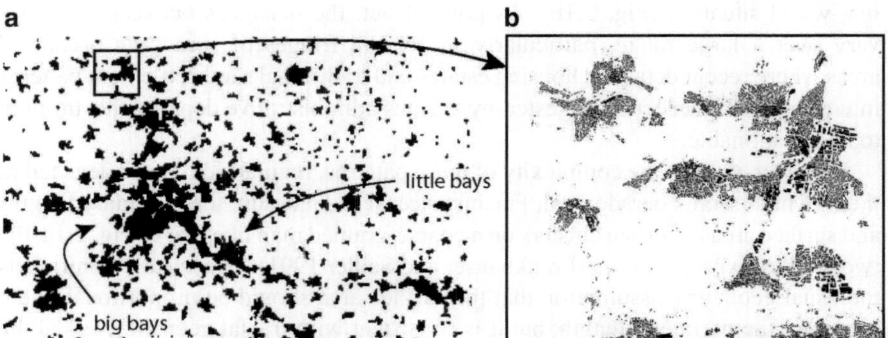

Fig. 2.1 (**a**) A simplified map of the metropolitan area of Stuttgart. (**b**) A detailed GIS-data base for one outskirt (Frankhauser 2005)

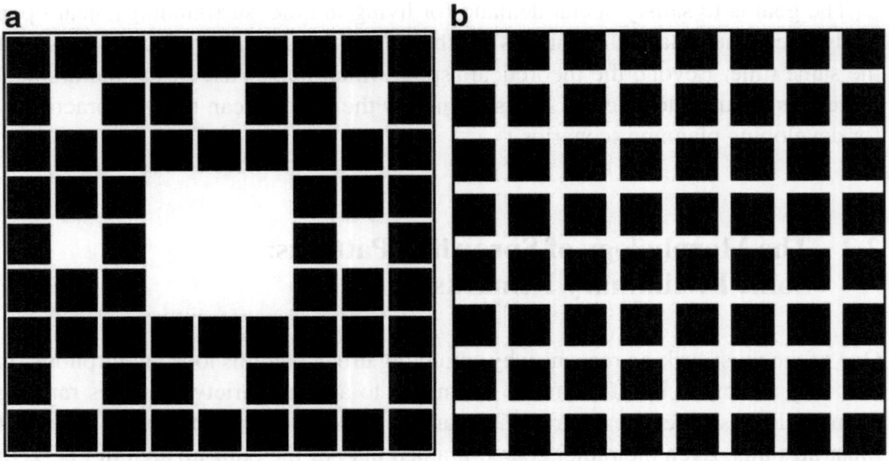

Fig. 2.2 64 blocks distributed in two different ways: the pattern on (**a**) obeys fractal logic and the pattern (**b**) is uniform (Thomas et al. 2008a)

of space. Accordingly, density is constant when the constituent parts of a structure, in our case buildings, are distributed uniformly across space, which does not seem to hold for urban patterns. Moreover, density does not really yield information about spatial distribution. Figure 2.2 shows two patterns in which 64 blocks of the same size are distributed in different ways within the same square: while the densities are the same, the first pattern is a fractal-like structure, unlike the second one.

Obviously, the nonuniform distribution of buildings makes it difficult to grasp the spatial organization of urban patterns. This is plain when looking for reliable criteria on which to define urban boundaries. Looking at the simplified map in Fig. 2.1a, it might be thought that the urban boundary can be identified easily, but not so in the real-world situation (Fig. 2.1b). As pointed out, the distances between buildings vary over a large range, particularly so for the fringes of sprawling urbanized areas where recent detached housing estates and traditional rural settlement patterns interdigitate. The criteria suggested by various administrative departments turn out to be questionable.

However, despite the complexity of these patterns, regularities can be detected in them, which seems paradoxical. For instance, if we measure the boundary lengths and surface areas of urban clusters on a coarse-grained map like that in Fig. 2.1a, the two tend to be *proportional* (Frankhauser and Sadler 1991). This clearly contradicts the usual geometric assumption that the surface area should be proportional to the square of the perimeter length, but it is consistent with fractal geometry, as will be seen next.

2.3 Fractal Models for Urban Patterns

We now illustrate by means of examples that fractals exhibit properties reminiscent of those discussed for urban patterns. As pointed out, urban patterns consist of elements, i.e., buildings forming what are often irregular clusters where the distances between the buildings vary over a large range. But even when boundaries are defined (e.g., by defining criteria for simplified maps), these boundaries are not smooth but display outgrowths and indents of various sizes.

Unlike Euclidian geometrical objects such as circles or squares, fractal geometry allows us to construct geometrical reference models consisting of elements distributed in a completely nonuniform way, forming clusters at different scales. It is then possible to illustrate several types of spatial pattern, which resemble specific aspects of urban patterns like fragmentation or the complex morphology of boundaries. These structures may look irregular; even so, the spatial distribution of the constituent parts obeys a powerful distribution law, which may be characterized by a single value. Thus, if urban patterns really do exhibit the particular features of fractal objects, it may be concluded that, despite their highly irregular aspect, they comply with a well-defined principle of spatial organization, which can be characterized quantitatively. The usual notion of "regularity" or "irregularity" then becomes meaningless.

Since we focus here on the question of how buildings are distributed within areas of settlement, we make use of particular types of fractals consisting of "black" elements distributed in space. We associate these elements with zones containing buildings, whereas empty, i.e., "white," zones are essentially unbuilt. Two kinds of fractal objects prove of interest, Sierpinski carpets and Fournier dust. These two approaches can be combined. In Fig. 2.3a–c, we show how a Sierpinski carpet is generated by iteration. Starting from the square-like initiator, a mapping procedure known as the generator is applied. Here the procedure reduces the initial figure by the factor $r = 1/3$, and $N = 5$ of these "elements" are assembled to form a cross. The same operation is repeated for each of the smaller squares. Hence, as this iteration procedure is repeated, an ever more filigree object appears consisting of

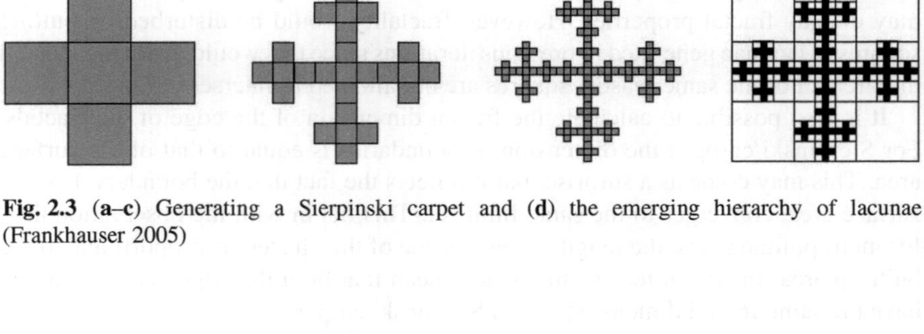

Fig. 2.3 (**a–c**) Generating a Sierpinski carpet and (**d**) the emerging hierarchy of lacunae (Frankhauser 2005)

an increasing number of smaller and smaller square-like elements. Indeed at each step, the number of elements is multiplied by N and their size is reduced by r so that at the nth step, we have N^n elements of length r^nL, where L is the length of the initial square. So, both the number of elements and their size follow geometric series. Moreover, at each step, an increasing number of smaller and smaller lacunae are generated, and hence, a hierarchical system of lacunae appears (Fig. 2.1d). This is one of the most striking features of these fractals. At the same time, the boundary line of the Sierpinski carpet is lengthened, since a growing number of smaller indents are generated. Hence, the boundary length tends to infinity, whereas the surface area tends to zero, and the limit geometrical object is neither two dimensional like a surface nor one-dimensional like a line. This is one of the most striking features of such fractals. Euclidian objects like squares or circles are compact and their border is smooth—except at some singular points, the corners—what never can be for fractals.

This led mathematicians to introduce the notion of fractal dimension in order to characterize these objects which are made up of elements that are distributed highly nonuniformly in space. The basic requirement is that there exists a measure M which remains constant throughout the iteration and is characteristic of the fractal. This measure is defined by computing the total "mass" of the fractal at each step, which is the product of the number of elements and their size weighted by the free parameter D, the fractal dimension

$$N^n \times (r^n L)^D = M \equiv L^D \tag{2.1}$$

This relation yields

$$N^n = (r^n)^{-D} \Rightarrow D = -\frac{\log N}{\log r} \tag{2.2}$$

so that D is indeed a constant over all iterations and, in the case of constructed fractals, is related to the two parameters N and r. Hence, for the Sierpinski carpet of Fig. 2.3, we obtain $D = 1.47$ (Mandelbrot 1982; Frankhauser 1994). Obviously, the dimension does not depend on the position of the elements within the square within which they were generated. Hence, we may randomly change their position without affecting fractal properties, and accordingly, irregular empirical structures may display fractal properties. However, fractality would be disturbed by putting squares in lacunae generated in previous iterations since this would affect the lacunal hierarchy. For the same reason, squares are not allowed to intersect.

It is also possible to calculate the fractal dimension of the edge of the fractals. For Sierpinski carpets, the dimension of boundaries is equal to that of the surface area. This may come as a surprise, but it reflects the fact that the boundary, like the surface area, converges to the same limit set. This fits in with the observation that, for metropolitan areas, the length of the outline of the clusters is proportional to the built-up area. In fractal terms, this would mean that both the edge and the surface have the same fractal dimension, like a Sierpinski carpet.

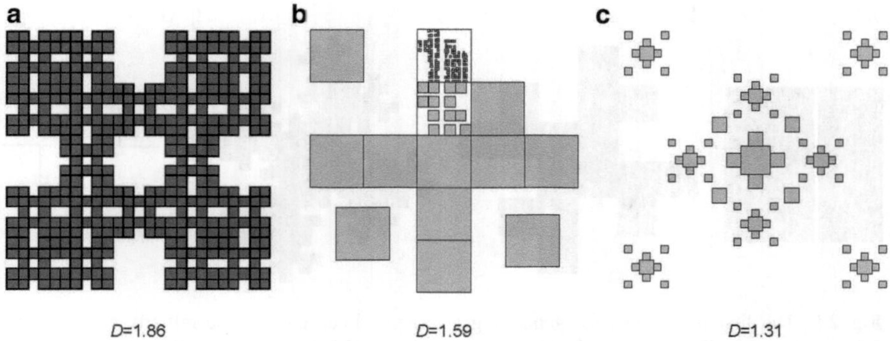

Fig. 2.4 Three different types of fractals. (**a**) A Sierpinski carpet where the outer border is a fractal, but the dimension of which differs from that of the carpet. (**b**) A fractal, the features of which remind urban patterns. Iteration has been developed up to the third step only in the *upper square* of the cross and up to the second step in the *square below*. (**c**) A fractal consisting of a hierarchical system of clusters reminding the logic of a central place system (Source: (**b**) Tannier et al. 2006)

Figure 2.4 shows other examples of similar fractals. For the Sierpinski carpet of Fig. 2.4a, the outer boundary is connected, and its complex shape is reminiscent of that of the urban pattern on coarse-grained maps as in Fig. 2.1a. In Sierpinski carpets, all the elements are connected and so the structure consists of a single cluster. Fournier dusts follow the same iterative logic, but the elements are disconnected. By combining both these logics, fractals like those shown in Fig. 2.4b, c can be connected. In Fig. 2.4b, the positions of the elements in the generator are less regular. In two squares, the second and third iterations are illustrated by changing the position of the elements each time respecting the hierarchy of lacunae. Hence, a more city-like shape is obtained at the microscale, with blocks of houses along the streets and inner courtyards. Finally, in Fig. 2.4c, two reduction factors, $r = 1/3$ and $r = 1/6$, are combined and thus a hierarchy of clusters occurs. Hence, a generalized version of Eq. (2.1) allows the dimension to be computed. This multifractal structure could be an illustration of the distribution of central places in an urban system like that of Fig. 2.1a.

Let us emphasize that the definition of fractal dimensions is entirely consistent with standard Euclidean geometry; for uniform distributions like that of Fig. 2.2b, the dimension is $D = 2$ and for lines $D = 1$. But furthermore the fractal dimension has a clear meaning. It measures the degree of concentration of the occupied sites across scales or, more specifically, the relative decrease in mass with increasing distance from any site where mass is concentrated. Accordingly, the more uniformly mass is distributed within a fractal structure, the closer the dimension will be to $D = 2$, and vice versa, if the mass is concentrated in one point, D is *zero*. Indeed, since mass is more uniformly distributed in Fig. 2.4a, the dimension value is higher for pattern 2.4a than for patterns 2.4b and 2.4c which are more contrasted. As pointed out, boundary and surface area have the same dimensions in Sierpinski carpets. However, in Fig. 2.4a, the outer boundary is itself a fractal consisting of $N = 7$ elements, and thus, the dimension of this fractal subset is $D_b^{(\mathrm{bord})} = 1.2$.

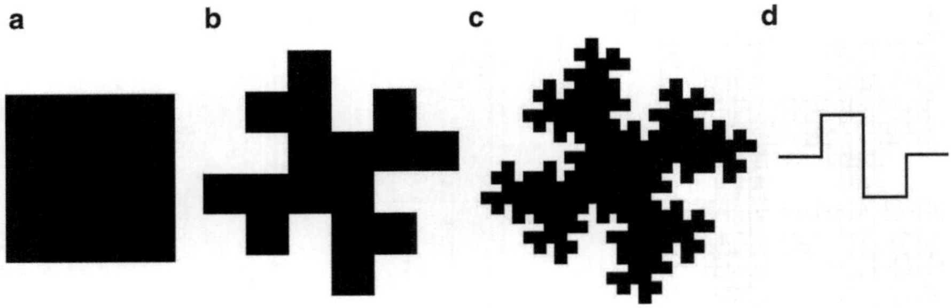

Fig. 2.5 The first three steps for generating a teragon. Each side of the initially given *square* (**a**) is replaced by the polygon (**d**) what generates figure (**b**). By reiterating this mapping, figure (**c**) is obtained (Mandelbrot 1982)

There are also fractals with compact surface areas of $D=2$ but which have just a fractal boundary, like the teragon of Fig. 2.5. Here, in the course of iteration, the surface area remains the same but is distributed differently at each step so that the border becomes increasingly filigree, its length tending to infinity, and its fractal dimension amounts to $D=1.5$.

We remind that fractal dimension should not be confused with density (Fig. 2.2). A suburban area consisting of big lots where individual houses are uniformly distributed in space has globally a fractal dimension of $D=2$, but low density. However, a neighborhood made of uniformly distributed row houses along the streets is certainly denser, but, since build-up area is uniformly distributed, too, the fractal dimension would also be $D=2$.

When referring to real-world situation, we usually stop the iteration at a certain step n. Mandelbrot called such objects *prefractals*. For instance, if the size of the initial figure corresponds to the area occupied by an agglomeration of a metropolitan area, we stop the iteration as soon as the size of the elements is comparable to the size of a building. Hence, we have a certain range for which we observe the scaling properties of fractals for which the distribution of the elements can be described by the fractal dimension. We now focus on the question of how to explore the fractal properties of empirical structures.

2.4 Measuring Fractality

2.4.1 General Methodology

Obviously, urban patterns are not generated by an iterative mapping procedure but result from complex interactions between different types of decision makers, such as politicians, planners, landowners, land developers, and so on. Hence, measuring methods must be used by which to verify to what extent the spatial organization of urban patterns exhibits scaling properties, as they are characteristic for fractals. To this end, several methods have been developed. We focus here on four methods referring to different types of information about the fractal behavior of empirical structures and to different properties of empirical textures like urban patterns.

It is clear in real-world textures such as urban patterns that the spatial organization, even if it can well be described by fractal measures, does not obey a fractal law exactly. This prompts us to speak often of "scaling behavior" which can change at certain scales, vary from one urban district to another, or be disturbed at certain scales. Mixing different scaling behaviors is reminiscent of multifractal geometry, where the fractal behavior varies from one site to another.

There are a number of methods used to estimate the fractal dimension of empirical structures. Not all of these methods obey the same logic, and if the empirical structures exhibit multifractal properties, the results obtained will not be the same. In this case, the methods provide complementary information.

The basic idea is to explore the texture under consideration at different scales. For this purpose, a distance ε is introduced and the number of elements $N(\varepsilon)$ lying within this range is determined. Then, the value ε is changed and the procedure repeated. For fractals, the following power-law relation between the number of elements $N(\varepsilon)$ and distance (ε) holds

$$N(\varepsilon) = a \times \varepsilon^{-D} \tag{2.3}$$

which is used to estimate the dimension D and the prefactor a. Hence, we assume that the analysis provides a sequence $N^{(obs)}(\varepsilon_i)$ of empirical data, the empirical curve, for a discrete series of ε-values called ε_i serving to estimate the parameters a and D by which the "theoretical curve" (2.3) is obtained. Theoretically, this prefactor corresponds to the measure M introduced previously. For empirical structures, however, we should expect deviations at different scales from a pure fractal law. Thus, for each ε_i value, we should introduce a local version of the fractal law which reads

$$N^{(obs)}(\varepsilon_i) = a_i \varepsilon_i^{-D_a} \tag{2.4}$$

In Thomas et al. (2012), it was shown that the global prefractor a is just the geometrical mean value of these scale-specific prefactor values and so a mean measure for the object across scales:

$$a = \left(\prod_{i=1}^{n} a_i \right)^{\frac{1}{n}} \tag{2.5}$$

Up to now, it has proved difficult to interpret the observed values directly. Recent tests have shown that the a-values are neither associated with the correlation between the empirical curve and theoretical curve, i.e., the quality of adjustment, nor with the confidence interval. Hence they provide, in principle, supplementary information about the object. Previous investigations (Thomas et al. 2012) showed that the more homogeneous the built-up area of a city, the lower the a and the higher the D.

2.4.2 Covering Methods

The most commonly used methods are based on the idea of covering the elements of the object analyzed by a minimum of elements like squares or circles of size ε. Hence, information is obtained about the number of elements lying within a distance less than or equal to ε. The most widely used method is "box counting," although we prefer to speak of it as "grid analysis." It consists of covering a chosen zone with a grid of a given mesh (grid square) size ε. The number of grid squares containing buildings is counted. At the next step, the mesh size is reduced and the procedure repeated. By repeating the procedure, we obtain an empirical relation between the number of grid squares containing buildings and the grid square size. Figure 2.6 illustrates the method for a Sierpinski carpet by choosing grid square sizes in accordance with the iteration procedure. The number $N(\varepsilon)$ as well as size ε of the grid squares corresponds here to the construction rule of the Sierpinski carpet, i.e., both follow the series of the mapping procedure and so the fractal law holds. For empirical structures, it is hence possible to verify whether the empirical data obey the fractal law. Deviations from this law, e.g., occurring for certain scales ε, can hence be identified. However, it is difficult to find a position of the grid which fulfills the requirement of minimal covering. Even for the constructed Sierpinski carpet, an inadequate position of the grid would falsify the results since we would obtain more grid squares containing elements of the fractal. In order to reduce this artifact, the

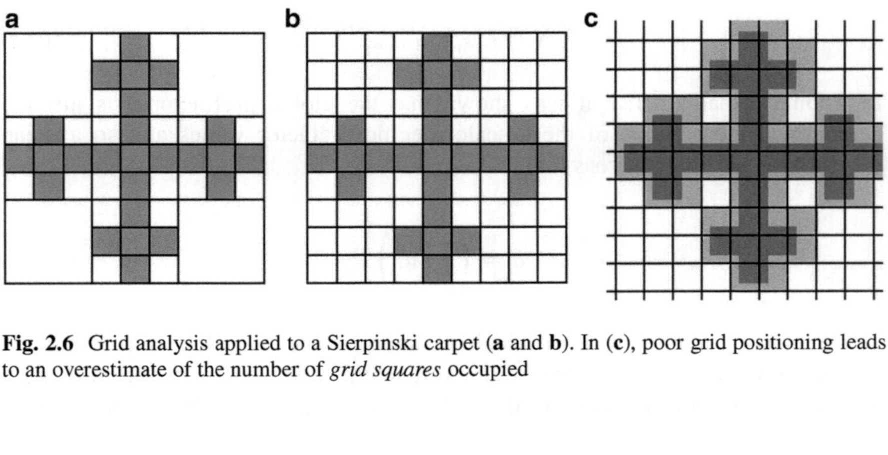

Fig. 2.6 Grid analysis applied to a Sierpinski carpet (**a** and **b**). In (**c**), poor grid positioning leads to an overestimate of the number of *grid squares* occupied

grid position is often subsequently moved and the minimal number of occupied grid squares is retained (see, e.g., Jiang and Liu 2012). We are currently testing a method allowing free positioning of the boxes for which the number necessary to cover the texture is optimized by means of a genetic algorithm.

A similar method is dilation analysis, based on the algorithm introduced by Minkowski and Bouligand. In dilation analysis, each occupied point is surrounded by a square of size ε, the surface area of which is considered to be completely occupied. The size of these squares is then gradually enlarged, and we measure the total surface area covered $A(\varepsilon)$ at each stage. As the squares are enlarged, any details smaller than ε are overlooked and we gradually obtain an approximation of the original shape. This is reminiscent of a gradual change in the degree of cartographic detail in drawing. Because more and more squares overlap, the total area occupied $A^{(\mathrm{dil})}(\varepsilon)$ for a particular value e is less than what it would be if the same number of occupied points that make up the original shape were surrounded individually. By dividing this total area by the area $A^{(\mathrm{dil})}(\varepsilon) = \varepsilon^2$ of a test square, we get the number of elements $N(\varepsilon)$ necessary to cover the whole and we obtain a relation corresponding to relation (2.3). The corresponding fractal dimension $D^{(\mathrm{dil})}$ is known as the Minkowski dimension or dilation dimension.

2.4.3 *Mass-Distance Relations*

A rather different method is radial analysis. It provides information about the spatial organization around a chosen counting point. A circle is drawn around this point, and the radius ε is gradually increased. At each step, the total number of occupied points $N(\varepsilon)$ inside the circle is counted. Here, the fractal law takes the form

$$N\ (\varepsilon) = a \times \varepsilon^D \tag{2.6}$$

As pointed out for more complex structures, different fractal behaviors may be mixed, the radial analysis provides specific information about local fractal behavior, and so the scaling exponent is known as the local fractal dimension.

It is possible to realize such analysis for each occupied, i.e., built-up, site within a given zone and to compute the mean number of occupied points observed for each distance value ε (Vicsek 1989). An equivalent fractal law is associated with the method. This method first proposed by Grassberger and Procaccia (1983) is known as correlation analysis. The information obtained provides mean information about the zone analyzed, as in grid analysis. However, the information is more detailed than for grid analysis, since the exact position of occupied sites is explored and not just the fact of lying within a box of a given size. Hence, other dimension values can be expected for more complex patterns. In fractal terms, this is referred to as a second-order dimension and refers thus to multifractal logic.

2.4.4 The Curve of Scaling Behavior

In previous work, it turned out that a particular way of representing the empirical results helps to provide more detailed information about the spatial organization of urban patterns (Frankhauser 1998; Thomas et al. 2010). This type of representation is known as the "curve of scaling behavior" (Frankhauser 1998; Thomas et al. 2010). Palmer (1988) called this a "fractogram," but that terminology is confined to ecology (Leduc et al. 1994). We start by taking the logarithm of relation Eq. (2.6):

$$\log N\ (\varepsilon) = \log a + D \log \varepsilon \tag{2.7}$$

which is a linear relation between log $N(\varepsilon)$ and log ε where the dimension is simply the slope value of the straight line.

We assume now that the prefactor a and the fractal dimension D may both depend on the distance parameter ε in accordance with the idea that real-world textures do not necessarily obey a strictly fractal law:

$$\log N\ (\varepsilon) = \log(a)\varepsilon + D\ (\varepsilon) \log \varepsilon \tag{2.8}$$

and thus the variation of log $N(\varepsilon)$ with respect to log ε becomes (Frankhauser 1998):

$$\frac{d \log N\ (\varepsilon)}{d \log \varepsilon} \equiv \alpha\ (\varepsilon) = \frac{d \log a\ (\varepsilon)}{d \log \varepsilon} + \frac{dD\ (\varepsilon)}{d \log \varepsilon} + D\ (\varepsilon) \tag{2.9}$$

This shows that if the dimension and the prefactor both depend on the scale parameter ε, the α-values are no longer constant and may even exceed the upper limit value of $D = 2$. Hence, representing the α-value as a function of distance ε informs us about deviations from fractal laws due to a variation in D or a. It turns out that the shape of these curves provides insight into the spatial organization of urban patterns and can distinguish different types of urban district (for details, see Thomas et al. 2010).

2.5 The Fractality of Urban Patterns

2.5.1 Toward a Typology of Urban Patterns

As pointed out above, a number of papers have been published about fractal analysis of urban patterns. We focus here on some striking results obtained at the Thema institute of Besançon (France) in cooperation with the CORE institute at Louvain-la-Neuve (Belgium). In order to realize fractal analyses of urban patterns, a computer program was developed which estimates the fractal dimension and the prefactor from the observed data and makes it possible to verify to what extent the empirical data $N(\varepsilon)$ obey a fractal law.[1] In general, the fractal approach is well suited for characterizing the spatial organization of urban fabrics. The quality of adjustment between empirical and theoretical curves is usually rather high.

Different authors emphasized that urban fabrics are generally a mixture of different scaling behavior, and hence, multifractal approach suits better for measuring their morphology than unifractal ones (e.g., Chen and Wang 2013). We agree to this point of view in that sense that considering an agglomeration as a whole, different town sections may have different scaling properties. Hence, considering large urban areas, multifractal behavior occurs since at this scale different types of patterns are mixed. We focus here essentially on the intraurban scale of wards for which we observed, e.g., by a recently developed special method we call multiradial analysis, a rather stable scaling behavior reminding that of unifractals. Ariza-Villaverde et al. (2013) used multifractal analysis on an intraurban scale for street networks, but the differences between the values of the different dimensions observed for two areas are very small, what confirms rather unifractal behavior on this scale.

However, since the town sections we analyzed are chosen within a square-like window and not according to local morphological properties, we do not exclude the potential presence of a certain mix of scaling behavior.

This incited us comparing different measuring methods. It turned out that correlation analysis provides the most robust results on a rather large range of scales. However, the dimension values differ for this method from results obtained with covering methods, which can be explained, as pointed out, by the fact that the method takes into account the position of built-up sites across scales in a precise way and follows a multifractal analysis logic.

All analyses presented were carried out with the open access software tool Fractalyse, developed at Thema institute by Gilles Vuidel under the supervision of Pierre Frankhauser and Cécile Tannier.

We first discuss some results obtained on the basis of rather coarse-grained data for metropolitan areas as a whole (Frankhauser 2004). For the agglomerations of Berlin and Stuttgart, surface dimensions are rather low, which expresses rather contrasted patterns. Indeed, in both cases, the urban fabric is dominated by axial growth along valleys for Stuttgart and alongside the suburban railway lines for Berlin. We also determined the fractal dimension of the boundaries of the central clusters of these agglomerations. The obtained value of Stuttgart is rather high, reflecting its rather sinuous appearance (Fig. 2.1). The most compact situation is observed for London, which can be explained by the green belt policy that attempted to contain urban growth. We shall return to this aspect later (Table 2.1).

We now come to the results obtained by using GIS-data which take into account the microstructure of the urban fabric, i.e., resolutions of 4 m minimum.

Several series of investigations have been realized for various European cities in Italy, Finland, France, Germany, and Switzerland, but essentially Belgium and

[1]The free software “fractalyse” was developed at the THEMA institute (Besançon) by Gilles Vuidel. A version allowing GIS-data to be used directly is currently under development.

Table 2.1 Some fractal dimensions of metropolitan areas

Metropolitan area	Surface dimension	Dimension of the central cluster boundary
Berlin	1.75	1.58
Stuttgart	1.75	1.88
London	1.86	1.41

Table 2.2 Fractal surface dimensions of some agglomerations

Agglomeration	*D*
Bayonne-Anglet-Biarritz	1.483
Bergamo	1.752
Besançon	1.638
Basel	1.723
Brussels	1.883
Cergy-Pontoise	1.695
Charleroi	1.857
Helsinki	1.708
Liège	1.914
Lille	1.683
Lyon	1.786
Montbéliard	1.558
Namur	1.526
Sarrebrücken	1.659
Strasbourg	1.785

France. We present here results obtained exclusively by correlation analysis. First, we refer to Thomas et al. (2012) which analyzed 18 cities. When considering the agglomerations as a whole, we may distinguish different ranges of dimension values. As Table 2.2 shows, Besançon, Strasbourg, Sarrebrücken, Basel, and Bergamo have dimensions of about 1.6–1.8. This holds even for Helsinki, which is influenced by its coastline, but also by a ribbon-like development on its periphery. These are cities of different sizes, in different countries and geographical situations, but they are dominated by a historical center and a gradual decline in density toward the periphery. A second group consisting of Bayonne-Anglet-Biaritz, Cergy-Pontoise, Lille, and Montbéliard has lower dimensions, lying between 1.5 and 1.7. All these towns are conurbations consisting of a rather disparate patchwork of zones with different morphologies where transient areas fill spaces between denser zones. Brussels and Lyon have high-dimension values; they are dominated by rather homogenous town sections dating from the nineteenth century. The highest values are observed for Charleroi and Liège, which were greatly affected by heavy industry constructions in the nineteenth century such as ribbons of working-class housing and an intricate urban structure (see, e.g., Vanneste et al. 2008 or Thomas et al. 2008a).

These examples, which are fully consistent with those obtained on the basis of more coarse-grained data, show that the historical context which contributed to urban development influences fractal dimension values. At the scale of agglomerations, axial development along valleys or transportation routes leads to more highly contrasted patterns.

This also means that some fractal values refer rather to physical factors (coast) or local historical contexts than to national planning rules, confirming Frankhauser (2003, 2008). More generally, this also confirms that nineteenth-century cities have higher fractal values than twentieth-century cities, and more generally, "historical cities are fractal, whereas the twentieth-century city is not" (Salingaros 2003) or they are multifractal (see, e.g., Batty 2005).

On the scale of urban districts, different investigations have shown that it is possible to identify different types of spatial organization which can be linked to peculiar historical or geographical contexts. In Thomas et al. (2012), a ward classification was applied to minimize intragroup variance. Four classes were identified. The first one with very high D values close to $D = 1.9$ corresponds to city centers with a high concentration of buildings, constructed in a continuous way along the streets where blocks are always very clearly recognizable. A second class comprises built-up neighborhoods composed of detached housing typical of pericentral areas. These districts have a low or middling density of urbanization, with quite a regular morphology, where houses are located alongside streets. Fractal dimensions are lower here since the patterns are less compact and hence more contrasted. Whereas in the previous patterns buildings follow the streets, in the third class dwellings and nonresidential, buildings are mixed and do not necessarily follow the street pattern. Such districts were often built during the period 1950–1980. Since buildings are located rather arbitrarily, dimensions are rather low (about $D = 1.67$). The most contrasted patterns are observed in French new towns like Cergy-Pontoise and districts constructed on the *Charte d'Athènes* (Le Corbusier 1971) principles. Here the spatial arrangement of buildings as well as their forms is varied, and green areas of various sizes separate the large isolated buildings.

2.5.2 Information Provided by the Curves of Scaling Behavior

Similar results were obtained in another way. In Thomas et al. (2010), we not only used the values of fractal dimensions to distinguish different types of urban patterns, but we looked at the shape of the curves of scaling behavior in order to distinguish different types of urban patterns. Indeed, for correlation analysis, it turns out that these curves often display a substantial fall before rising again for small distances corresponding to the size of buildings, courtyards, etc. Here, fractality does not come into play yet.

Figure 2.7 shows an example in which the lowest point corresponds to 28 m; the stable situation is reached for 68 m. But for other districts, the shape of the curves is quite different, e.g., smoother from the very beginning. Hence, we used the k-medoid algorithm (Bishop 2006) for classifying curves by their shape.

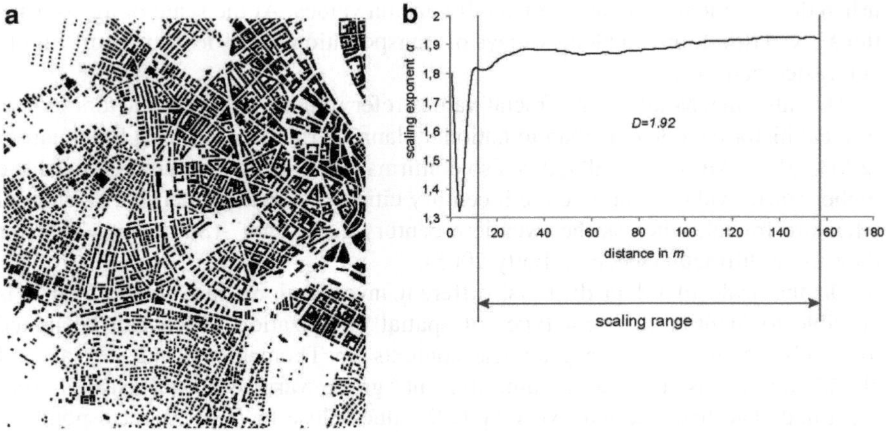

Fig. 2.7 (**a**) A town section of Basel and (**b**) the curve of scaling behavior of the correlation analysis (Source: digitized topographic maps)

This algorithm avoids the artifact of the k-means algorithm, which is based on the centroid of experimental data, which is often not one of the experimental data, whereas in the algorithm used, the representative of each cluster is necessarily one of the initial data forming the cluster. In this project, we analyzed a set of 49 town sections from the data set for nine European cities: Besançon, Cergy, Lille, Lyon, and Montbéliard in France, Brussels and Charleroi in Belgium, and Stuttgart and the Ruhr area in Germany.

Finally, five different morphological classes were distinguished. The first class may be associated with districts with ribbon-like semidetached housing typical of not too densely urbanized centers or pericentral zones as in, say, cities belonging to the Flemish part of Belgium. Whereas for these curves the "root-like" shape (Fig. 2.7b) of the curve is very significant, it is less pronounced for a second group of patterns which are typical for dense historical city centers. The town sections with the big "Corbusier" buildings again form their own morphological class, and the new towns also exhibit peculiar shapes of their curves of scaling behavior. This is due to a very typical mix of individual detached housing and apartment blocks, within large green areas and public places. Finally, very particular shapes are observed for commercial and industrial zones dominated by huge buildings, where intra-building distances are considerable, erasing the usually observed deviations for small distances.

While we find globally similar information to that given by fractal dimension ranges, closer analysis of the curves gives details about distance ranges for which substantial changes in spatial organization occur, or alternatively, for which the parameters are stable or not. Hence, the information contained in these curves turns out to be complementary to that of the fractal dimension D which remains a useful, albeit rather synthetic indicator.

The results show that the spatial organization of neighborhoods is influenced rather by the historical context or underlying urban planning concepts which contributed to its emergence. These conditions may be comparable for neighborhoods belonging to different cities or countries. Hence, the resemblances between neighborhoods of different cities may be greater than those observed within a city.

2.5.3 The Multiradial Approach

Very recently, we have tested a specific method in order to segment town sections on the basis of their spatial organization. For this purpose, we realized radial analyses for each built-up site and estimated the fractal dimension within a fixed distance range, e.g., 400 m around the counting point. We then introduced classes of fractal dimensions. Figure 2.8 shows an example for a city near Lyon. It clearly turns out that the entire core of the old center has the same high-dimension values, in keeping with their rather homogenous shape. Around this, the more recent town sections are more contrasted and have lower dimensions. The fringes where non-built areas come into play have lower dimensions still. The quality of adjustment is here checked by means of the confidence interval. This indicator is rather sensitive to local changes in spatial organization. The range becomes larger, e.g., in the vicinity of a big building

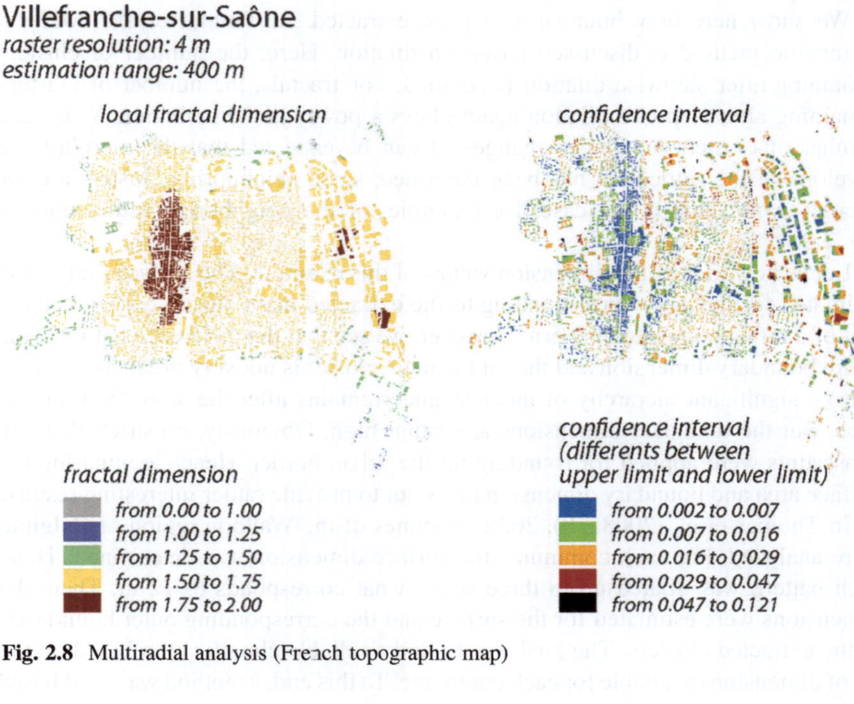

Fig. 2.8 Multiradial analysis (French topographic map)

which is surrounded by smaller ones. In general, the interval becomes larger for sites lying close to sites where spatial organization changes. This type of analysis seems rather promising for exploring the spatial organization of urban patterns.

2.5.4 Including Boundary Analysis

We now discuss results where urban borders have been extracted and where their morphology has been compared to that of built-up areas. As pointed out, in Sierpinski carpets, the built-up space and the entire boundary should have the same dimension. However, there may be subsets, e.g., outer boundaries (Fig. 2.4a) or subclusters, with different fractal dimensions.

Moreover, we recall the peculiar feature of teragons with compact areas but fractal boundaries. Hence, we may seek out the kind of fractal that urban patterns resemble most. To this end, we dilated stepwise urban patterns. Buildings were seen to merge increasingly and form clusters. Investigations show that large clusters emerge after just a few dilation steps, when courtyards and small streets are filled in, which usually occurs at distances of about 8–16 m (De Keersmaecker et al. 2003; Thomas et al. 2008a). Here again, we find the distance range for which we often observed non-fractal behavior when looking at the curves of scaling behavior (Fig. 2.7b). Hence, we could say that, beyond this threshold, urban fabrics may be considered to be prefractal structures.

We show here how boundaries can be extracted. In Tannier et al. (2011), a systematic method is discussed based on dilation. Here, the number of clusters remaining after stepwise dilation is counted. For fractals, the number of clusters remaining after stepwise dilation again obeys a power-law distribution. When the number of clusters no longer changes, it can be expected that the morphologic envelope of the settlement has been identified, since neighboring clusters are far enough away. This is the case, for example, of the neighboring settlements in Fig. 2.9.

Let us now look at the dimension values of this example. The surface dimension estimated for the zone corresponding to the extracted main cluster corresponds to that of a rather contrasted pattern. However, we see that the difference between the entire boundary dimension and that of the main cluster is not very great. This shows that no significant hierarchy of inner lacunae remains after the smoothing procedure. But the boundary dimensions are rather high. Obviously, no strict planning constraints were applied for rounding up the urban border. Hence, comparing the surface area and boundary dimension turns out to provide rather interesting results.

In Thomas et al. (2008a, b), 262 communes of the Walloon region of Belgium were analyzed. For each commune, the surface dimension was determined. Then, each pattern was dilated up to three steps, what corresponds to 12 m. Then, the dimensions were estimated for the surface and the corresponding outer boundaries of the extracted clusters. The goal was to find a reliable classification based on the set of dimensions available for each commune. To this end, a method was used based

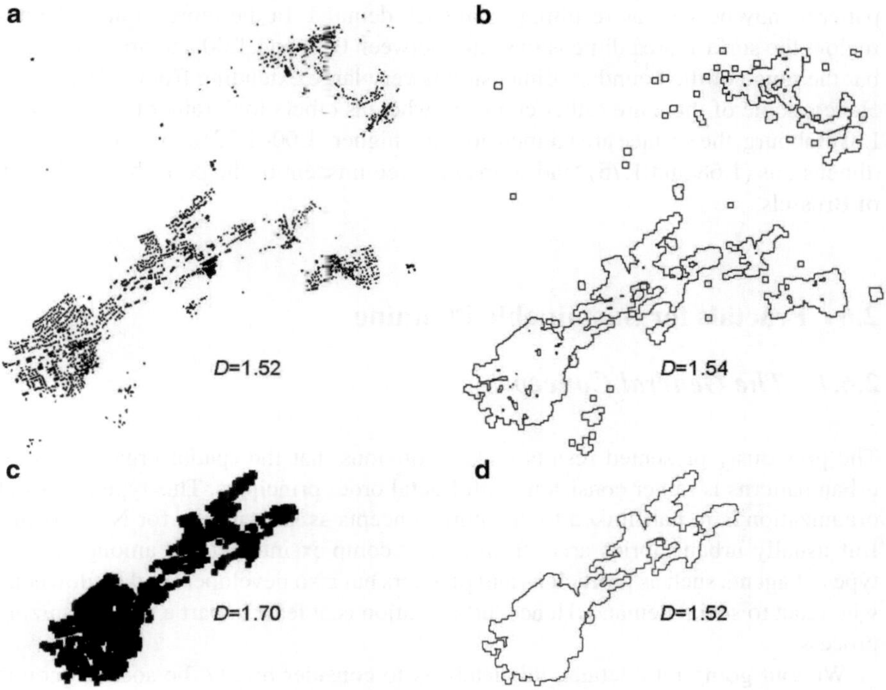

Fig. 2.9 A periurban village of Besançon, the original pattern (**a**), seven-step dilation isolates the settlement from neighboring ones (**c**), corresponding to a distance of 28 m (French topographic map), (**b**) and (**d**) show the extracted corresponding borders

on a statistical analysis of the cloud of points combining the different dimension values for each commune. The method is based on bivariate Gaussian distributions; such a model provides a powerful and nowadays standard tool for clustering (see references in McLachlan and Peel 2000). Details are given in Thomas et al. (2008a). Finally, six morphological classes were retained. On average, it turned out that the morphology of the built-up surfaces in Wallonia is strongly influenced by the history of urbanization and the underlying processes: the history of the urban network, the nineteenth-century urbanization leading to the Sambre-Meuse industrial corridor, and the twentieth-century suburbanization spreading from Brussels and other cities (including Luxembourg). Hence, surface area dimensions of old industrial areas lie within the range of 1.58–1.84, whereas the boundary dimensions of the dilated patterns lie between 1.68 and 1.75. Recent periurbanization has generated patterns with low surface area dimensions of 1.39–1.57 and high boundary dimensions lying between 1.76 and 1.87. These patterns are rather fragmented with respect to old industrial cities and confirm the increasing preference of households for living in semirural areas offering peace and quiet with pleasant landscapes, while still within reach of urban amenities. In some sense, the morphological properties of periurban patterns may be seen as resulting from such demand. In the more rural Ardennes region, the surface area dimensions vary between 0.54 and 1.00 and are rather low, but the range of the boundary dimension is very large extending from 1.33 to 1.75. Hence, some of them are rather compact, whereas others look rather tattered. Near Luxembourg, the surface area dimensions are higher (1.00–1.23) as are the boundary dimensions (1.68 and 1.75), and so they are reminiscent of the periurban periphery of Brussels.

2.6 Fractals for Sustainable Planning

2.6.1 The General Concept

The previously presented results make it obvious that the spatial organization of urban patterns is rather consistent with fractal order principles. This type of spatial organization is in part linked to planning concepts as shown, say, for New Towns. But usually urban fabrics arise from highly complex interactions among various types of agents such as politicians and planners but also developers and landowners, who react to social demand. Hence, urbanization is at least in part a self-organizing process.

Without going into details, this leads us to consider briefly the socioeconomic processes contributing to what is usually called urban sprawl. We should be aware that this phenomenon is not just due to lower lot prices in the periphery than in city centers. As already pointed out, households choose these areas since they want to flee urban density and prefer to live in individual houses surrounded by a garden and to enjoy a green and quiet environment. In France, for example, a survey conducted in 2007 by the *Département Stratégies d'Opinion de l'institut d'études marketing et d'opinion TNS Sofres* (Gault and Bedeau 2007) revealed that 56 % of French households want to live in detached individual houses surrounded by a large garden. Individual houses belonging to housing estates are preferred by 20 % and semidetached housing in an urban environment by 11 %. It is obvious that this residential choice behavior tends to generate diffused settlement patterns where residential areas are localized far away from jobs as well as from retail centers and services. This increases the number and the length of commuting trips and hence energy consumption and pollution. The negative impact of this evolution has essentially been made evident by Newman and Kenworthy (1989). Moreover, natural areas risk becoming ever more fragmented, so lowering biodiversity.

This prompted numerous authors to recommend a return to compact cities in order to limit urban sprawl. However, even if this lifestyle contributes to urban sprawl (Von Hoffman and Felkner 2002) such households will reject densifying (Breheny 1997; Gordon and Richardson 1997; Fouchier 1995).

Investigations showed that households integrate a couple of criteria concerning accessibility to different types of sites frequented when choosing their residence. Brun and Fagnani (1994) and McDowell (1997) showed that they try to minimize the distance or the travel time for acceding to their jobs, but also to retail centers (Lerman 1976) and, too, to leisure areas (Guo and Bhat 2002). Hence, densifying residential zones may prompt households to move to lower density areas (Schwanen et al. 2004). Moreover, policies favoring the compact city have turned out to be less efficient than expected. They induce an increase in housing costs and traffic congestion and reduce the accessibility of leisure areas (Breheny 1997). As observed by Levinson and Kumar (1994), trip time does decrease with population density but increases again when a certain threshold is exceeded.

Investigations for Walloon communes have shown that, where substantial proportions of households are dissatisfied with their immediate environment, D values are commonly close to 2 (Thomas et al. 2008b), whereas communes with large percentages of very satisfied households have a high variety of D values (from 0.5 to 1.8). Of course, there is no simple relation between the fractal shape of the urban fabric and household satisfaction since a large number of criteria come into play and there is considerable diversity in housing environments and tastes. However, in two papers on economic residential choice modeling, we explored the impact of Sierpinski carpet-like urban patterns on household satisfaction. It turned out that, for households which have preferences for both "urban amenities" and "green amenities," such a model provides advantages that can be measured by market-induced lot prices (Cavailhès et al. 2004).

We showed, too, for a teragon, that the mean minimal distance from all sites for acceding to the boundary falls to 56 % when passing from a square to the first iteration step and is reduced to 91.5 % for the transition from the first to the second step. The mean distance to the center increases, of course, but for the first step, we obtain 112.5 % with respect to the initial square and 104 % for second step with respect to the first one (Frankhauser 2000). Hence, the "loss of centrality" is less important than the gain in acceding to the urban boundary.

These reflections prompted us to ask whether the emerging fractal order principle cannot be made operational for managing urban sprawl, which means that instead of rejecting the non-compact shape of urban patterns, we may use this approach to structure urban fabrics. The goal is to develop a planning concept that takes account of household preferences but that reduces energy consumption and prevents fragmentation of both built-up and open landscape. Hence, one of the most crucial objectives is to improve accessibility to the different kinds of sites residents frequent more or less regularly. On a local scale, we saw that a fractal shape of an urbanized area can provide leisure areas in the neighborhood of residential zones thereby satisfying residents' desired quality of life. But one of the most criticized impacts of urban sprawl is that urbanization increasingly affects rural areas far from jobs and shopping facilities and so generates large traffic flows.

Returning to urban history, we must be aware that, after the industrial revolution, market-determined growth generated irregular urban patterns. However, during the "trolley period," i.e., before motorization, urban growth remained restricted to ribbon-like development beside public transportation routes. Particularly in Northern metropolitan areas, planning strategies deliberately oriented urbanization in this direction on the scale of metropolitan areas. Let us cite as examples the

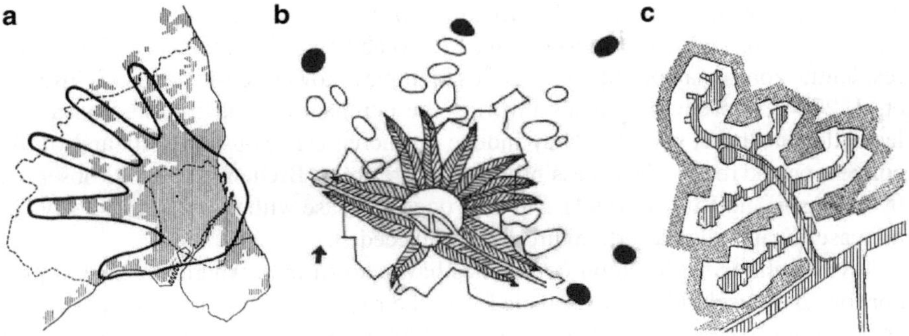

Fig. 2.10 (**a**) The Copenhagen finger plan (Source: Städtbauliches Institut Universität Stuttgart). (**b**) Schumacher's palm plan for Hamburg (Source: Güldner 1968). (**c**) A proposal by the architect Schöfl for a "fractal" town outline (Schöfl 1986)

Table 2.3 Change in the fractal surface area dimension of Basel

Basel	Central city cluster	Agglomeration
1880	1.42	–
1950	1.68	1.64
1990	1.7	1.74

famous Copenhagen finger plan (Fig. 2.10a) or the Schumacher plan for Hamburg (Fig. 2.10b). The same logic inspired already Eberstadt et al. (1910) for developing the Berlin metropolitan area.

But in many cases, we observe that in the course of urbanization, the fractal dimension increases on the scale of metropolitan areas (see the example of Basel, Table 2.3). This means that interstitial space between the main transportation routes is increasingly filled since motorization provides good accessibility even to more isolated villages.

This contributes not only to lengthening the daily commuting trips but is one of the reasons for the deterioration of natural resources. Moreover, improving the street network contributes largely to land consumption. For example, on the outskirts of the urbanized area of the Paris metropolitan area, within the period of 1987–1997, only 1.4 % of space was consumed by new residential and mixed use constructions, whereas 50 % of space was used for improving the road network (Tourneux 2006).

Let us remind that dimensions close to $D = 2$ refer to a uniform distribution of build-up space whatever the density, whereas lower fractal dimensions correspond to higher local concentrations of build-up space across scales like in the example of Figs. 2.2 and 2.4. Hence, we may summarize that on the scale of agglomerations, lower fractal dimensions of build-up space correspond to a lower degree of sprawl since buildings are locally more concentrated and do not fill up space in a diffuse more or less uniform way.

Chen and Feng (2010) showed that in Chinese agglomerations, fractal dimensions increases sometimes rather dramatically since the 1980th, and they link this phenomenon to the rather top-down planning strategy in China in opposition to urban dynamics in Western countries more seen as resulting from self-organization processes. The discussed empirical results show that weaker planning strategies tend, too, to generate more uniform patterns, but in Western countries, this homogeneity is linked to low density, contrarily to Chinese cities. This shows again that fractal dimension provides different and indeed complementary information about spatial organization to density.

Indeed, the hierarchy of lacunae, typical of fractals, at the metropolitan scale saves large zones from urbanization by concentrating urbanization along development routes, thus recalling the Northern countries' planning concept. However, the fractal planning concept goes beyond this, as shown in Fig. 2.11, where we illustrate the concept by comparing the Sierpinski carpet already introduced to a scheme resembling the finger plan. We assume that in both systems the axes correspond to a transportation network and that urbanization follows these routes.

In both the patterns b and c, the total network lengths are the same. Even if both the networks follow a radial-concentric logic, the density of network (Fig. 2.11b) decreases continuously when moving away from the center. Network (Fig. 2.11c) is a Sierpinski gasket which follows the same fractal logic as that of the Sierpinski carpet of Fig. 2.3c. It fills space in a diversified and more contrasted way than network (Fig. 2.11b). Zones around nodes, even in periphery, are densely covered by network branches, whereas other zones, of different size, are not served. This corresponds to the underlying hierarchical organization. Hence, the large square-like zones on the edges of the figure can be thought of as large natural reserves or rural areas. As illustrated for the upper left part of scheme (Fig. 2.11c), these large zones are connected to smaller squares, so forming a connected spatial system of green areas penetrating into urbanized zones. These green corridors have the double function of providing leisure areas close to residential areas and of guaranteeing good "ventilation" of the city center. But the concept avoids the damaging effects of landscape fragmentation for biodiversity. It should be emphasized that the scaling principle of fractals extends the idea of articulating green space and urbanized areas to a local scale. Indeed, the observed lengthening of urban boundary, one of the

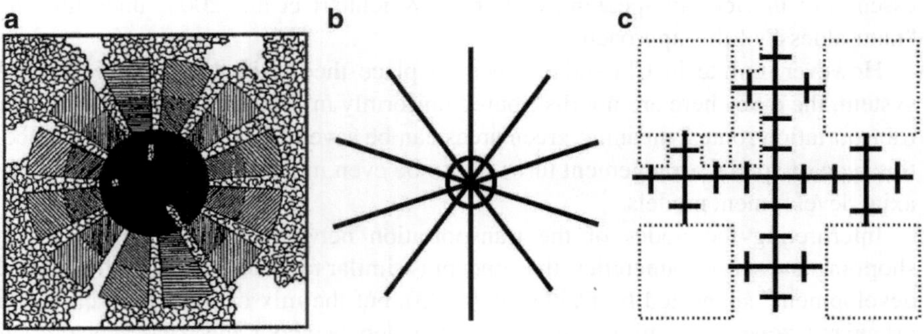

Fig. 2.11 (**a**) The Eberstadt-Möhring-Petersen plan for Berlin (Source: Eberstadt et al. 1910) and (**b**) a non-fractal axial plan, compared to a fractal one (**c**), both having the same total axis length

fractal features of the "sprawling" urban patterns, makes for easy access to open landscape, since green pockets enter into urbanized areas and may be a useful goal for planning (Fig. 2.10c), unlike the usual tendency to round off urban borders as was made evident for the teragon!

Looking at the dimension values for agglomerations (Tables 2.1 and 2.2), we see that fractal dimension of about 1.5–1.7 corresponds to urban fabrics with rather contrasted patterns resembling the constructed fractal of Fig. 2.4b. In Walloon examples, dimensions were even lower.

Comparing schema, Fig. 2.1b–c shows not only that fractal patterns save larger patches from urbanization but illustrates another important feature about accessibility. We may imagine that the nodes of the fractal networks correspond to service centers of various ranks. Hence, all nodes can be considered as providing facilities for daily needs; the four larger peripheral nodes can be interpreted as centers also offering shopping and services amenities for weekly frequency of recourse. Finally, the main center concentrates all types of amenities, also those called on more rarely. Indeed, the less often amenities are frequented; the better users accept long trips. This type of reasoning is reminiscent, of course, of central place theory. Different authors emphasize the importance of developing secondary centers or polycentric urban networks (Fouchier 1995). This is, e.g., discussed in the frame of Calthorpes' concept of "New Urbanism" (Calthorpe 1993) in order to minimize trip lengths. Of course, investigations have shown that consumers choose not always closest shopping facilities (Clark 1968) but this is also due to the fact that car accessibility has been improved over a long time and fuel costs were low. This is why we introduce in the following a poorly linked public transportation network avoiding direct concurrence between shopping areas providing the same type of offer. Of course, street network should be conceived in the same sense avoiding high-quality links between subcenters of same level.

Central place theory has also been criticized for other reasons often referring to real-world situation which seems in many cases not in coherence with this model (Berry and Pred 1961). But this does not exclude that for planning, the basic principle of a planning model based on a hierarchical principle taking into account the frequency of recourse can be useful and remains a reference in planning essentially in German speaking countries. Weichhart et al. (2005) underline the "naturalness" of this approach.

However, unlike in Christallers' central place theory, in the proposed spatial system, the cities here are not distributed uniformly in space but concentrated near transportation routes, meaning green areas can be saved from urbanization. Hence, this type of spatial arrangement turns out to be even more efficient than the purely axial development models.

Interpreting the nodes of the transportation network as sites concentrating shopping and service amenities, the concept is similar to that of the "transit-oriented development" suggested by Calthorpe (1993), but the mix of open landscape and urbanized zones is reminiscent also of the debate about the *Zwischenstadt* of Sieverts (1997) or the reflections of Dubois-Taine and Chalas (1997) about the *ville émergente*.

2.6.2 Applying the Concept

Of course, the Sierpinski carpets presented are not directly suitable for concrete applications. We have explored different approaches which all take advantage of the possibility of changing the position of elements in a fractal without modifying its fractal properties. Let us emphasize that the objective is not to radically change the urban fabric but to provide support for further development in a given situation. For this aim, planning rules are introduced on various topics:

- Fractal standards and supplementary morphological standards like the articulation of built-up space and open landscape across scales
- The evaluation of accessibility of urban amenities (retail centers, services) and green amenities (leisure areas)
- The potential of city size development (rank-size distribution of cities)
- Natural and environmental constraints or recommendations

Two main approaches have been developed and are currently being tested. We present first the approach implemented in the MUP-city simulation tool for developing scenarios for further urbanization but which is more suitable for the more local scale of periurban communes. Then, we discuss the Fractalopolis approach which is better adapted when considering the scale of metropolitan areas.

2.6.2.1 MUP-city

This concept has been developed in cooperation with Cécile Tannier and Hélène Houot (both members of the Théma research institute) as part of a research project of the PREDIT 3 program financed by the French Ministry of the Environment and Sustainable Development and the Environment and Energy Management Agency (ADEME) (Tannier et al. 2010; Frankhauser et al. 2011; Frankhauser 2013). A software tool has been developed by Gilles Vuidel for developing scenarios for further urbanization.

It has been applied to several periurban zones of Besançon, a medium-sized city in the east of France, and modified versions are currently being tested for the Luxembourg area. Moreover, it is used for simulating local scenarios in the framework developed with the enlarged model, Fractalopolis, presented below.

The approach is also inspired both by the grid analysis mentioned in Sect. 2.4.1 and by the iterative mapping procedure used for generating the Sierpinski carpet in Fig. 2.3. We assume that a square-like zone is selected for exploring possible future urbanization. We apply a procedure we call "fractal decomposition." The zone is covered by nine grid squares of equal sizes. The user now sets, as with a generator, how many grid squares in which building is authorized. Let us assume he chooses as fractal standard $N_{\text{norm}} = 6$. Now, the grid squares already containing buildings are identified. Let us assume N_{emp} grid squares contain buildings. If $N_{\text{emp}} \geq N_{\text{norm}}$, then no additional cells can be opened to urbanization. We may now either exclude $N_{emp} - N_{norm}$ grid squares from further urbanization or accept urbanization, but, in that case, applying the defined standard again for the following steps. It is obvious that for shrinking cities, we can define which and how many grid squares can be reconverted into green areas. However, if $N_{emp} < N_{norm}$ an amount of $N_{norm} - N_{emp}$, grid squares can be chosen for urbanization. In the next step, each of these grid squares is divided into nine smaller squares, with sides one-third of the length of the size of the initial ones (Fig. 2.12). In each of the grid squares retained for urbanization in the first step, we look again among the nine smaller grid squares to see which of them contain buildings. Of course, since our grid squares are smaller, we will again find empty ones within the larger grid squares occupied at the previous step.

The grid squares excluded from urbanization at a certain step are never again considered in further steps, so stringently respecting the lacunal hierarchy of fractals. By choosing a standard N_{norm}, we define, of course, a fractal dimension since the reduction factor is set at $r = 1/3$. Hence, the fractal dimension D_{norm} becomes a *multiscale land occupation index* and hence a planning standard. It describes how the built-up area is concentrated in space across scales: if D tends to two, the built-up mass is uniformly distributed, but the lower the value of D, the more the built-up area is locally concentrated (Table 2.4).

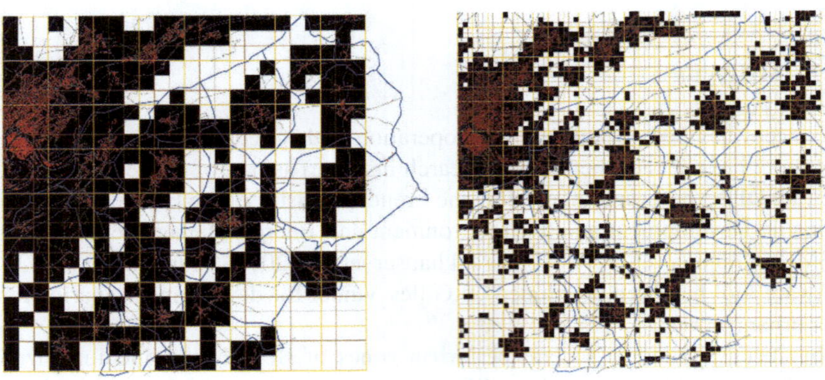

Fig. 2.12 Two successive steps of decomposition used for developing planning scenarios with MUP-city (Source: Frankhauser et al. 2008)

Table 2.4 The relation between occupied grid squares and fractal dimension in MUP-city

Number N of occupied grid squares at each scale	Fractal dimension D
4	1.26
5	1.46
6	1.63
7	1.77

The fractal dimension is not the only criterion used for designing development scenarios. We check also that

- Green corridors remain connected up at the microscale.
- Urbanized grid squares remain connected in order to avoid fragmentation.
- Each urbanized grid square has adjacent non-urbanized grid squares.

In the MUP-city development tool, these rules are strict and the software does not allow them to be broken.

2.6.2.2 Fractalopolis

We now come to the second approach used for planning purposes named Fractalopolis. The geometric concept, the underlying central place system with its assigned coding system, as well as the population model were developed as part of the "Vilmodes" project of the PREDIT 4 program financed by the French Ministry of the Environment and Sustainable Development and by the Environment and Energy Management Agency (ADEME) (Frankhauser 2012). The model was tested for the first time for the Vienna-Bratislava metropolitan region in the frame of the PhD thesis written as part of a collaboration project with the Technische Universität Wien (Czerkauer-Yamu 2012). For this aim, a computer application was developed by Gilles Vuidel (Thema) which can be used to design scenarios for real-world applications. Based on previous work (Tannier et al. 2012), the rules for acceding to facilities have been enlarged and adapted to the Vienna case, and specific morphological rules have been introduced. The Fractalopolis approach is currently being adapted to the Besançon agglomeration and the Lyon metropolitan area.

The spatial modeling approach is based on a multifractal model, and it too refers to an iterative mapping procedure, but combines different reduction factors, e.g., r_0 and r_1 in Fig. 2.13. This allows the creation of urban zones of different sizes which

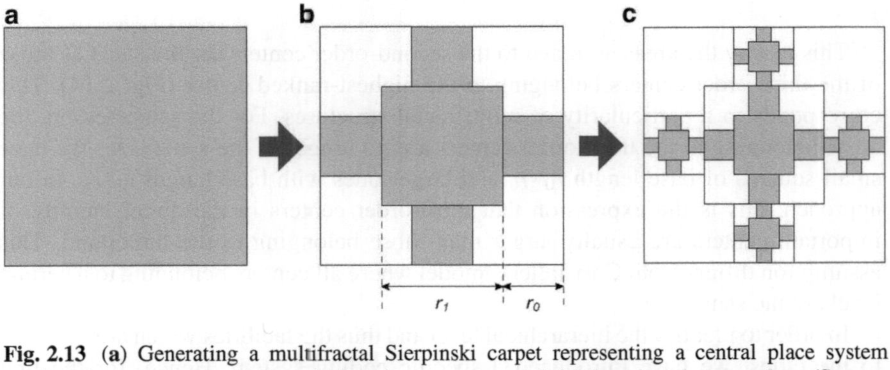

Fig. 2.13 (**a**) Generating a multifractal Sierpinski carpet representing a central place system starting from a *square*, (**b**) introducing the generator, and (**c**) iterating the procedure (Source: Frankhauser 2012)

can be used to generate a hierarchical city system in accordance with the central places approach (Frankhauser 2012).

Starting from a square-like area, we introduce a generator as illustrated in Fig. 2.13. Of course, the reduction factors can be chosen arbitrarily allowing the biggest square to be adjusted to the size of the largest center offering the whole range of facilities. The smaller squares correspond to second-order cities which we assume do not provide the highest level of facilities. Depending on the real-world situation, more or fewer than four centers may be introduced. Moreover, the position of all the squares can be chosen freely, the only restriction is that the squares are not allowed to intersect and they must lie within the initially given square. Hence, the squares tend to be centered on already existing cities. Moreover, natural and environmental constraints can be respected which generally condition urban development (Mohajeri et al. 2013).

We see that, in this concept, the logic is not to cover zones containing built-up space but to define from the outset the areas we wish to develop. By this logic, we accept that settlements lie in the residual zones, i.e., the "lacunae." These zones are interpreted as rural zones. The iteration proceeds by replacing each square by a smaller replication of the generator.

Hence, areas for urbanization are even more concentrated in zones which we assume are served by public transportation networks. This prompts us to admit that, in the zones cut off at this step, a low level of development is possible, thus weakening the strong fractal model.

We see that by iteration, the reduction factors r_1 and r_0 are now combined according to all possible permutations, which yields, for example, for the second step:

$$r_1 \cdot r_1, \quad r_1 \cdot r_0, \quad r_1 \cdot r_0, \quad r_0 \cdot r_0 \tag{2.10}$$

Since permutations are allowed, we have a degree of "degeneration," since

$$r_0 \cdot r_1 = r_1 \cdot r_0 \tag{2.11}$$

This is why the areas assigned to the second-order centers are the same as those of the third-order centers belonging to the highest-ranked center (Fig. 2.14). This corresponds to a particularity of multifractal structures. For the same reason, the areas belonging to the third-order centers are no longer of the same size. We have small squares of base length $r_0 \cdot r_0$ and larger ones with base length $r_0 \cdot r_1$. In our approach, this is the expression that third-order centers in the direct vicinity of important centers are usually larger than those belonging to the hinterland. This assumption differs from Christallers' model where all centers belonging to a certain level are the same size.

In order to identify the hierarchical level and thus the facilities which are assigned to the cities, we have introduced a specific coding system. Hence, for the first iteration, we distinguish the large central square which we denote by the digit 1

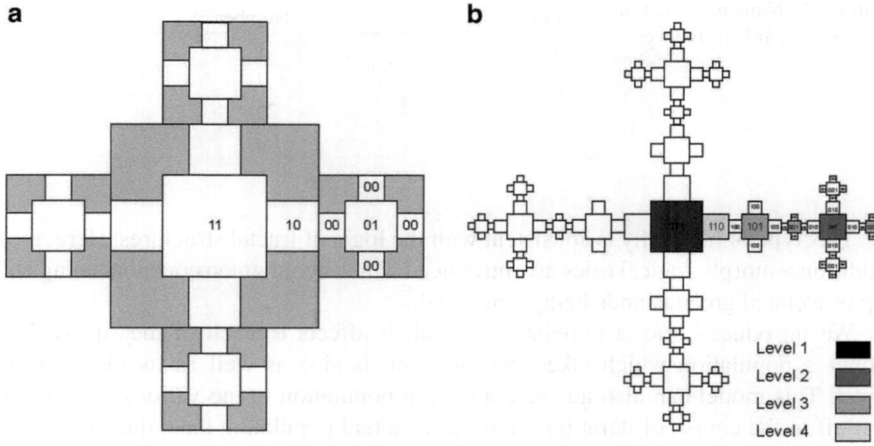

Fig. 2.14 The coding system of the central place system. (**a**) Shows the system at first iteration (*gray squares*) and the second one (*white squares*) (**b**) corresponds to the third step and illustrates the underlying hierarchy of service levels (for details, cf. text) (Source: Frankhauser 2012)

and the four smaller peripheral squares denoted by 0. In each following step, we add another digit to the right of each digit, by the same logic. Hence, the hierarchy is created just by combining two factors. In the next step, the highest-order central square is now called 11 and the four smaller adjacent ones 10. The four peripheral squares generated in the previous step are replaced, too, by the generator. The central place is called 01 and the four peripheral ones 00 (Fig. 2.14a). This procedure is reiterated at the third step (Fig. 2.14b). We then obtain a set of eight different codes, each consisting of three digits. The first level center with the highest facility level $m = 1$ has the code 111. The four directly adjacent squares of level $m = 2$ have the codes 110. They correspond to suburban areas of the main center. The four centers 011 correspond to the four centers of level $m = 2$ generated at the second iteration step and correspond to centers of the facility level $m = 3$. According to the logic of iteration, we assume that higher level facilities are provided by the center 111 for the 101 centers, whereas the second level centers 011 provide the same type of facilities for the centers 001. The small elements 100 and 000, adjacent to these third level centers, are all low level centers $m = 4$ (Fig. 2.14b). By introducing these codes, we have given up the previously discussed commutativity. Indeed, in the system introduced, the codes 101 and 110 or 011 are not equivalent, even if the areas are of the same size. Hence, the code introduces a noncommutative operation. Consequently, the system displays certain properties corresponding rather to properties of unifractals than to multifractals, which corresponds rather to the hierarchical city system assumed.

Hence, making abstraction of their size, we confirm that the total number of centers belonging to the different levels obeys a geometrical series, except for the transition from the highest to the next level (Table 2.5).

Table 2.5 Number of central places by their facility level

Level	Number
1	1
2	4
3	20
4	100

This type of hierarchy is consistent with the logic of fractal structures. Here, too, additional morphological rules are introduced which avoid green corridors being cut up or isolated green islands being generated.

We introduced, too, a population model. It affects to each of the square-like zones a population which takes into account its size as well as its hierarchical level. This model can also attribute a certain population to the various rural areas cut off in the course of iteration. Starting from real population data, the simulation tool identifies the population in the various zones defined by the simulation and computes the mean population for each hierarchal level. The simulator shows for each city the surplus or deficit with respect to this mean number. Moreover, the model can define distribution laws according to the underlying hierarchical logic and hence propose population development scenarios. Details are presented in a working paper (Frankhauser 2012).

Figure 2.15 shows the different steps for a simulation of the metropolitan area of Lyon. Here, the colors refer to the population surplus or the deficit with respect to the mean values for the different hierarchical levels.

2.6.3 The Accessibility Rules

For both planning concepts presented here, the accessibility to different types of amenities frequented by residents is one of the fundamental aspects. Indeed, as pointed out, the accessibility to different types of amenities offered to the inhabitants is fundamental in order to propose planning scenarios that can reduce the total length of trips. Moreover, it is possible to clearly identify accessible sites which could be developed in the future. Both "urban" (goods and services) and "rural" amenities (green leisure areas) are considered. By the underlying central place logic, which appears explicitly in the Fractalopolis coding system, different levels of amenities are distinguished depending on how often residents use them. We present here the methodology developed for measuring accessibility to retail and service centers (Frankhauser et al. 2011). For a medium-sized agglomeration, we distinguish three service levels:

- *Level 1* (monthly use of less occasionally attended places such as medium and large specialized stores, town halls, banks/insurance companies, restaurants, cinemas, theaters, etc.). Since it is unlikely that new centers of this level will be

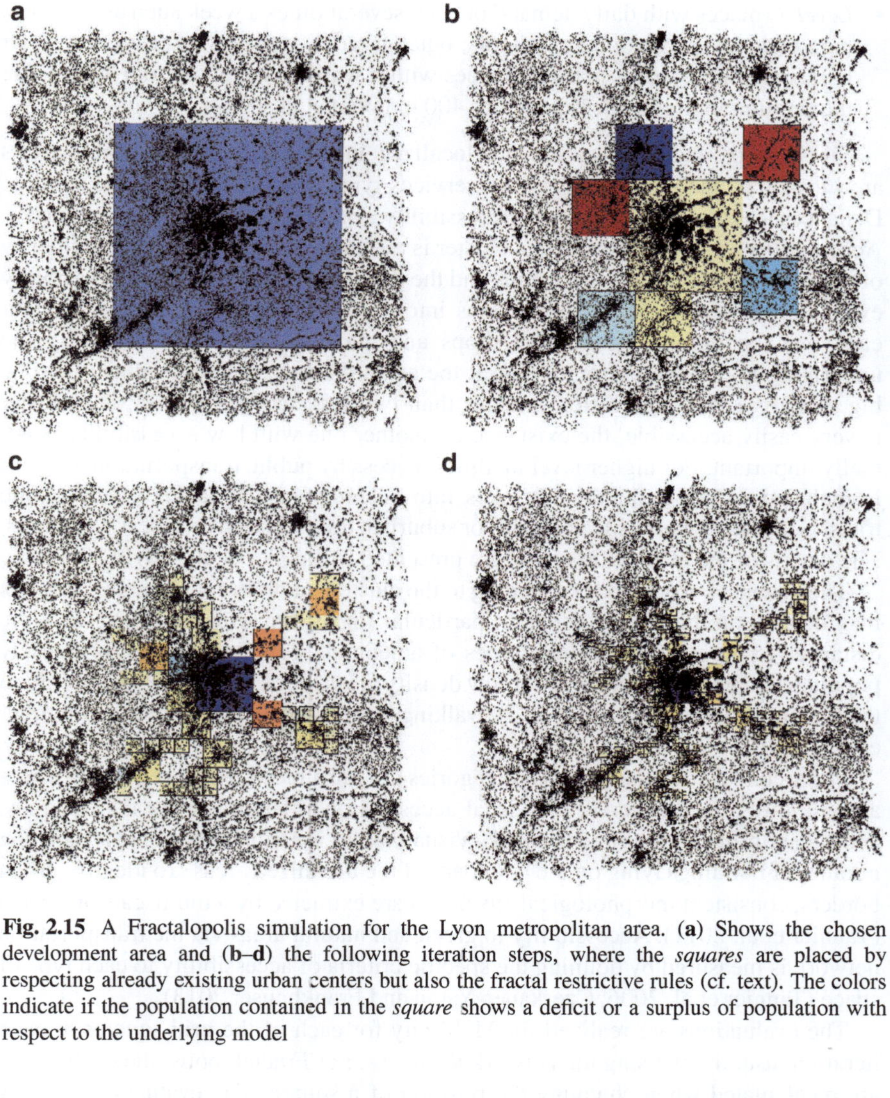

Fig. 2.15 A Fractalopolis simulation for the Lyon metropolitan area. (**a**) Shows the chosen development area and (**b–d**) the following iteration steps, where the *squares* are placed by respecting already existing urban centers but also the fractal restrictive rules (cf. text). The colors indicate if the population contained in the *square* shows a deficit or a surplus of population with respect to the underlying model

created, the distance to these centers is crucial and limits sprawl; however, until now no precise limiting distance has been introduced. For metropolitan areas, level 3 will be split into two subclasses.

- *Level 2* (weekly use of places such as markets, automobile repair shops, gas stations, cafés, hypermarkets, family doctors, etc.). Since the literature provides sparse information about distances to these services, the maximum distance within the given area is identified for each type of service which is used to define the range of distances accepted for that type of amenity.

- *Level 3* (places with daily demand or used several times a week such as bakeries, cigarette/newspaper outlets, schools, butcher's/delicatessens, grocery stores). It is assumed here that it is the distances within walking distance from home. The maximum range considered is set to 400 m (Wiel et al. 1997).

Since shops and services are not localized in the same place, retail clusters are introduced including all shops or services lying within 200 m of one another. Distances less than 200 m improve accessibility, whereas greater distances reduce it. Accessibility from a given site to a cluster is evaluated by combining the frequency of use of the different services/shops and the distance on the network. Accessibility evaluation, based on fuzzy logic, takes into account all existing clusters and for each one the number of services/shops and the diversity of the offer. By the underlying logic, two clusters lying at the same distance are better than one; one big, readily accessible cluster is better than two less accessible ones; if a cluster is very easily accessible, the existence of another one with low accessibility is not really important. For higher level facilities, access by public transportation network is included in the evaluation. It takes into account the timetables of the public transportation networks such as buses or suburban trains as well as access to stations. The details of the evaluation model are presented in Tannier et al. (2010, 2012).

The distances accepted for acceding to the clusters of different hierarchical levels may depend on the local context, in particular for higher level facilities, since the distances to cities offering these types of amenities are not the same for densely populated metropolitan areas as for low density areas. For evaluating accessibility to the facilities for daily needs (level 3), walking or cycling distances are used (Tannier et al. 2006).

For green leisure areas, similar categories are introduced. However, two types of accessibility are considered, the visual access to open landscape for the buildings and accessibility by the road network. Visual access is evaluated by considering the number of buildings lying on the boundary of the urbanized areas. To identify urban borders, consistent morphological envelopes are extracted by a multiscale approach (Tannier et al. 2011). Accessibility to green and natural areas via the transportation network is measured by defining the specific criteria of accessibility to each type of space (Tannier et al. 2006; Czerkauer-Yamu and Frankhauser 2013).

The evaluations are realized for MUP-city for each of the grid squares at each iteration, usually by using the network distances. For Fractalopolis, the evaluations are recalculated when changing the position of a square. The evaluation result is coded by a color which is assigned to the square.

2.7 Conclusion and Outlook

We presented a couple of results about how fractal geometry allows better understanding spatial organization of urban fabrics. We saw that fractal approach suits indeed rather well for describing spatial organization of urban fabrics even if their shapes seem to be irregular. That such patterns are perceived as irregular is rather due to the fact that they follow another type of spatial organization principles as the usual Euclidean references like smoothness of borders, uniform distribution of constitutive elements like buildings, etc. Urban fabrics seem in many cases made up of rather clearly organized town sections which form a complex patchwork. Historical contexts and particular planning concepts may highly contribute to the type of fractal behavior observed. Hence, planning strategies as well as self-organization processes issued from social interactions or trends in society contribute to an emerging spatial order principle.

This led us to reflect about the interest of using fractal approach for developing planning strategies useful in the context of Western countries in order to manage urban sprawl by respecting nevertheless social demand often favoring living in the countryside. Simulation tools have been developing allowing to apply the concept to real-world cities.

Further research will focus on deepening the comparison between various methods of measuring fractal properties of urban patterns including 3D data about buildings in the analyses. Different types of visualization of the results will also be tested with a view to making the results accessible to a wider audience.

For the planning concept, the question of the catchment areas of different types of services necessary to survive will be considered in more detail and included in the evaluation as well as the localization of jobs.

References

Ariza-Villaverde AB, Jimenez-Hornero FJ, De Rave EG (2013) Multifractal analysis of axial maps applied to the study of urban morphology. Comput Environ Urban Syst 38:1–10

Bairoch P (1985) De Jericho à Mexico. Gallimard, Paris

Batty M (2005) Cities and complexity: understanding cities with cellular automata, agent-based models, and fractals. MIT Press, Cambridge, MA

Batty M, Kim SK (1992) Form follows function: reformulating urban population density functions. Urban Stud 29(7):1043–1070

Batty M, Longley P (1986) The fractal simulation of urban structure. Environ Plan A 18: 1143–1179

Batty M, Longley P (1994) Fractal cities. A geometry of form and function. Academic, London

Batty M, Xie Y (1996) Preliminary evidence for a theory of the fractal city. Environ Plan A 28:1745–1762

Benguigui L, Czamanski D, Marinov M, Portugali Y (2000) When and where is a city fractal? Environ Plan B 27(4):507–519

Berry BJ, Pred A (1961) Central place studies: a bibliography of theory and applications. Regional Science Research Institute, Philadelphia

Bishop C (2006) Pattern recognition and machine learning. Springer, Berlin

Breheny MJ (1997) Urban compaction: feasible and acceptable? Cities 14:209–217

Brun J, Fagnani J (1994) Lifestyles and locational choices—trade-offs and compromises: a case-study of middle-class couples living in the Ile-de-France region. Urban Stud 31(6):921–934

Calthorpe P (1993) The next American metropolis. Princeton Architectural Press, Princeton

Cavailhès J, Frankhauser P, Peeters D, Thomas I (2004) Where Alonso meets Sierpinski: "an urban economic model of fractal metropolitan area". Environ Plan A 36:550–578

Chen Y (2009) A new model of urban population density indicating latent fractal structure. Int J Urban Sustain Dev 1(1–2):89–110

Chen Y, Feng J (2010) Spatiotemporal evolution of urban form and landuse structure in Hangzhou, China: evidence from fractals. Environ Plan B 37:838–856

Chen Y, Wang J (2013) Multifractal characterization of urban form and growth: the case of Beijing. Environ Plan B 40:884–904

Clark WAV (1968) Consumer travel patterns and the concept of Range. Ann Assoc Am Geogr 58(2):386–396

Czerkauer C (2012) Strategic planning for developing sustainable metropolitan areas with a multiscale decision support system. The Vienna case. PhD thesis, Université de Franche-Comté

Czerkauer-Yamu C, Frankhauser P (2013) Development of sustainable metropolitan areas using a multiscale decision support system. Working paper, HALSHS: hal-00837493

De Keersmaecker ML, Frankhauser P, Thomas I (2003) Using fractal dimensions for characterizing intra-urban diversity: the example of Brussels. Geogr Anal 35(4):310–328

Dubois-Taine G, Chalas Y (eds) (1997) La ville émergente. La Tour d'Aigue, Edition de l'Aube

Eberstadt R, Möhring B, Petersen R (1910) Gross-Berlin. Ein Programm für die Planung der neuzeitlichen Grosstadt. Wasmuth, Berlin

Fouchier V (1995) La densification: une comparaison internationale entre politiques contrastées. Les Annales de la Recherche Urbaine 67:95–108

Frankhauser P (1994) La fractalité des structures urbaines. Anthropos, Paris

Frankhauser P (1998) The fractal approach. A new tool for the spatial analysis of urban agglomerations. Population: an English selection, Special issue New Methodological Approaches in Social Sciences, pp 205–240

Frankhauser P (2000) La fragmentation des espaces urbains et périurbains—une approche fractale. In: Derycke PH (ed) Structures des villes, entreprise et marchés urbains. L'Harmattan, Paris, pp 25–54

Frankhauser P (ed) (2003) Morphologie des "villes émergentes" en Europe à travers les analyses fractales. Report, Université de Franche-Comté, UMR 6049 Théma

Frankhauser P (2004) Comparing the morphology of urban patterns in Europe: a fractal approach. In: Borsdorf A, Zembri P (eds) European cities structures insights on outskirts: structures. European Cooperation in the Field of Scientific and Technical Research, Bruxelles, pp 93–103

Frankhauser P (2005) La morphologie des tissus urbains et périurbains à travers une lecture fractale. Revue Géographique de l'Est XLV(3/4):145–160

Frankhauser P (2008) Fractal geometry for measuring and modelling urban patterns. In: Albeverio S, Andrey D, Giordano P, Vancheri A (eds) The dynamics of complex urban systems—an interdisciplinary approach. Springer, Heidelberg, pp 241–243

Frankhauser P (2012) The Fractalopolis model—a sustainable approach for a central place system. Working paper HALSHS: hal-00758864

Frankhauser P (2013) La ville fractale—un concept d'aménagement multi-échelle. In: Brun G (ed) Ville et mobilité—Nouveaux regards de la recherche. Documentation française, Paris, pp 85–99

Frankhauser P, Sadler R (1991) Fractal analysis of agglomerations. In: Hilliges M (ed) Natural structures: principles, strategies, and models in architecture and nature. University of Stuttgart, Stuttgart, pp 57–65

Frankhauser P, Tannier C, Vuidel G, Houot H (2008) Une approche multi-échelle de l'accessibilité pour maîtriser l'étalement urbain, actes en ligne du Colloque mobil. TUM, Munich, 19 p. http://www.bfhz.uni-muenchen.de/cms/upload/downloads/Tagung-TUM-2008-Frankhauser-etal.pdf

Frankhauser P, Tannier C, Vuidel G, Houot H (2011) Une approche multi-échelle pour le développement résidentiel des nouveaux espaces urbains. In: Antoni JP (ed) Modéliser la ville. Forme urbaine et politiques de transport. Economica, Paris, pp 306–332

Gault G, Bedeau L (eds) (2007) Les Français et leur habitat—Perception de la densité et des formes d'habitat, Principaux enseignements du sondage réalisé pour l'Observatoire de la Ville du 10 au 12 janvier 2007. TNS Sofres, Département Stratégies d'Opinion/Société, http://www.observatoiredelaville.com/pdf/Synthesesondage.pdf

Goodchild M, Mark D (1987) The fractal nature of geographic phenomena. Ann Assoc Am Geogr 77(2):265–278. doi:10.1111/j.1467-8306.1987.tb00158.x
Gordon P, Richardson HW (1997) Are compact cities a desirable planning goal? J Am Plan Assoc 63:95–106
Grassberger P, Procaccia I (1983) Measuring the strangeness of a strange attractor. Phys D 9: 189–208
Güldner H (1968) Unsere Stadt – Tragödie einer Spätkultur. Beig-Verlag, Pinneberg
Guo J, Bhat C (2002) Residential location modeling: accommodating sociodemographic, school quality and accessibility effects. University of Texas, Austin
Jiang S, Liu D (2012) Box-counting dimension of fractal urban form: stability issues and measurement design. Int J Artif Life Res 3(3):521–525
Lam N, de Cola L (2002) Fractals in geography. The Blackburn Press, Caldwell
Le Corbusier (1971) La Charte d'Athènes. Entretien avec les étudiants des écoles d'architecture. Éditions du Seuil, Paris, 190 p
Leduc A, Prairie Y, Bergeron Y (1994) Fractal dimension estimates of a fragmented landscape: sources of variability. Landsc Ecol 9(4):279–286
Lerman SR (1976) Location, housing, automobile ownership, and mode to work: a joint choice model. Transp Res Rec 610:5–11
Levinson DM, Kumar A (1994) The rational locator: why travel times have remained stable. J Am Plan Assoc 60:319–332
Mandelbrot B (1982) The fractal geometry of nature. Freeman, San Francisco
McDowell L (1997) The new service class: housing, consumption and lifestyle among London bankers in the 1990s. Environ Plan A 29:2061–2078
McLachlan G, Peel D (2000) Finite mixture models. Wiley, New York
Newman PWG, Kenworthy JR (1989) Cities and automobile dependence: an international sourcebook. Gower, Aldershot/Brookfield
Mohajeri N, French J, Batty M (2013) Evolution and entropy in the organisation of urban street patterns. Ann GIS 19:1–16
Palmer M (1988) Fractal geometry: a tool for describing spatial patterns of plant communities. Vegetatio 75:91–102
Rodrigue J-P (2013) The geography of transport systems. Routledge, New York
Salingaros N (2003) Connecting the fractal city. Key note speech, Biennial of Towns and Town planners in Europe, Barcelona. http://zeta.math.utsa.edu/~yxk833/connecting.html
Schöfl G (1986) Minimalnetze. ARCUS 2
Schwanen T, Dijst M, Dieleman FM (2004) Policies for urban form and their impact on travel: the Netherlands experience. Urban Stud 41(3):579–603
Shen G (2002) Fractal dimension and fractal growth of urbanized areas. Int J Geogr Inf Sci 16(5):519–437
Sievert T (1997) Zwischenstadt. Bauwelt Fundamente 118
Tannier C, Frankhauser P, Houot H, Vuidel G (2006) Optimisation de l'accessibilité aux aménités urbaines et rurales à travers le développement de modèles fractals d'urbanisation. In: XIIème colloque de l'ASRDLF, XIIème colloque du Grerbam, Sfax, Conference proceedings on CD
Tannier C, Vuidel G, Frankhauser P, Houot H (2010) Simulation fractale d'urbanisation – MUP-city, un modèle multi-échelle pour localiser de nouvelles implantations résidentielles. Revue internationale de géomatique 20(3):303–329
Tannier C, Thomas I, Vuidel G, Frankhauser P (2011) A fractal approach to identifying urban boundaries. Geogr Anal 43(2):211–227
Tannier C, Vuidel G, Houot H, Frankhauser P (2012) Spatial accessibility to amenities in fractal and nonfractal urban patterns. Environ Plan B 39:801–819. doi:10.1068/b37132
Thomas I, Frankhauser P, Biernacki C (2008a) The morphology of built-up landscapes in Wallonia (Belgium): a classification using fractal indices. Landsc Urban Plan 84:99–115
Thomas I, Tannier C, Frankhauser P (2008b) Is there a link between fractal dimensions and other indicators of the built-up environment at a regional level. Cybergeo 413:24. http://www.cybergeo.eu/index16283.html

Thomas I, Frankhauser P, Frenay B, Verleysen M (2010) Clustering patterns of urban built-up areas with curves of fractal scaling behavior. Environ Plan B 37(5):942–954

Thomas I, Frankhauser P, Badariotti D (2012) Comparing the fractality urban districts: do national processes matter in Europe? J Geogr Syst 14(2):189–208. doi:10.1007/s10109-010-0142-4

Tourneux FP (2006) L'évolution de l'occupation du sol dans les franges franciliennes: des artificialisations concentrées plus qu'un étalement urbain? In: Larceneux A, Boiteux-Orain C (eds) Paris et ses franges: étalement urbain et polycentrisme. Editions universitaires de Dijon, Dijon, p 101

Vanneste D, Thomas I, Vanderstraeten L (2008) The spatial structure(s) of the Belgian housing stock. J Hous Built Environ 23(3):173–198

Vicsek T (1989) Fractal growth phenomena. World Scientific, Singapore

Von Hoffman A, Felkner J (2002) The historical origins and causes of urban decentralization in the United States. Joint Center for Housing Studies, Harvard University, Cambridge, MA (WP 02–1)

Weichhart P, Fassmann H, Hesina W (2005) Zentralität und Raumentwicklung, Nr. 167. Schriftenreihe der Österreichischen Raumordnungskonferenz, Wien

White R, Engelen G (1994) Urban systems dynamics and cellular automata: fractal structures between order and chaos. Chaos Solitons Fractals 4(4):563–583

Wiel M, Tauty S, Rollier Y, Morvan A, Le Guirriec P, Desse RP, Barthélémy JP (1997) Comportement de mobilité et évolution de l'organisation urbaine (région urbaine de Brest). Etude pour la DRAST, la DTT, l'UTP, le Plan urbain, Agence de développement et d'urbanisme du pays de Brest 2

Chapter 3
Knowledge Discovery in Spatial Planning Data: A Concept for Cluster Understanding

Martin Behnisch and Alfred Ultsch

Abstract The objective of this paper is to present a methodology for discovering comprehensible, valid, potentially innovative, and useful patterns, i.e., new knowledge, in multidimensional spatial data. Techniques from statistics, machine learning, and data mining are applied in consecutive logical steps to allow the visualization of results and the application of validation procedures at each stage. However, the approach does not end with a data cluster; rather, if such a valid cluster has been achieved, then the question is posed: "What do the clusters mean?". Symbolic machine learning methods are employed to produce an explanation of the clusters in terms of rules employing an understandable subset of the high-dimensional data variables. This combined with canonical representatives of a cluster and consideration of the spatial distribution of the clusters lead to hypothesis on emergent data structures, that is, potential new knowledge. The approach is demonstrated on an exemplary data set of German urban districts featuring seven dimensions of land use.

Keywords Knowledge discovery • Data mining • Cluster • Spatial planning

3.1 Introduction

The rapid growth of freely available spatial data and advances in information technology have made an application of the techniques of data mining and knowledge discovery in databases (KDD) (Ultsch 2013; Laube 2011; Guo 2009; Fayyad et al. 1996) both possible and necessary. The goal of this chapter is to present a methodology for applying knowledge discovery to spatial planning data. Here our

M. Behnisch (✉)
Leibniz Institute of Ecological Urban and Regional Development, Weberplatz 1, D-01217 Dresden, Germany
e-mail: m.behnisch@ioer.de

A. Ultsch
Department of Mathematics and Computer Science, Philipps-University of Marburg, Hans-Meerwein-Straße, D-35032 Marburg, Germany
e-mail: ultsch@informatik.uni-marburg.de

M. Helbich et al. (eds.), *Computational Approaches for Urban Environments*, Geotechnologies and the Environment 13, DOI 10.1007/978-3-319-11469-9_3

initial assumption is that data is available without any concrete notion of inherent structure (cf. process of KDD Miller and Han 2009, p. 4).

In many instances when knowledge discovery is applied to spatial planning data, a clustering is more or less the final result of analysis, intended to answer a specific research question (European Spatial Planning Observation Network 2011; Aumayr 2007; Blume and Sack 2010; Demsar 2009; Kronthaler 2005; Hietel et al. 2004; Rasul et al. 2004; Thompson et al. 2002; Qu 2000). The objects of interest can be regions, municipalities, settlement blocks, or raster cells, described by a number of attributes and gathered into uniform clusters. A small number of clusters are extracted from data sets that may contain thousands of individual data objects. The aim is to identify important shared features of the target objects from such huge pools of data in order to provide concrete findings to assist in planning decisions. In the most recent approaches, clusters are described by measures of central tendency, variability, or discriminant analysis (Geyler et al. 2008; Frenkel 2004; Bätzing and Dickhörner 2001; Siedentop et al. 2003).

Here we propose to go a step further. In the presented approach, clustering is merely the starting point for the actual generation of knowledge. Useful clusters are ones that help spatial planners, politicians, and decision-makers in their actions. Therefore, the question "What do the clusters mean?" is addressed using several different approaches involving interaction with a human expert. A special class of classifier generation algorithm from machine learning is applied, with the aim of producing human-understandable characterizations of the classes in the form of decision rules (Alpaydin 2008; Hastie et al. 2009; Izenman 2008; Kuncheva 2004). It should be emphasized that this technique can generate knowledge by investigating the variables previously used for the clustering partitions (intrinsic explanation) or by exploring other variables (extrinsic explanation).

This chapter is structured as follows: Section 3.2 introduces the sample spatial planning data set to provide the basic framework. In Sect. 3.3, the individual steps of the knowledge discovery approach are explained. Finally, section "Conclusions and Future Challenges" concludes with some remarks and addresses future challenges.

3.2 Sample Spatial Data Set

Our approach to knowledge discovery for spatial planning data is demonstrated on a data set describing land use in 111 urban districts (UDs) as a subset of all German districts (Kreise) ($n = 412$). The land use in each UD is specified by seven variables, measured in the year 2010. The data is compiled from the Monitor of Settlement and Open Space Development (Krüger et al. 2013). This is a scientific service operated by the Leibniz Institute of Ecological Urban and Regional Development to provide information on land use trends in Germany (IOER Monitor, http://www.ioer-monitor.de). The advantage of data supplied by the IOER Monitor is the provided classification featuring a wide range of variables (see Land Use Classification) and the explicit spatial reference of the land use categories.

Table 3.1 Variables used to describe the UD data

Label	Measurement	Description	Literature
OpenSpaceMeshSize	Effective mesh size (modified) of open space areas (%)	The larger the effective mesh size in a reference area, the lower the degree of fragmentation of the landscape	European Environment Agency (2011), Moser et al. (2007), and Jaeger (2000)
BuildingArea	Area per resident (m^2/resident)	High indicator values indicate high land consumption for buildings per resident (low efficiency of land use)	Behnisch et al. (2013) and Zentrale Stelle Hausumringe und Hauskoordinaten (2013)
SettlementDensity	No. of residents per settlement area (residents/km^2)	The indicator describes the urban density, based on the settlement area and transport area of a reference area	Siedentop and Fina (2010) and Siedentop et al. (2007)
SealedSurface	Ratio of sealed surface to total area (%)	This is a general ecological indicator to describe various forms of environmental loads associated with settlements, available for the first time in Germany in this form. The base data is a European data set that has only been available for a short period of time and which makes use of ortho-rectified satellite images, analyzed into classes	European Environment Agency (2013) and Siedentop et al. (2007)
LandConsumption	Ratio of settlement and transport area to total reference area (%)	Land consumption encompasses both the loss of agricultural land and natural habitats as well as the expansion of settlement and transport areas	Krüger et al. (2013), Siedentop and Fina (2010), Storch and Schmidt (2008), Siedentop et al. (2007), and Dosch (2001)
ProtectedAreas	Ratio of protected areas to total reference area (%)	This indicator describes the proportion of land in an urban district (excluding municipal coastal waters) designated as protected	Krüger et al. (2013)
HemerobyIndex	Spatially weighted mean of the hemeroby steps of all forms of land use in a reference area [nondimensional]	The index provides a measure of the cultural influence of humans on the natural environment, or an inverse measure of naturalness (1 = no cultural influence, . . . , 6 = strong cultural influence)	Walz and Stein (2014) and Steinhardt et al. (1999)

In Table 3.1, you can find a brief label of the variables, the form of measurement, units used, and a short description. For further information on the variables, see also the literature references on the right. The indicators of the IOER Monitor, derived from standard geo-topographic data sets, are more often reliable than official land use statistics. The latter are currently derived from property registries and are frequently out-of-date, and updated data is only available upon payment of a fee (Meinel 2013).

3.3 Systematic Approach to Knowledge Discovery in Spatial Planning Data

As already mentioned, exploratory techniques such as clustering are often employed in the analysis of spatial planning data. Here we propose to follow a systematic step-by-step approach, starting from the raw data and ending hopefully with the discovery of some new data structures, which can then be subjected to rigorous statistical testing. It should be noted that although the steps are presented here in succession, in practice it is often the case that insights gained at some step of the knowledge discovery process will be used to revise the procedures of previous steps. Therefore, in practice, the process is generally circular or spiral in form, as shown in Fig. 3.1 (Ultsch 2013; Behnisch 2009).

The various steps of the current approach to knowledge discovery for spatial planning data are as follows:

1. **Descriptions: Modeling the distribution of each variable separately**
   - Initial data inspection
   - Exploring the distributions of the individual variables
2. **Structures: Finding structures in high-dimensional space**
   - Looking for correlations of variables

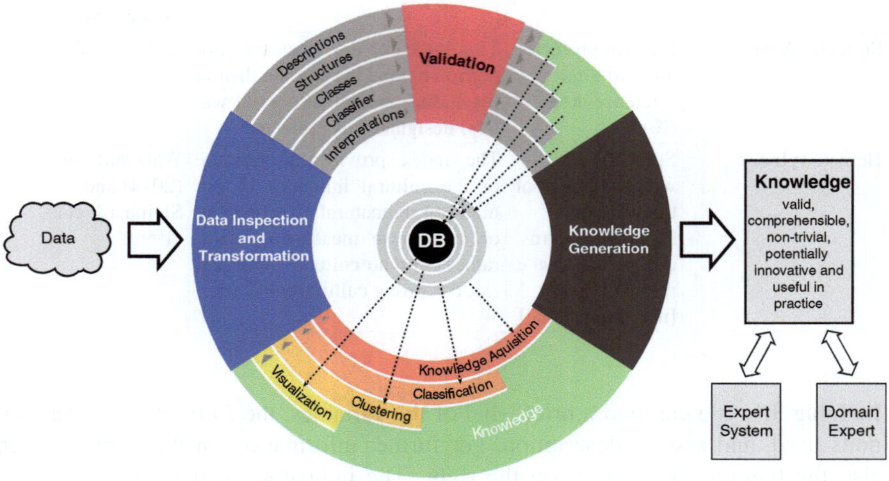

Fig. 3.1 A systematic approach to knowledge discovery

– Definition of a distance measure for the high-dimensional data
– Projection of the data from the high-dimensional data space onto a visualizable space

3. **Classes: Finding intrinsic groups, called clusters, in a data set**
   – Data clustering/unsupervised classification
4. **Classifier: Symbolic classifiers to assist human skills of comprehension**
   – Machine-generated explanations in form of rules or decision trees
5. **Interpretations: Human understanding of clusters**
   – Analysis of the spatial distribution of clusters
   – Mindful translation of machine-generated explanations
   – Cluster labeling/finding spatial abstractions
   – Knowledge generation/domain experts gain new insights
6. **Testing new insights**

We will demonstrate this approach on the UD data set described above. Although this is intended simply as an illustrative example, nevertheless, some results have been obtained. For example, potential subclasses of the variable SealedSurface have been identified. Another piece of knowledge discovered by the presented approach is the identification of two types of German coastal urban districts. Furthermore, it was possible to rediscover a predicted cluster of urban districts characterized by a dense building structure, fragmented open space, and a high degree of sealed surface. Most urban districts of this cluster belong to the official type highly central of the spatial monitoring system of the Federal Institute for Research on Building, Urban Affairs and Spatial Development (2013). The following semantic is suggested for urban districts in this cluster: urban districts, regarding density, ecological impacts of soil sealing, and fragmentation of the urban area.

3.3.1 Initial Data Inspection

The first and most important step in a knowledge discovery process is to gain an initial overview by inspecting the data set as a whole and closely reviewing each variable individually. To gain an overview of the data, a heat map of the entire data set can be made (Wilkinson and Friendly 2009). A heat map displays each data point as an area of colored pixels in a matrix. The presented colors reflect data values. In particular, missing values can be identified by a unique color (white) so that the number and distribution of data gaps can be clearly seen. Figure 3.2 gives such an overview of the UD data scaled to percent and ordered by the official district key (01001 ... 16056). In this data set, there are no missing values, normally coded as "NaN" (IEEE 754-1985). No obvious structures can be identified in this heat map, for example, in the ordering of data.

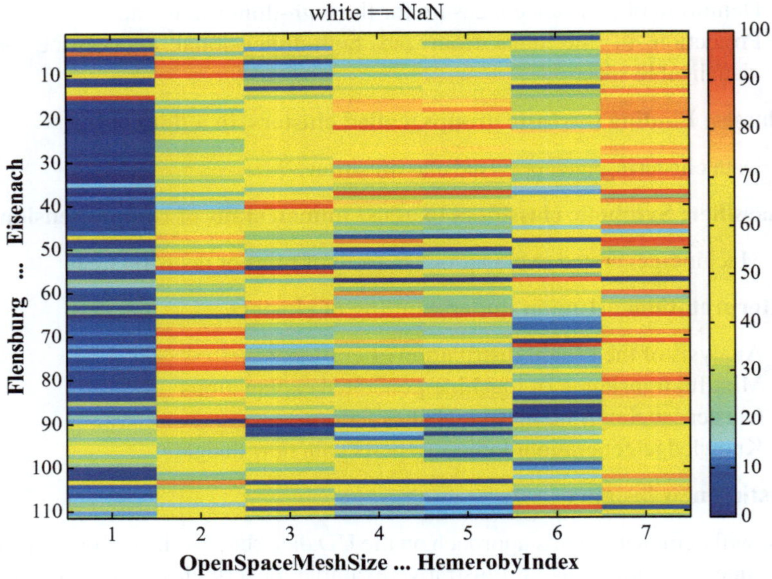

Fig. 3.2 Heat map of the scaled UD data

3.3.2 Exploring the Distributions of the Individual Variables

The next step of data inspection is to determine the distribution of the individual variables. Important tools for this inspection are the quantile-quantile plot (QQ-plot) and kernel estimators for the probability density function (pdf). Here we use the PDE method for pdf estimation (Ultsch 2003) as it is specially designed to uncover subsets in the variables. Consider, for example, the variable SealedSurface. The graph on the left of Fig. 3.3 presents the empirically measured pdf as a blue curve. One can see that the degree of sealed surface in UD data appears to have several subsets: a small proportion of sealed surface vs. medium and higher percentages of sealed surfaces. The black lines show a mixture of Gaussians to model these subset distributions (Bilmes 1998; Dempster et al. 1977). In this way, a close inspection of each variable separately offers some initial insights into the data set. The right panel of Fig. 3.3 gives an indication of the quality of the GMM model. Here the straight line confirms a good data fit to the model.

The main goal of the inspection of the individual variables is to find out how they are distributed in comparison to standard distributions. A QQ-plot of the variable HemerobyIndex versus a Gaussian (N(0,1)) shows that this variable has an approximately normal (i.e., Gaussian) distribution (cf. Fig. 3.4).

The QQ-plots of other variables reveal different types of distribution. While the variable OpenSpaceMeshSize is clearly non-normally distributed, the logarithm of OpenSpaceMeshSize shows nearly a normal distribution (cf. Fig. 3.5).

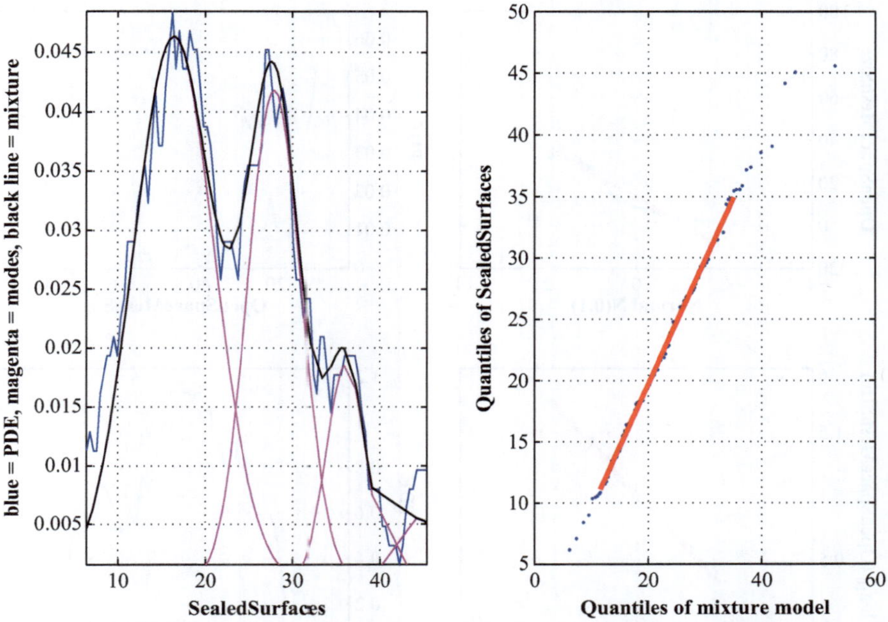

Fig. 3.3 Inspection of the variable SealedSurface using PDE and a suitable GMM (*left*) and a QQ-plot of model vs. data (*right*)

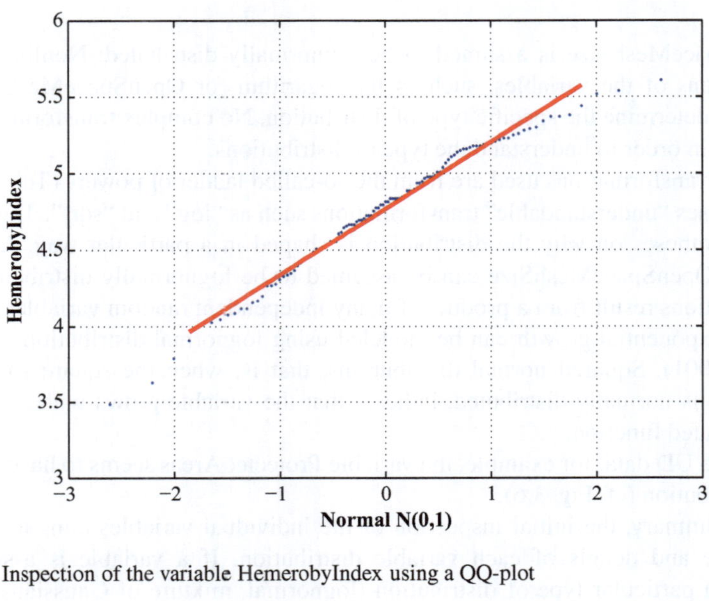

Fig. 3.4 Inspection of the variable HemerobyIndex using a QQ-plot

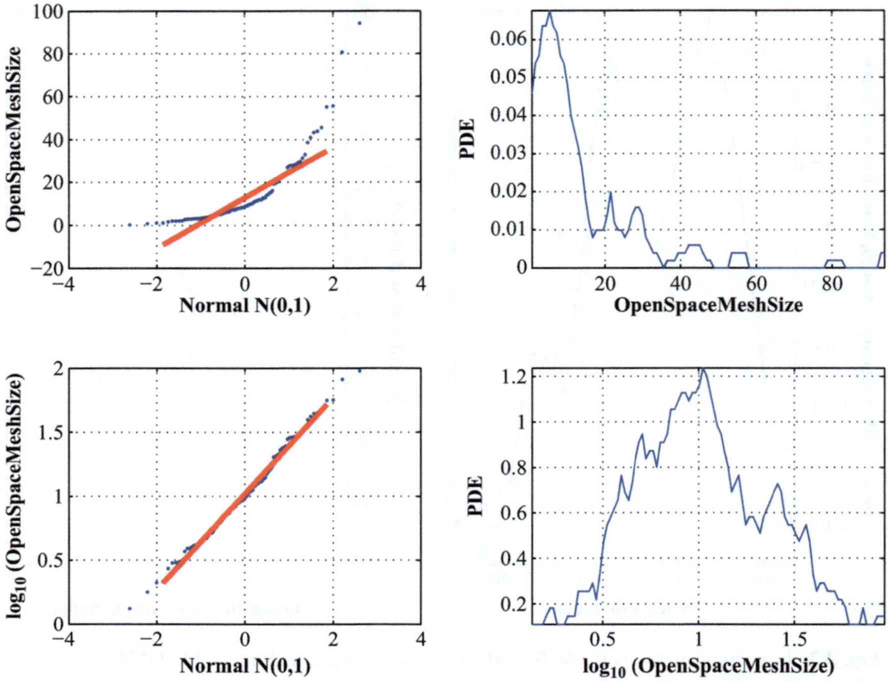

Fig. 3.5 Inspection of the variable OpenSpaceMeshSize using QQ-plot and PDE

OpenSpaceMeshSize is assumed to be lognormally distributed. Nonlinear transformations of the variables, such as the logarithm for OpenSpaceMeshSize, are used to determine the specific type of distribution. No complex transformations are applied in order to understand the type of distributions.

The transformations used are from the so-called ladder of powers (Tukey 1977), which uses "understandable" transformations such as "log" and "sqrt". These allow for hypotheses on why the distribution is shaped in a particular way. As shown above, OpenSpaceMeshSize can be assumed to be lognormally distributed. Such distributions result from a product of many independent random variables; furthermore, exponential growth can be modeled using lognormal distributions (Limpert et al. 2001). Squared normal distributions, that is, when the square root of the variable is normally distributed, indicate that the variable grows with a quadratic area-related function.

In the UD data, for example, the variable ProtectedAreas seems to have this type of distribution (cf. Fig. 3.6).

In summary, the initial inspection of the individual variables aims to discover the type and details of each variable distribution. If a variable is assumed to follow a particular type of distribution (lognormal, mixture of Gaussians, etc.), a first validation should be attempted by applying statistical tests for distributions

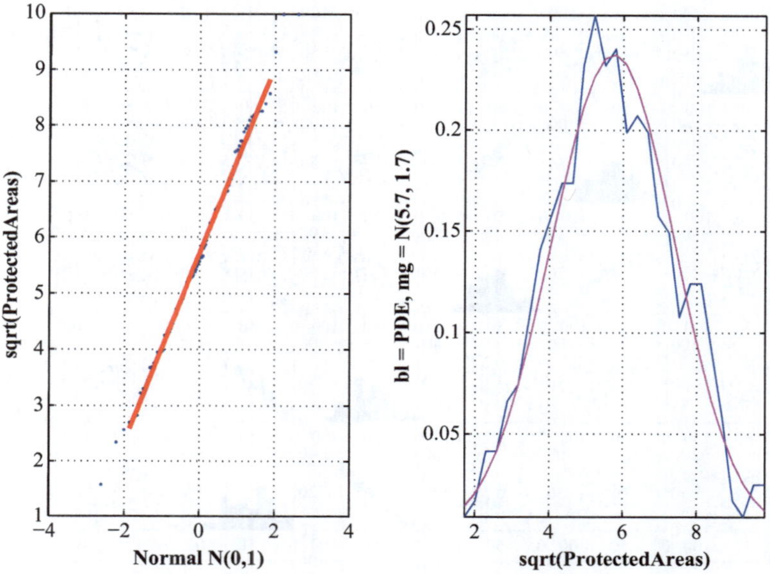

Fig. 3.6 Inspection of the variable ProtectedAreas using QQ-plot and PDE

(Kolmogorov-Smirnov, Chi2, Jarque-Bera, etc.) and/or visual checks such as the QQ-plots (cf. Fig. 3.1, where the validation step is highlighted in red).

Sometimes initial classifications may be discovered by this description step (cf. Fig. 3.1). One example is the detection of possible subclasses of SealedSurface as described above. Another important aim is to identify useful nonlinear transformations, such as log or sqrt in order to enable the comparison of distributions of variables (cf. Fig. 3.9 below). The nonlinear transformations applied to the UD data are given in Appendix 1.

3.3.3 *Looking for Correlation Structures*

After selecting a normalizing nonlinear transformation, it is useful to identify correlations among the variables. Two typical methods for this are scatterplots and the calculation of correlation measures.

If nonlinear transformations are first applied to the data, then linear correlation measures such as the Pearson correlation coefficient can be used. Otherwise rank-based correlation measures such as the Spearman correlation coefficient or Kendall's Tau must be used. Figure 3.7 shows a matrix of all pairwise scatterplots. It can be seen that some variables are highly correlated. For example, SealedSurface and LandConsumption are strongly positively correlated, while BuildingArea and SettlementDensity show strong negative correlation. Figure 3.8 visualizes the Pearson correlation coefficient of the transformed data.

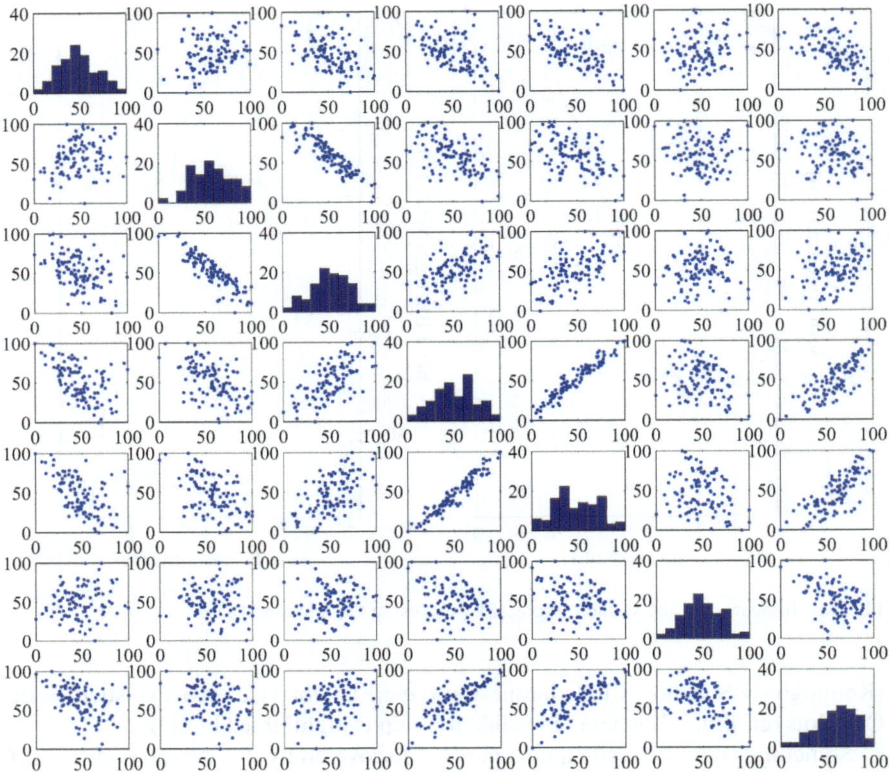

Fig. 3.7 Matrix of all pairwise scatterplots on the transformed UD data

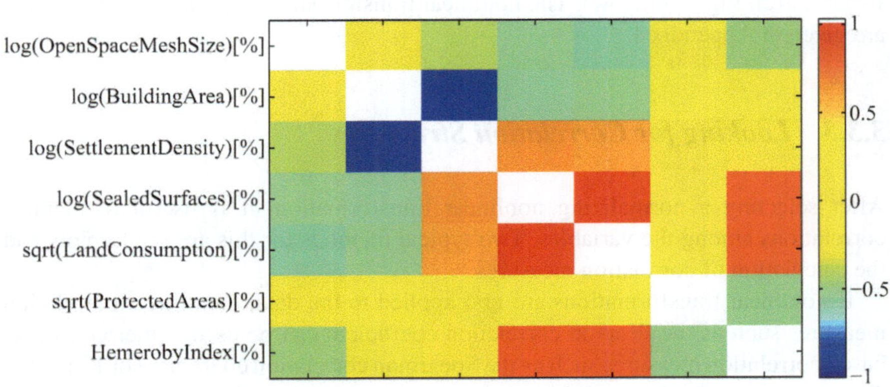

Fig. 3.8 Pearson correlation of the nonlinear transformed UD data

3.3.4 *Definition of a Distance Measure for the High-Dimensional Data*

In order to project the high-dimensional data onto a space which can be visualized or to identify clusters in the data, a meaningful (dis-) similarity measure (data distance) must be defined. The distance measure must be such that similar data are close and differing data distant. Each variable should be well captured by this measure. Two variables with a high correlation represent basically the same information. Thus, if both variables are included in a data distance, the same information is weighted by a factor of 2. One simple approach to address this effect is to remove highly correlated data from the definition of a meaningful data distance.

For the UD data, it makes sense to include only the four variables OpenSpaceMeshSize, BuildingArea, SealedSurface, and ProtectedAreas as the other variables are highly correlated to this subset, so that their information is already contained in the selected variables. Comparison is of the transformed variables. Otherwise, the differences between two data points within a variable would not be comparable (cf. Fig. 3.9). In order to adjust the scaling for the data, all data was rescaled to percent.

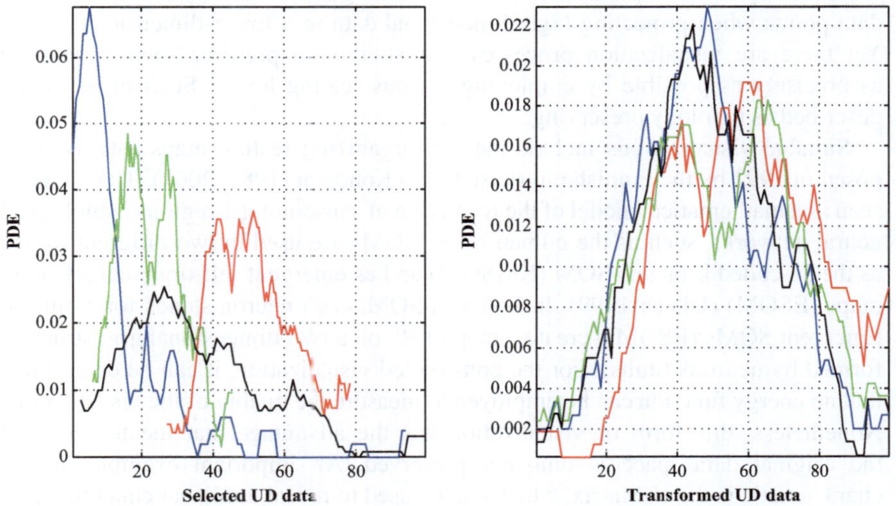

Fig. 3.9 Comparison of the distributions of original (*left*) and transformed variables (*right*). Blue = OpenSpaceMeshSize, red = BuildingArea, green = SealedSurfaces, black = ProtectedAreas

3.3.5 Projection of Data from the High-Dimensional Data Space to a Visualizable Space

In order to detect structures in data, in particular clusters, it is helpful to project structures in the high-dimensional data space (R^n) onto a visualizable space of only two or three dimensions (R^{viz}). Any projection of a high-dimensional data space onto a lower dimensional space cannot preserve all the spatial relationships of the original space. Nonetheless, the projection onto R^{viz} should reflect as closely as possible the distances and clustering of data points in R^n. The visual projection must certainly enable an estimation of relative distances between data points. An overview of projection methods can be found in Hand et al. (2001). The so-called nonlinear projection methods do not strive to preserve linear relationships between projected data and the original data space. Instead, a projection is sought that provides the optimal visualization of the structural characteristics of data. Such structural characteristics can be spatial relationships, as, for example, the occurrence of dense clusters of data. A further important consideration is that nonlinear projection methods should enable a precise illustration of neighborhood relationships existing in the high-dimensional space in the form of similar or identical neighborhood relationships in the projection space.

In principle it is impossible to preserve all neighborhood relationships between data points when projecting high-dimensional data to a lower-dimensional space. Yet there are visualization processes that attempt to preserve such structuring as precisely as possible by employing various scaling levels. Such methods are described as topology preserving.

Suitable visualizations include the self-organizing feature maps (SOMs) proposed in 1992 by the Finnish physicist Teovo Kohonen (1982, 2001). These can be seen as a mathematical model of the formation of sensomotoric regions in biological neural networks, such as the human brain. SOMs are used in two different forms: as the so-called k-means-SOM (KMSOM) and as emergent self-organizing feature maps (ESOM) (Ultsch 1999). In k-means-SOM, each neuron stands for a cluster. Emergent SOMs (ESOMs) create a map of R^n on a two-dimensional grid structure formed by neurons (units). For the constructed visualization, it can be determined that no energy function can be employed to measure the quality of the visualization. Nonetheless, this form of visualization has the advantage that the topology of the original data space is uniquely preserved. An important example of this characteristic is the U-Matrix, which can be used to reveal nonlinear entanglements within Kohonen maps. A typical example is the case of two intertwined toroidal data sets (cf. Figs. 3.10 and 3.11). The Emergent Self-Organizing Map also enhances the investigation of multidimensional spatial objects. An ESOM with 50×82 neurons is trained with the inspected and preprocessed data.

The corresponding U*-Map (Ultsch 2003) (island view, cf. Fig. 3.12) delivers a geographical landscape of the UD data on a projected map (imaginary axis). A clear structure can be easily recognized. The structures are expressed by mountains (displayed on the z-axis), the height of which define the distance between different

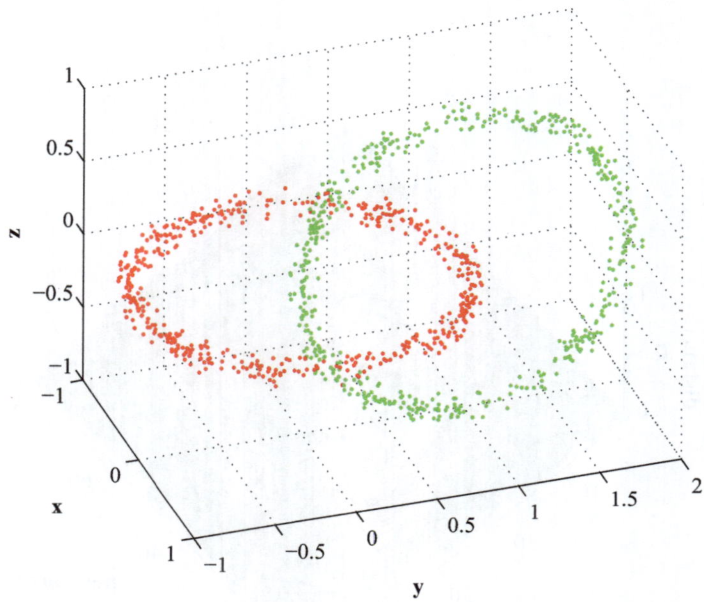

Fig. 3.10 Chain-link dataset

objects. A valley describes similar objects, characterized by small U-heights on the U*-Map. Data points found in coherent regions are assigned to one structure. All local regions lying in the same structure have nearly the same properties. Outlier groups can also be clearly discerned, characterized by small valleys and very prominent boundaries (large U-heights). Close inspection of the U-matrix reveals that two "structures" can in fact be regarded as outlier, having only two members each. The first group of outlier consists of the major cities Munich and Berlin, both of which display a large value for sealed surfaces, a small value for building area per inhabitant, and low value for protected areas. The second group of outlier consists of Suhl and Baden-Baden, two cities which are remarkable for their extraordinarily large proportion of protected areas and low value for sealed surfaces.

3.3.6 Data Clustering

After an initial inspection of the multidimensional data space, the next important step is to undertake a partitioning into coherent data structures. Groups must be formed of similar data objects which can be easily distinguished from the remaining data. This task is termed clustering. Depending on the type of resulting groups, cluster processes can be subdivided into hierarchical, partitional, or overlapping clustering (Hand et al. 2001).

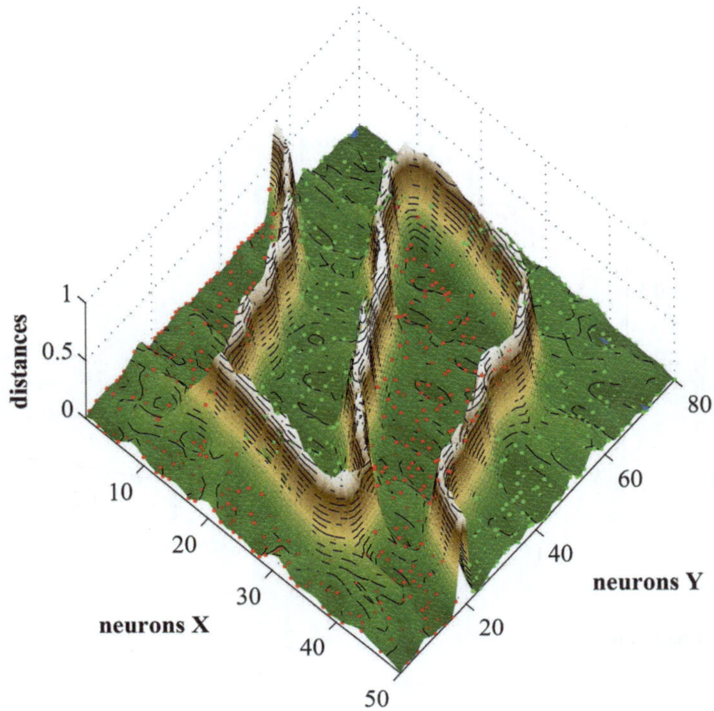

Fig. 3.11 Emergent SOM of the chain-link dataset

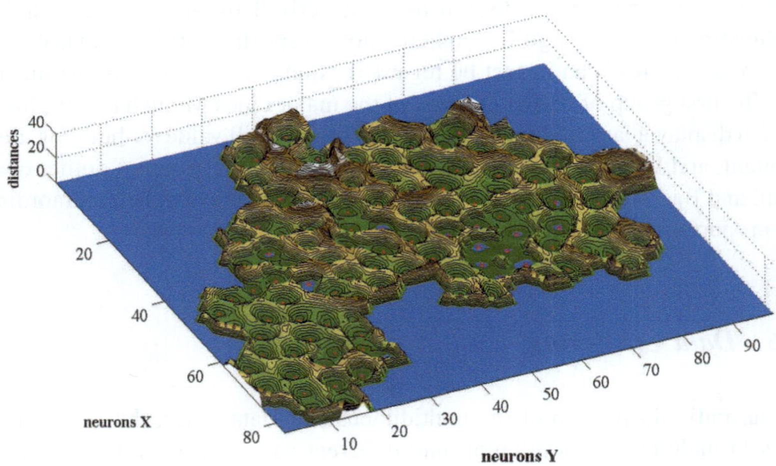

Fig. 3.12 U*-Map (island view) of the UD data

Frequently the number of partitions must be defined beforehand. Here we adopt the approach of partitional clustering, building on the emergent self-organizing feature map. The structures of the U-matrix are used to define the clusters, that is, when the projections of data points (bestmatches) are found in a common valley. The neurons of an ESOM can also be clustered using the clustering algorithm U*C, which is based on grid projections and makes use of distance and density information (Ultsch and Herrmann 2006). In our case, this approach leads to nine U-matrix cluster (UC).

Clustering methods partition the data into clusters. The cluster structures critically depend, first, on the definition of a meaningful measure of distance (see above) and, second, on the details of the clustering algorithm. If a known pre-classification is at hand, then this may be used to evaluate the clustering. However, in most real knowledge discovery cases, no such pre-classification is given. The question arises as to which form of clustering is optimal. For the purposes of knowledge discovery, the quality of any data clustering is determined by whether the resulting classes offer some useful interpretation; in particular, whether these data classes reveal unsuspected structures and correlations in the original data space. Hand et al. (2001) emphasize that the numerical size of clusters should not be accorded too great importance, as it is precisely the unexpected something that goes against the rules which is being sought.

> Generally speaking, however, the validity of a clustering is often in the eye of the beholder; for example, if a cluster produces an interesting scientific insight, we can judge it to be useful. (Hand et al. 2001, p. 292)

In such cases where new structures are detected, other non-supervised approaches should be adopted to validate the clustering results. One such approach is to cluster the data using a different cluster algorithm. Another is to calculate some cluster immanent measure. Finally, the approach which best meets the aims of knowledge discovery is to seek a semantic interpretation of the detected clusters. This means determining whether a cluster makes sense through the application of knowledge generation methods (see next section).

Figure 3.13 shows a hierarchical clustering of the data using Ward clustering (Ward 1963) to produce a dendrogram (Carlsson and Mémoli 2010). The user has to define either a threshold distance or the number of clusters in order to define the clustering in a hierarchical algorithm. In our case, a threshold distance of 100 was used, giving 8 Ward Clusters.

The results of different clustering algorithms can be compared using contingency tables (Fienberg 2007). In our case, the two methods have produced rather similar clustering partitions (cf. Table 3.2). One of the outlier clusters, that is, number UC9 in the U-matrix clustering, has been subsumed to Ward Cluster WC6. In this case, the Ward clustering basically confirms the U-matrix clustering and vice versa.

The silhouettes proposed by Rousseeuw (1987) are a useful graphical display for the interpretation and validation of data partitioning. The values in a silhouette range from -1 to $+1$ for each data point. Large positive values indicate that a data

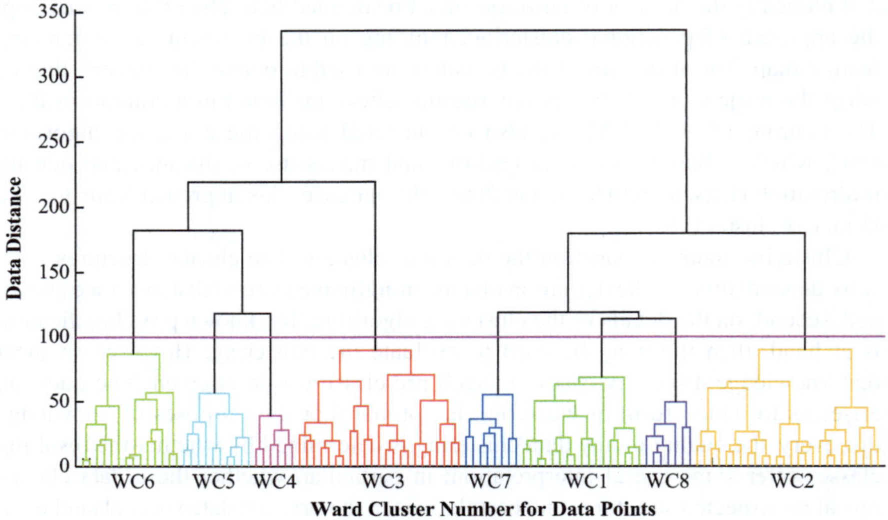

Fig. 3.13 Ward clustering of the UD data

Table 3.2 Comparison of two different cluster partitions (own source)

	U-matrix cluster								
Ward cluster	UC1	UC2	UC3	UC4	UC5	UC6	UC7	UC8	UC9
WC1	15	0	0	1	0	0	0	1	0
WC2	10	15	0	0	0	1	0	0	0
WC3	0	7	15	0	0	1	0	0	0
WC4	0	0	0	6	0	0	0	0	0
WC5	1	0	0	2	7	0	0	0	0
WC6	0	4	0	0	0	5	3	0	2
WC7	4	3	0	0	1	0	0	0	0
WC8	5	1	0	0	0	0	0	1	0

point is at the center of a cluster. Zero or negative values indicate that this data point is on the periphery of the cluster or may even belong to a different cluster. However, it is very important to bear in mind that the classical silhouette values assume a hypersphere as the shape of the cluster.

The silhouette plot of Fig. 3.14 shows that all clusters, apart from numbers UC2 and UC6, can be well modeled as hyperspheres. Some points in cluster UC1 may be outliers. The data points with the largest silhouette values can be used as good representatives for each cluster. The table in Appendix 2 contains a list of cluster representatives determined in this way. The process of knowledge generation produces a clear semantic for each cluster.

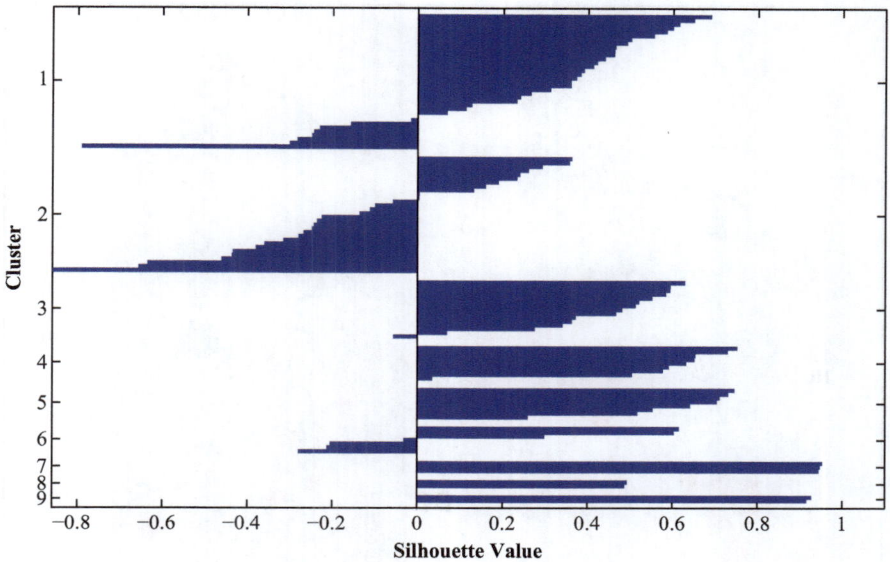

Fig. 3.14 Silhouette plot of the clustering partition

3.3.7 Explaining the Clusters/Knowledge Generation

Having obtained a clustering of the data as described above, the next important step in the process of knowledge discovery is to ask "What do the clusters mean?". This is the layer termed "knowledge generation" in Fig. 3.1. In order to answer this question, we use algorithms from machine learning to produce "symbolic classifiers". These algorithms take a given classification of the data, such as the clustering calculated above (cf. Fig. 3.15: Spatial distribution of the U-matrix clustering), and construct from this decision trees (CART Breiman et al. 1984; C4.5 Quinlan 1993; C5.0 Quinlan 2013; Random Forest Breiman 2001 etc.) or decision rules (such as sig* Ultsch 1991 or Ripper Cohen 1995 etc.).

In the case at hand, we applied rule extraction from a CART decision tree on the UD data. The generated rules are listed in Appendix 3. The rule generation is steered so that a spatial planning expert could easily understand the rules (e.g., low, medium, high values). The application of these rules to the unclassified data assigns a data point to a cluster. The quality of the rules can be reviewed by drawing up a contingency table of the clustering vs. the assignment of cluster labels by the rules (cf. Table 3.3). The two outlier cluster UC8 and UC9 are not taken into account for a rule-based explanation.

As the rules assign almost all data to the correct clusters, it can be assumed that they are sufficiently precise. The rules can be read and understood by a spatial planning expert in order to assign meaning to a particular cluster.

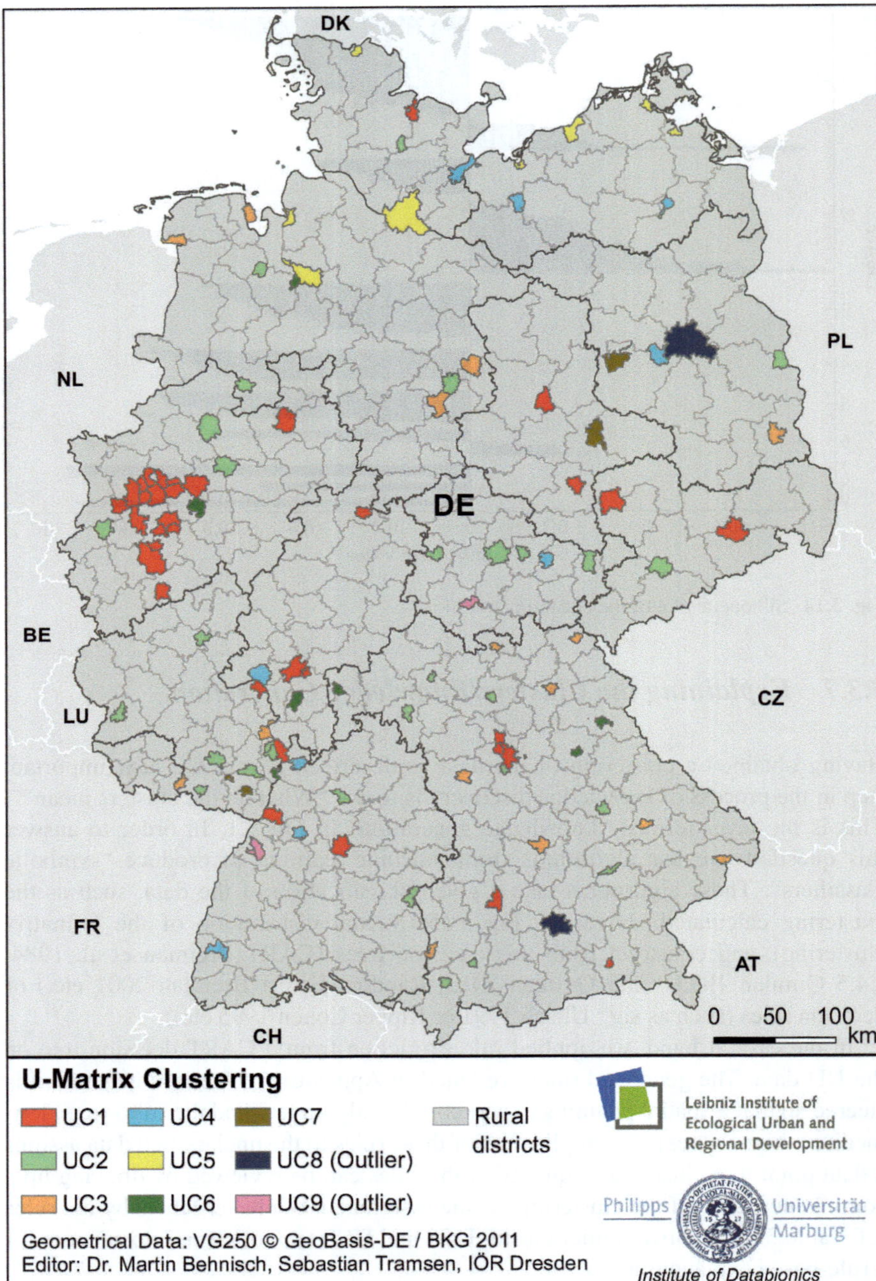

Fig. 3.15 Spatial distribution of the U-matrix clustering: "What do the clusters mean?"

Table 3.3 Performance of the generated rules (own source)

	U-matrix cluster						
Rules	UC1	UC2	UC3	UC4	UC5	UC6	UC7
R1	35	4	0	0	1	0	0
R2	0	23	2	1	0	0	0
R3	0	0	12	0	0	0	0
R4	0	1	0	8	0	2	0
R5	0	0	0	0	7	0	0
R6	0	0	1	0	0	3	0
R7	0	2	0	0	0	2	3

3.3.7.1 Rediscovering Known Structures

If the procedures above are applied correctly to a data set, it should be possible to rediscover structures which are already known. For example, the spatial monitoring system of the Federal Institute for Research on Building, Urban Affairs and Spatial Development (Federal Institute for Research on Building, Urban Affairs and Spatial Development 2013) captures the attribute *geographical position* (German: räumliche Lage), of which there are four types: *highly peripheral, peripheral, central, and highly central*. The BBSR spatial typology classifies 51 % of the UD data as *highly central*. The U-matrix cluster UC1 contains 35 data points, of which 18 (51 %) should, statistically speaking, belong to the class *highly central*. In fact 31 of the 35 urban districts in Cluster UC1 are labeled *highly central* (i.e., 89 %). An urn model can be used to calculate the probability of Cluster UC1 containing at least 31 *highly central* districts by pure chance (p-value). Applying the hypergeometric distribution (Rice 2007,p. 42ff.), we calculate a p-value of 2.8441*10-9. Thus, it can be safely assumed that Cluster UC1 largely consists of *highly central* districts. The city of Leverkusen was found to be the most representative urban district for this cluster. The generated rule for Cluster UC1 is as follows:

UD data belongs to Cluster UC1, if
$log(SealedSurfaces) \geq 48.3179$ and
$log(OpenSpaceMeshSize) \leq 72.6255$ and
$log(BuildingArea) < 65.4549$

This enables a possible characterization of Cluster UC1: "*highly central districts with a large proportion of sealed surfaces, substantial fragmentation, and dense building areas*". In summary, Cluster UC1 basically coincides with the known type of the prior classification (Federal Institute for Research on Building, Urban Affairs and Spatial Development 2013) *highly central* district. The proposed clustering has defined a particular subset of this type of urban district. This can eventually indicate paths for further research into these urban districts, for example, regarding effective land usage, ecological impacts of soil sealing, and fragmentation of the urban area.

3.3.7.2 Generation of New Knowledge

As elaborated in the previous section, one initial expected result of clustering is to rediscover previously known structures in the data. The experience from many knowledge discovery tasks (Loetsch and Ultsch 2013; Behnisch and Ultsch 2009; Moerchen et al. 2006; Kupas et al. 2004) is that about 80 % of clusters coincide with known processes. Typically about 10 % may be attributed to erroneous data, while the remaining 10 % may generate entirely new knowledge.

This latter situation can be sketched out in the case of Cluster UC5 in the U-matrix clustering. The members of this cluster are Flensburg, Hamburg, Bremen, Bremerhaven, Greifswald, Rostock, Stralsund, and Wismar. Bremen was found to be the most representative object of this cluster (maximum value in the silhouettes). In Fig. 3.14, the members of Cluster UC5 are highlighted in yellow. It can be seen that these cluster objects are all coastal urban districts. So a first observation in regard to Cluster UC5 is that it represents a subset of Germany's coastal urban districts. Other coastal urban districts are shown in different color.

Note that information on whether an urban district is located on the coast or is a harbor city is not included in the variables. Some coastal urban districts are not grouped in Cluster UC5, implying that this is more than just a collection of seaports. The rules for the cluster can be examined in order to gain greater insight into the particular meaning of the cluster. The rule describing Cluster UC5 is:

UD data belongs to Cluster UC5, if
$\mathit{OpenSpaceMeshSize} \geq 72.6255$ and
$\mathit{SealedSurface} \geq 48.3179$

This means that Cluster UC5 is the subset of the UD data with large values in OpenSpaceMeshSize and large values in SealedSurface. This rule assigns all seven districts correctly to the cluster. Other coastal urban districts are not included in Cluster UC5. In the case of Kiel, for example, although the city possesses a fairly high degree of sealed surface, the fragmentation of open space is substantially higher than in the coastal urban districts of Cluster UC5. The larger the effective mesh size in an urban district, the lower the landscape fragmentation. In our case, the regional transport network of roads and railway lines was adopted as a measure of landscape fragmentation. The procedure developed by Moser et al. (2007) was applied in order take account of target areas truncated by the borders of administrative units (cf. Table 3.1). The interpretation of the cluster properties leads to a hypothesis regarding UDs: there are two types of coastal urban districts in Germany, those with high and those with low fragmentation of open space. Thus, a potential meaning of Cluster UC5 is "coastal urban districts with less landscape fragmentation due to linear transport infrastructure". A statistically testable hypothesis could be formulated as follows: the coastal urban districts in Cluster UC5 are substantially different from all other such districts in Germany. If this hypothesis cannot be refuted, then it may be worthwhile examining the reasons behind these differences in landscape fragmentation. For example, one can assume that several other properties will influence the characteristics of a coastal urban district, for example, the historical settlement development, the shape of the coastline, shape of the bay, size of the harbor area, development of traffic infrastructure, municipal environmental protection, proximity effects of the surrounding area, natural and administrative boundaries, as well as the location of settlement functions. In some cases, coastal waters can affect the development of built-up areas and traffic infrastructure in coastal urban districts, which in turn will have an impact on open space fragmentation. Such information could be used for further spatial investigations of coastal urban district types.

Another interesting thing was, for example, the identification of a cluster that was characterized by many protected areas and a compact building utilization (e.g., small value of building area per inhabitant). The cluster objects also belong to the first discovered group of urban districts with lower values of sealed surfaces (see Fig. 3.3: Exploring the distributions of the individual variables). The rule describing Cluster UC4 is:

UD data belongs to Cluster UC4, if
$sqrt(ProtectedAreas) \geq 60.8555$ and
$log(BuildingArea) < 75.0076$ and
$log(SealedSurfaces) < 48.3179$

Freiburg im Breisgau was found to be representative for this cluster. This may lead to another hypothesis: There is a type of urban districts in Germany the ones with an assumed efficient and compact building structure and a prominent influence of natural protection laws on the settlement development. The following label is suggested: "Urban Protected Areas". The details and in particular the causes of the observed land use properties (e.g., lower values of sealed surfaces, small value of building area per inhabitant) can presumably not be exhaustively ascertained on the basis of the presented approach. In addition to the quantitative approach, a qualitative examination of selected results might be necessary (e.g., integration of case studies, interviews, local inspections). One should ask to what degree the settlement structure or the type of buildings can be traced back to conscious planning decisions of local decision-makers. Moreover, the question as to the extent to which local construction activity is regulated by regional spatial planning is of interest. However, this is not an issue of the presented quantitative investigation.

In summary, an investigation of clusters in the ways sketched above may lead to a fresh interpretation of what features of the data are of interest. In this way, the application of exploratory techniques can eventually produce knowledge that is entirely new.

3.4 Conclusions and Future Challenges

In view of the continual growth in geodata and the heterogeneity of datasets, there is an urgent need for alternatives to traditional spatial analysis within geographic information systems (GIS) (Miller and Han 2009, p. 13). Techniques of (spatial-temporal) data mining, in particular knowledge discovery in databases (KDD), provide one such approach to automated knowledge extraction (Streich 2009, p. 252). This chapter has shown how methods of data mining and knowledge discovery can be applied to a spatial data set, in this particular case, 111 urban districts in Germany classified by seven dimensions of land use.

Any evaluation on this basis can best be realized through an interdisciplinary collaboration of computer science (data mining) and the spatial sciences. In the example given here, the raw data was comprehensively reviewed and pre-processed to allow the application of methods of data mining and knowledge discovery. A quantifiable measure was determined to enable the multidimensional comparison of urban districts. The data was structured using projection, clustering, and machine learning algorithms. By visualizing results, it was possible to discern prominent structures and characteristics of the data set, such as correlations, spatial outliers, and potential clusters (groups). The automated clustering led to the discovery of clusters of urban districts sharing common features. Representative districts were selected from each cluster. Any classification of data should, ideally, also support spatial planning decisions. We can speak of knowledge discovery if the extracted knowledge is entirely new and is nonobvious and also if this knowledge can be of use at the practical level (e.g., by spatial planners, politicians, decision-makers). The authors believe that methods are required to explicitly elucidate the relevance of classes produced by such automated data processing. In some instances, the wide range of possible interpretations has been explored in rather unsystematic fashion by simple inspection of the characteristics of variables (e.g., measures of central tendency, variability). Approaches which use machine learning can provide a more systematic review of the complete range of hypotheses. This can produce decision rules or trees that can then be applied to discover useful and previously unknown correlations in the data set. Such methods have been presented in this chapter using the example of a small data set (111 urban districts in Germany) with only seven dimensions (variables). Presentation of this exploratory approach should show how the applied processes can help generate hypotheses as well as extract important correlations from the data set. Methods of KDD can be used to produce a much larger number of hypotheses than would be possible manually. Yet this powerful approach has at the same time a particular drawback, which should also be emphasized in regard to the results of this chapter: the hypotheses derived from many data mining procedures must not be interpreted as statistically validated truths. Rather they should be understood as suggestions for discussion, which should be assessed by suitable methods in follow-up spatial investigations.

For example, it is possible to find data subsets that are perhaps irrelevant, which have already been discovered (rediscovered, not new for the target/application domain) or are highly uncertain. Some subsets require a more detailed investigation (e.g., "*coastal urban districts with a large mesh size due to linear transport infrastructure*", "Urban Protected Areas", *subclasses of the variable SealedSurface*). It should be emphasized that the extraction of knowledge for spatial planning generally requires further investigations, validations, and tests. In particular, it can be helpful to apply other methods to confirm specific hypothesis extracted by the knowledge discovery approach. Furthermore, it is necessary to review the results at different spatial scales (i.e., nationally, regionally, and locally). However, the step of hypothesis testing was not part of this chapter. Relevant methods will already be known to readers.

The data on settlement and open space development employed here was borrowed from the publicly available data sets of the IOER Monitor (http://www.ioer-monitor.de/). In future, depending on the research question and the availability of additional variables, the presented illustrative data set on land use (static perspective) can be expanded, while changes in land use can be characterized in a multidimensional approach (dynamic perspective). According to the explanatory power of the parameter land as a barometer for sustainability (Siedentop et al. 2007), the static perspective can generate knowledge for the high-level political evaluation of land use, whereas the dynamic perspective primarily produces practical information for operative political action at the level of concrete, individual cases of planning and project-based decisions. In the years to come, the methods outlined here will find a wide range of application in diverse fields. For example, they are suited to the machine-based analysis and evaluation of the impact of spatial programs and measures for urban development imposed at the political level. Automated spatial survey systems (automated monitoring, classifiers) could assist in the periodic automatic or semiautomatic estimation and comparison of spatial objects (reference areas).

Acknowledgements The authors acknowledge the yearly data provided by the Federal Agency for Cartography and Geodesy, which was crucial for the development of the land use monitoring. The Federal Institute for Research on Building, Urban Affairs and Spatial Development (BBSR) makes several spatial typologies available. The authors would like to thank the colleagues of the Leibniz Institute of Ecological Urban and Regional Development (IOER) for the indicator computation and the fruitful cooperation. Further, we cordially appreciate the remarks of the reviewers and editors for giving constructive and helpful comments to improve the quality of this chapter.

Appendix 1

Nonlinear Transformations of the Variables in the UD Data

$OpenSpaceMeshSize : log$
$BuildingArea : log$
$SettlementDensity : log$
$SealedSurface : sqrt$
$LandConsumption : sqrt$
$ProtectedAreas : sqrt$
$HemerobyIndex : identity(notransformation)$

Appendix 2

Cluster Representatives

U-matrix cluster	Urban district
UC1	Leverkusen
UC2	Heilbronn
UC3	Zweibrücken
UC4	Freiburg im Breisgau
UC5	Bremen
UC6	Aschaffenburg
UC7	Landau in der Pfalz
UC8	Berlin
UC9	Suhl

Appendix 3

Rules Explaining the U-Matrix Clustering

UD data belongs to Cluster UC1, if
$log(SealedSurfaces) \geq 48.3179$ and
$log(OpenSpaceMeshSize) \leq 72.6255$ and
$log(BuildingArea) < 65.4549$

UD data belongs to Cluster UC2, if
$log(SealedSurfaces) < 48.3179$ and
$log(BuildingArea) < 75.0076$ and
$sqrt(ProtectedAreas) < 60.8555$ or
$log(SealedSurfaces) \geq 48.3179$ and
$log(BuildingArea) \geq 65.4549$ and
$log(OpenSpaceMeshSize) < 72.6255$

UD data belongs to Cluster UC3, if
$log(BuildingArea) \geq 75.0076$ and
$log(SealedSurfaces) < 48.3179$

UD data belongs to Cluster UC4, if
$sqrt(ProtectedAreas) \geq 60.8555$ and
$log(BuildingArea) < 75.0076$ and
$log(SealedSurfaces) < 48.3179$

UD data belongs to Cluster UC5, if
$log(OpenSpaceMeshSize) \geq 72.6255$ and
$log(SealedSurfaces) \geq 48.3179$

References

Alpaydin E (2008) Introduction to machine learning, 2nd edn. MIT, Cambridge

Aumayr ChM (2007) European region types in EU-25. Eur J Comp Econ 4(2):109–147

Bätzing W, Dickhörner Y (2001) Die Typisierungen der Alpengemeinden nach Entwicklungsverlaufsklassen für den Zeitraum 1870–1990. Mitteilungen der Fränkischen Geographischen Gesellschaft 48:273303

Behnisch M (2009) Urban data mining. KIT Scientific Publishing, Karlsruhe

Behnisch M, Ultsch A (2009) Urban data mining: spatiotemporal exploration of multidimensional data. Build Res Inf. doi:10.1080/09613210903189343

Behnisch M, Hagemann U, Meinel G (2013) Analysergebnisse zum Gebäudebestand in Deutschland auf der Grundlage von Geobasisdaten. In: Meinel G, Schumacher U, Behnisch M (eds) Flächennutzungsmonitoring V Methodik Analyseergebnisse Flächenmanagement (IR Schriftenreihe, Bd. 61) Rhombos, Berlin

Bilmes J (1998) A gentle tutorial on the EM algorithm and its application to parameter estimation for Gaussian mixture and hidden Markov models. Technical report. Available via DIALOG. http://crow.ee.washington.edu/people/bulyko/papers/em.pdf. Cited 25 Oct 2013

Blume L, Sack D (2010) Patterns of social capital in West German regions. Eur Urban Reg Stud. doi:10.1177/0969776408090416

Breiman L (2001) Random forests. Mach Learn. doi:10.1023/A:1010933404324

Breiman L, Friedman J, Olshen RA, Stone CJ (1984) Classification and regression trees. Wadsworth, Belmont

Carlsson G, Mémoli F (2010) Characterization, stability and convergence of hierarchical clustering methods. J Mach Learn Res 11:1425–1470

Cohen WW (1995) Fast effective rule induction. In: Prieditis A, Russel S (eds) Machine learning. Proceedings of the twelfth international conference (ML 95), Lake Tahoe. Morgan Kaufmann, San Francisco

Dempster AP, Laird NM, Rubin DB (1977) Maximum likelihood from incomplete data via the EM algorithm. J R Stat Soc Ser B (Methodol) 39(1):1–38

Demsar U (2009) Data mining of geospatial data: combining visual and automatic methods. Royal Institute of Technology (PhD thesis), Stockholm

Dosch F (2001) Flächenverbrauch in Deutschland und Mitteleuropa. Struktur, Trends und Steuerungsoptionen durch das Boden-Bündnis. Terra-Tech 6:19–23

European Environment Agency (2011) Landscape fragmentation in Europe. Joint EEA-FOEN report, EEA report No. 2. European Environment Agency (EEA). Available via DIALOG. http://www.eea.europa.eu/publications/landscape-fragmentation-in-europe. Cited 25 Oct 2013

European Environment Agency (2013) Technical note on HR imperviousness layer product specification. European Environment Agency (EEA). Available via DIALOG. http://www.gmes-geoland.info/fileadmin/geoland2/redakteur/pdf/Project_Documentation/Service_Specification/TechnicalProductSpecification_HR_Imperviousness_Layer_I1-01.pdf. Cited 25 Oct 2013

European Spatial Planning Observation Network (ESPON) (2011) Climate change and territorial effects on regions and local economies. Available via DIALOG. http://www.espon.eu/main/Menu_Projects/Menu_AppliedResearch/climate.html. Cited 25 Oct 2013

Fayyad UM, Piatetsky-Shapiro G, Smyth P (1996) From data mining to knowledge discovery in databases. doi:http://dx.doi.org/10.1609/aimag.v17i3.1230

Federal Institute for Research on Building, Urban Affairs and Spatial Development (ed) (2013) Spatial typologies of the Federal Institute for Research on Building, Urban Affairs and Spatial Development. Technical report. Available via DIALOG. http://www.bbsr.bund.de/BBSR/DE/Raumbeobachtung/Raumabgrenzungen/raumabgrenzungen_node.html. Cited 25 Oct 2013

Fienberg SE (2007) The analysis of cross-classified categorical data, 2nd edn. Springer, New York

Frenkel A (2004) Land-use patterns in the classification of cities: the Israeli case. Environ Plan B Plan Des 31(5):711–730

Geyler S, Warner B, Brandl A, Kuntze M (2008) Clusteranalyse der Gemeinden in der Kernregion Mitteldeutschland. Eine Typisierung der Region nach Entwicklungsparametern und Rahmenbedingungen. In: Foschungsverbund KoReMi (eds) Schriftenreihe des Forschungsverbundes KoReMi (Band 2), Leipzig

Guo D (2009) Multivariate spatial clustering and geovisualization. In: Miller HJ, Han J (eds) Geographic data mining and knowledge discovery, 2nd edn. Chapman & Hall, Boca Raton

Hand D, Mannila H, Smyth P (2001) Principles of data mining. MIT, Cambridge

Hastie T, Tibshirani R, Friedman J (2009) The elements of statistical learning. Data mining, inference, and prediction, 2nd edn. Springer, New York

Hietel E, Waldhardt R, Otte A (2004) Analysing land-cover changes in relation to environmental variables in Hesse, Germany. Landsc Ecol. doi:10.1023/B:LAND.0000036138.82213.80

Izenman AJ (2008) Modern multivariate statistical techniques. Springer, New York

Jaeger J (2000) Landscape division, splitting index, and effective mesh size: new measures of landscape fragmentation. Landsc Ecol. doi:10.1023/A:1008129329289

Kohonen T (1982) Self-organized formation of topologically correct feature maps. Biol Cybern. doi:10.1007/BF00337288

Kohonen T (2001) Self-organizing maps, 3rd edn. Springer, New York

Kronthaler F (2005) Economic capability of East German regions: results of a cluster analysis. Reg Stud. doi:10.1080/00343400500213630

Krüger T, Meinel G, Schumacher U (2013) Land-use monitoring by topographic data analysis. Cartogr Geogr Inf Sci. doi:10.1080/15230406.2013.809232

Kuncheva L (2004) Combining pattern classifiers: methods and algorithms. Wiley, Hoboken

Kupas K, Klebe G, Ultsch A (2004) Comparison of substructural epitopes in enzyme active sites using self-organizing maps. J Comput Aided Mol Des. doi:10.1007/s10822-004-6553-x

Laube P (2011) Raumzeitliches data mining. In: Schilcher M (ed) Geoinformationssysteme. abc, Heidelberg

Limpert E, Stahel WA, Abbt M (2001) Log-normal distributions across the sciences: keys and clues. BioScience 51(5):343–352

Loetsch J, Ultsch A (2013) A machine-learned knowledge discovery method for associating complex phenotypes with complex genotypes. Application to pain. J Biomed Inform. doi:10.1016/j.jbi.2013.07.010

Meinel G (2013) Auf dem Weg zu einer besseren Flächenstatistik. Raumforschung Raumordnung. doi:10.1007/s13147-013-0256-5

Miller HJ, Han J (2009) Geographic data mining and knowledge discovery, 2nd edn. Chapman & Hall, London

Moerchen F, Ultsch A, Hoos O (2006) Extracting interpretable muscle activation patterns with time series knowledge mining. Int J Knowl Based Intell Eng Syst 9(3):197–208

Moser B, Jaeger JAG, Tappeiner U, Tasser E, Eiselt B (2007) Modification of the effective mesh size for measuring landscape fragmentation to solve the boundary problem. Landsc Ecol. doi:10.1007/s10980-006-9023-0.07.010

Qu W (2000) Zur Anwendung der Fuzzy-Clusteranalyse in der Grundstückswertermittlung. Univ. Hannover, Fachbereich Bauingenieur- und Vermessungswesen, Hannover

Quinlan R (1993) C4.5 – programs for machine learning. Mach Learn. doi:10.1007/BF00993309

Quinlan R (2013) C5.0 and see 5: illustrative examples. Available via DIALOG. http://www.rulequest.com/. Cited 25 Oct 2013

Rasul G, Thapa GB, Zoebisch MA (2004) Determinants of land-use changes in the Chittagong hill tracts of Bangladesh. Appl Geogr 24(3):217–240

Rice JA (2007) Mathematical statistics and data analysis, 3rd edn. Duxbury Press, Pacific Grove
Rousseeuw PJ (1987) Silhouettes: a graphical aid to the interpretation and validation of cluster analysis. Comput Appl Math. doi:10.1016/0377-0427(87)90125-7
Siedentop S, Fina S (2010) Monitoring urban sprawl in Germany: towards a GIS-based measurement and assessment approach. J Land Use Sci. doi:10.1080/1747423X.2010.481075
Siedentop S, Kausch S, Einig K, Gssel J (2003) Siedlungsstrukturelle Veränderungen im Umland der Agglomerationsräume. In: Bundesamt für Bauwesen und Raumordnung (ed) Forschungen (Band 114), Selbstverlag, Bonn
Siedentop S, Heiland S, Lehmann I, Schauerte-Lüke N (2007) Regionale Schlüsselindikatoren nachhaltiger Flächennutzung fr die Fortschrittsberichte der Nationalen Nachhaltigkeitsstrategie Flächenziele (Nachhaltigkeitsbarometer Fläche). In: Bundesamt für Bauwesen und Raumordnung (ed) Forschungen (Band 130), Bonn
Steinhardt U, Herzog F, Lausch A, Müller E, Lehmann S (1999) The hemeroby index for landscape monitoring and evaluation. In: Hyatt DE, Lenz R, Pykh YA (eds) Environmental indices systems analysis approach. Advances in sustainable development. Proceedings of the first international conference on environmental indices systems analysis approach, St. Petersburg, 7–11 July 1997. EOLSS Publishers, Oxford
Storch H, Schmidt M (2008) Spatial planning: indicators to assess the efficiency of land consumption and land-use. In: Schmidt M, Knopp L (eds) Standards and thresholds for impact assessment. Environmental protection in the European union series. Springer, Berlin
Streich B (2009) Stadtplanung in der Wissensgesellschaft, 2nd edn. VS, Wiesbaden
Thompson DM, Serneels S, Lambin EF (2002) Land use strategies in the Mara ecosystem: a spatial analysis linking socio-economic data with landscape variables. In: Walsh SJ, Crews-Meyer KA (eds) Linking people, place and policy: a GIScience approach. Kluwer Academic, Norwell
Tukey JW (1977) Exploratory data analysis. Pearson, London
Ultsch A (1991) Konnektionistische Modelle und ihre Integration mit wissensbasierten Systemen. Dekanat Informatik, Dortmund
Ultsch A (1999) Data mining and knowledge discovery with emergent self-organizing feature maps for multivariate time series. In: Oja E, Kaski S (eds) Kohonen maps. Elsevier, Amsterdam
Ultsch A (2003) Pareto density estimation: a density estimation for knowledge discovery. In: Baier D, Wernecke K D (eds) Innovations in classification, data science, and information systems. Proceedings 27th annual conference of the German classification society. Springer, Berlin
Ultsch A (2013) Databionic knowledge discovery. Lecture notes, Department of Mathematics and Computer Science, Philipps-University of Marburg. Available via DIALOG. http://www.uni-marburg.de/fb12/informatik/arbeitsgebiete/bioinf/profalfredultsch. Cited 25 Oct 2013
Ultsch A, Herrmann L (2006) Automatic clustering with U*C. Technical report, Department of Mathematics and Computer Science, Philipps-University of Marburg. Available via DIALOG. http://www.uni-marburg.de/fb12/forschung/berichte/berichteinformtk/autom_clust. Cited 25 Oct 2013
Walz U, Stein C (2014) Indicators of hemeroby for the monitoring of landscapes in Germany. J Nat Conserv. doi:10.1016/j.jnc.2014.01.007
Ward JH (1963) Hierarchical grouping to optimize an objective function. J Am Stat Assoc 58(301):236–244
Wilkinson L, Friendly M (2009) The history of the cluster heat map. Am Stat. doi:10.1198/tas.2009.0033
Zentrale Stelle Hausumringe und Hauskoordinaten (2013) Produktbeschreibungen Hausumringe und Hauskoordinaten. Bezirksregierung Köln. Available via DIALOG. http://www.bezreg-koeln.nrw.de/brk_internet/organisation/abteilung07/dezernat_74/zshh/index.html. Cited 25 Oct 2013

Chapter 4
Clustering Contextual Neural Gas: A New Approach for Spatial Planning and Analysis Tasks

Julian Hagenauer

Abstract Spatial clustering is a method that can reveal structures and identify groupings in large spatial data sets, which is in particular useful for spatial planning and analysis tasks. A recent and powerful clustering algorithm for spatial data is contextual neural gas (CNG). The CNG algorithm is closely related to the basic self-organizing map algorithm but additionally takes spatial dependence into account. However, like most clustering algorithms, CNG requires the analyst to specify the number of clusters beforehand. Even though the chosen number of clusters critically affects the results of the clustering, it is unclear how to determine it. This study introduces a new method which combines CNG, the learning of the CNG's topology, and graph clustering. It can be used to cluster spatial data without any prior knowledge of present clusters in the data. The proposed method is in particular useful for spatial planning and analysis tasks, because it provides means to find groupings in the data and identify homogeneous regions. To evaluate the method, this study draws from two experiments which are based on a synthetic and a real-world data set. The results of the synthetic data set show that it can correctly identify clusters in a predefined setting. The results of the real-world data set demonstrate that the proposed method outlines meaningful and theoretically sound regions.

Keywords Artificial neural networks • Cluster analysis • Spatial planning

4.1 Introduction

Clustering is the task of organizing observations into clusters such that the similarity of observations within a cluster is maximized and the dissimilarity between the clusters is maximized. It is particularly useful if no categorization or labeling of the observations is available, but some structural organization is needed. Many different clustering algorithms have been developed in the past, mainly in the fields

---

J. Hagenauer (✉)
Institute of Geography, Heidelberg University, Berliner Straße 48, D-69120 Heidelberg, Germany
e-mail: hagenauer@uni-heidelberg.de

M. Helbich et al. (eds.), *Computational Approaches for Urban Environments*,
Geotechnologies and the Environment 13, DOI 10.1007/978-3-319-11469-9_4

of statistics and machine learning. These clustering algorithms can be broadly classified by the paradigm they use. One of the most prominent and widely used clustering paradigms is partitioning clustering. Partitioning clustering algorithms, such as k-Means or neural gas (NG; Martinetz and Schulten 1991), divide a set of observations into a nonoverlapping set of clusters. Each observation is assigned to the cluster which it is closest to. For large data sets, partitioning clustering algorithms are typically more computationally effective than, e.g., hierarchical clustering algorithms (Jain et al. 1999). However, a severe disadvantage of them is that they require the analyst to choose the number of desired clusters beforehand.

There are several important special cases of clustering. One such case is spatial clustering, which deals with the clustering of spatially located observations. A basic property of such observations is that they are likely to be spatially dependent. Spatial dependence states that observations that are spatially located close to each other tend to have similar characteristics. This property is essential to spatial sciences because without it variation of phenomena would be independent of location and thus the notion of region would be totally meaningless (Goodchild 1986). The presence of spatial dependence has been traditionally regarded as problematic for statistical analysis, which typically requires sample independence (Bailey and Gatrell 1995). However, it can also serve as a valuable source of information about spatial processes, because it provides evidence of causality (Miller 2004). Therefore, it is generally useful for spatial clustering algorithms to take spatial dependence into account in order to utilize the full range of available information for discovering spatial patterns.

Spatial clustering is of special importance for spatial planning tasks: Administrative areas typically have their roots in historic administrative divisions of space, which disregard the nonspatial characteristics of place. As a consequence, administrative divisions often intersect contiguous regions and are often inhomogeneous. Decisions made concerning the planning, distribution, and allocation of resources among such administrative areas are likely to be ineffective and meaningless (Amedeo 1969). In fact, it has been shown by Van Der Laan and Schalke (2001) that local policies are more effective for homogeneous regions. These concerns are very closely related to the modifiable areal unit problem (MAUP; Openshaw 1984). Spatial analysis typically requires manageable discrete descriptions of spatial processes, which are continuous. For this purpose, it is necessary to aggregate observations over areal units. The MAUP states that the outline of these units and the scale of aggregation critically affect the results of any spatial analysis. In general, it is useful if the observations that are aggregated over the same areal unit are similar to each other. Consequently, since spatial clustering outlines mostly coherent and homogeneous areas, it has potential to serve as a valuable tool for spatial planning and analysis tasks (e.g., Helbich et al. 2013).

Various spatial clustering algorithms have been developed in the past (see, e.g., Han et al. 2001). Most of these methods are based on general-purpose clustering algorithms that have limited capabilities in recognizing spatial patterns that involve neighbors or cannot deal with high-dimensional data sets (Guo et al. 2003). Contextual neural gas (CNG; Hagenauer and Helbich 2013) is a recently developed algorithm for clustering spatial data that is specially designed for spatial data mining. The CNG algorithm combines the concepts of the NG algorithm with the GeoSOM (Bação et al. 2005), a variant of the famous self-organizing map algorithm (SOM; Kohonen 1982, 2001), in order to take spatial dependence into account. A particular advantage of the CNG is that it quantizes the data space better than the GeoSOM, because the adaptation of the CNG's neurons, in contrast to the SOM, does not depend on some predefined and fixed topology (Hagenauer and Helbich 2013). However, the topology of the SOM facilitates the analysis of the SOM and hence supports the understanding of the properties of the data (e.g., Hagenauer et al. 2011; Skupin and Esperbé 2011; Arribas-Bel and Schmidt 2013). In particular, it is useful for determining the actual number of clusters in the data, either computationally (e.g., Murtagh 1995; Costa and De Andrade Netto 1999) or by visualizing it (see Flexer 2001).

Another important special case of clustering is graph clustering. A graph is a set of vertices and edges that are connections between pairs of vertices. The edges can have a weight assigned which indicates the strength of the connection and can be directed or undirected. The task of spatial clustering is organizing the vertices of a graph into clusters such that the vertices within a cluster are better connected than the vertices within different clusters. The ability to find and analyze clusters is useful for understanding and visualizing the structure of networks, which is of great importance in many research areas that deal with social, technological, or information systems. Many different algorithms have been developed in the past for this purpose (see Schaeffer 2007). From the large set of available graph clustering algorithms, the heuristic multilevel modularity optimization algorithm (MLMO; Blondel et al. 2008) is particularly promising, because it is exceptionally fast even for very large graphs and automatically determines the number of clusters in the graph by optimizing its quality score.

This study introduces a new method that combines CNG, topology learning, and graph clustering algorithms to outline clusters in CNG. The method consists of the following steps: First, a CNG consisting of a sufficient number of neurons is trained. Second, a topology of the neurons is learned and the resulting topology is considered a weighted graph. Finally, this graph is clustered using advanced graph clustering algorithms, which do not require to specify the desired number of clusters. The resulting clusters represent homogeneous regions in the input data. Since the number of clusters is automatically determined depending on the topological patterns of the graph, the method is especially useful for outlining spatial clusters when no prior knowledge about the actual number of clusters is available.

This workflow is closely related to the clustering approach using the GeoSOM algorithm. In this approach, a GeoSOM consisting of a sufficiently large number of neurons is trained to project the input data onto a two-dimensional map. Subsequently, the map is visualized, usually by means of a U-matrix (Ultsch and Siemon 1990). Clusters appear on the U-matrix as valleys, cluster boundaries as ridges. However, for complex and high-dimensional data sets, U-matrices often show no clear patterns so that it is difficult or even impossible to determine clusters, in particular when using computational methods.

The proposed method is also closely related to the approach by Costa and Oliveira (2007). In their approach, they train a growing neural gas (GNG; Fritzke 1995) to obtain a topology. The main differences between the basic NG and the GNG algorithm are that the GNG does not require to specify the number of neurons beforehand and that it forms a topology in the process of training the network. However, the GNG also introduces numerous additional parameters, which must be set appropriately to obtain reasonable results. Then, in a post-processing step, the authors modify the topology of the GNG by heuristically removing connections between neurons; disjunctive sections of the topology are considered clusters. However, which connections are removed depends on arbitrary chosen threshold levels and critically affects the results. Additionally, complex structural properties of the topology are totally disregarded. Moreover, their approach is not appropriate for clustering spatial data, because it merely uses a basic NG algorithm, which does not take spatial dependency into account.

This study is structured as follows: Sect. 4.2 introduces the algorithms that this study utilizes, while Sect. 4.3 briefly explains the consecutive steps of which the proposed method consists of. The usefulness of the method is demonstrated with two different experiments (Sect. 4.4). Finally, the last section concludes with some remarks and identifies future work.

4.2 Methodical Background

4.2.1 Contextual Neural Gas

Contextual Neural Gas (CNG; Hagenauer and Helbich 2013) is a spatial clustering algorithm that combines the concepts of the GeoSOM with the NG algorithm. Like basic NG, CNG consists of an arbitrary number of neurons, which are not subject to any topological restrictions and provides a nonlinear mapping in high-dimensional data space. In each step of the training process, an input vector is selected from the input data and each neuron is moved into its direction. The strength of the movement depends on the neurons' ranking order with respect to the distance to the input vector, the adaptation rate, and the neighborhood range. Both the neighborhood range and the adaptation rate are typically chosen to decrease with time.

CNG differs from basic NG in the determination of the neurons' ranking order, which CNG accomplishes in a two-phase procedure to incorporate spatial dependence. In the first step, neurons are ordered by spatial closeness. In the second step, the first l neurons of the resulting spatial ordering are reordered within their ranks with respect to the similarity of attributes.

The parameter l determines the strength of spatial dependence which is incorporated into the mapping. If $l = 1$, the ordering in the second step has no effect on the final ordering at all. As a consequence, the adaption of the neurons depends solely on spatial closeness. The attributes of the input data are ignored. If l is increased, the ordering of the l spatially closest neurons depends on attribute similarity. Hence, spatial closeness is less important for the final ordering and less spatial dependence is being incorporated. If l equals the total number of neurons, the spatial ordering does not matter for the final ordering, because all neurons are totally reordered in the second step by similarity of attributes. Consequently, no spatial dependence is incorporated at all.

CNG has several advantages over other spatial clustering algorithms: Like the GeoSOM, CNG enforces spatial proximity between observations and neurons by means of neural distance, defined by either the map's topology or the rank ordering or neurons. Consequently, it is not necessary to weigh or scale spatial proximity and attribute similarity in the data space. Furthermore, the neurons are basically local averages. Thus, the process of incorporating spatial dependence is less sensitive to random variations in the input data. Finally, the parameter l restricts the mapping of observations; all observations are always mapped to one of its l spatially closest neurons. Hence, the mapping maintains a certain degree of spatial closeness, even for observations whose attributes are very different from those of their spatial neighbors (spatial outliers).

4.2.2 Competitive Hebbian Learning

Competitive Hebbian Learning (CHL; Martinetz and Schulten 1991; Martinetz 1993) forms a topology on a set of neurons by creating a number of connections between neighboring neurons. More specifically, the algorithm can be described as follows: For each input vector, the two closest neurons are determined and a connection between these is added to the total set of connections. Thereby, closeness is typically measured by Euclidean distance. After all input vectors have been presented, the set of connections represents the topology of the underlying data.

The resulting graph optimally preserves the topology in a very general sense (Martinetz 1993). In particular, each connection between two neurons belongs to the Delaunay triangulation corresponding to the neurons in data space. The theoretical foundations of CHL in terms of topology preservation have been provided by Edelsbrunner and Shah (1997).

CHL is especially useful for NG and other vector quantization algorithms which do not define a topological structure. It can be applied concurrently to the training of NG or as a post-processing step. However, in the first case, the movement of neurons during the training may make previously learned connections invalid. Therefore, it is necessary to constantly adapt the topology to these movements, e.g., by removing outdated connections (Martinetz and Schulten 1991). In the latter case, NG is trained before CHL is applied, and hence, the topology is not affected by the movement of the neurons. For simplicity, this study applies CHL as a post-processing step.

Since its introduction, numerous extensions and variants of the CHL algorithm have been proposed. For example, De Silva and Carlsson (2004) presented a generalization of CHL which produces a simplicial complex instead of a graph. An alternative to CHL was presented by Aupetit (2005). In this approach, each edge and vertex of the Delaunay triangulation is the basis of a generative model so that the triangulation generates a mixture of Gaussian density functions; the likelihood of the set of model parameters is maximized using the expectation-maximization algorithm.

4.2.3 Multilevel Modularity Optimization

The multilevel modularity optimization algorithm (MLMO; Blondel et al. 2008) is a heuristic method which seeks to find a clustering of a graph with maximum modularity. Modularity is a quality measure that evaluates the density of connections inside a cluster as compared with the connection between different clusters (Newman and Girvan 2004). Because optimizing modularity is a problem that is computationally hard (Brandes et al. 2008), heuristic algorithms are inevitable for practical applications.

The MLMO algorithm consists of two phases: Initially, each vertex of a graph is assigned to a single cluster. Then, in the first phase, each vertex is assigned to the cluster of the neighboring vertex which yields the largest increase of modularity, as long as it is positive. In the second phase, the original graph is replaced by a newly built graph whose vertices are the clusters found during the first phase. Connections between the new graphs' vertices exist if there is at least one connection between vertices of the corresponding clusters in the original graph. The two phases are iteratively repeated until there are no more changes to the graph and a maximum of modularity is reached.

The MLMO algorithm is computationally efficient and scales very well, because the number of clusters dramatically reduces with each pass. In particular, computer simulations on large graphs indicated that its complexity is linear on typical and sparse data (Blondel et al. 2008). A limitation of most modularity optimizing clustering algorithms is that they fail to detect small clusters in very large graphs (Fortunato and Barthelemy 2006). However, the MLMO algorithm seems to be unaffected by this limitation because of its multilevel nature (Blondel et al. 2008). In fact, it has been shown by Fortunato (2010) and Lancichinetti and Fortunato (2009) that the quality of the MLMO's results is superior to that of many other graph clustering algorithms.

4.3 Workflow

The proposed method consists of three major steps that are typically executed in sequential order:

1. *Contextual neural gas:* The CNG algorithm clusters the data set into *n* spatial clusters, where *n* is the number of neurons. The actual number of clusters in the data is typically unknown; *n* must be chosen large enough so that a reasonable cluster structure can be detected in the subsequent steps. However, if *n* is too large, some of the CNG's neurons may not map any data at all. These neurons must not be removed, because the rank ordering of CNG depends on the number of neurons (Hagenauer and Helbich 2013).
2. *Topology learning:* A topology of the CNG's neurons is learned with a modification of the CHL algorithm. The algorithm can be described as follows: For each input vector, the ranking order of neurons is determined according to the two-phase procedure of the CNG, and a connection between the two highest ranked neurons is added to the connection set. Additionally, the number of times a connection has been added to the set is stored for each connection. This number finally indicates the strength of a connection and is of use in the next step.
3. *Graph clustering:* Before clustering the resulting graph, single vertices that are not connected to any other vertex are removed because the neurons that these vertices represent do not map any data and bear no valuable topological information. Then the graph is clustered based on its structural properties using the MLMO algorithm.

4.4 Experiments

To evaluate the proposed method, two experiments on different data sets are conducted. In both experiments, a CNG with 25 neurons is applied. The neurons are randomly initialized and the training time is set to 100,000 iterations. The neighborhood range and the adaptation rate are chosen as proposed by Martinetz et al. (1993).

4.4.1 Synthetic Data

In this experiment, a synthetic data set is constructed whose properties are clearly determined. Consequently, the results of the proposed method can be easily evaluated. The data set consists of five clusters: one large cluster in the middle with low point density and four smaller clusters in the corners with higher point density (see Fig. 4.1). Each cluster contains 200 random data points and each point has three attributes: the *x* and *y* coordinates and a synthetic attribute, whose value is zero for the middle cluster and otherwise one.

The main challenge when clustering this data set is to differentiate between the spatial clusters in the corners of the data set, because their borders are defined by spatial point density. Spatial clustering algorithms which solely consider the spatial distances between points and/or the similarity of the points' attribute value are likely

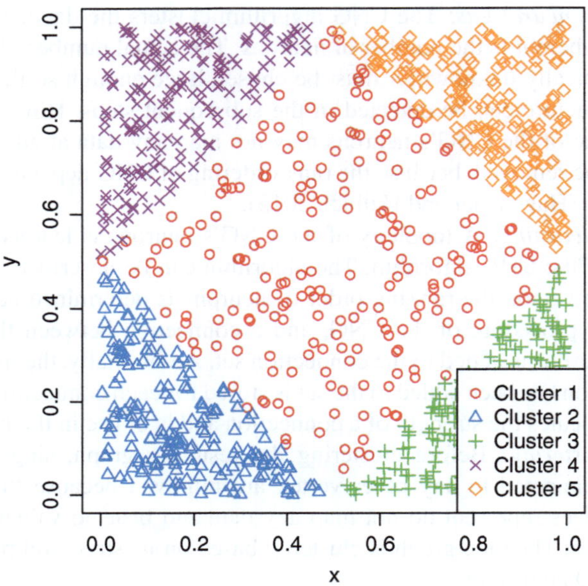

Fig. 4.1 Synthetic data set. The attribute values of the data points of cluster 1 are 0, the attribute values of the other clusters 1

to fail to correctly identify the spatial clusters. Additionally, the clustering of the data set becomes much more difficult, if the actual number of clusters is not known beforehand.

The results for the synthetic data set depend on parameter l. If l is set too low, differences in the observations' attribute values are neglected and the resulting graph therefore exhibits no distinct clusters. Otherwise, if l is set too high, the clustering does not consider the spatial configuration of the data set, resulting in a graph with only two clusters that represent the data points with the values 0 and 1. However, because of the clearly defined cluster structure of the data set, it can be assumed that l is chosen correctly, when the modularity of the resulting graph is maximal.

Figure 4.2 plots the mean modularity scores of 100 runs for different settings of l. For $l > 16$, the mean modularity score is basically constant and at its minimum. Hence, for large l values, the spatial configuration of the data set has no significant effect on the clustering of the resulting graph. Furthermore, the plot shows multiple local maxima; the highest mean modularity score (0.724) is achieved with $l = 7$.

Figure 4.3 exemplarily shows the graph resulting from a CNG that has been trained with $l = 7$ and its clustering, indicated by the colored vertices. At first, it is notable that the depicted graph consists of only 17 vertices, although CNG consists of 25 neurons. Generally, the number of neurons which map no data at all increases rapidly if l is increased, because of the simple clustering structure of the data set. The large number of vertices present indicates that the incorporated degree of spatial dependence is significant. Moreover, the figure reveals that the MLMO algorithm

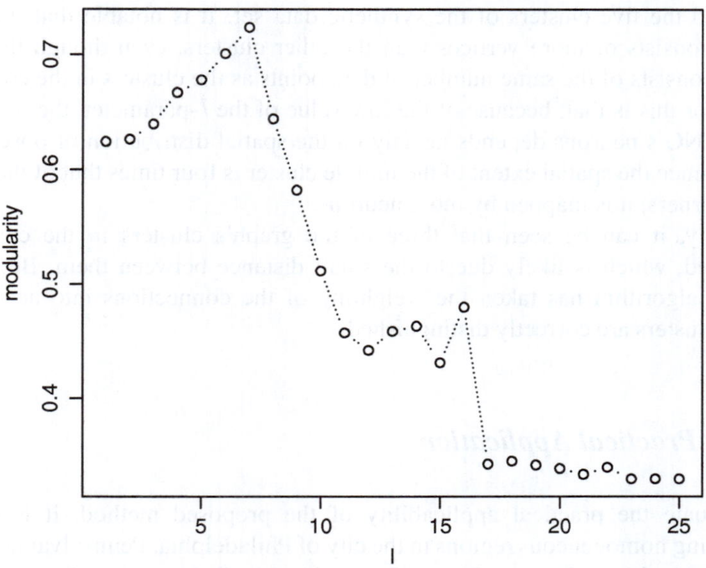

Fig. 4.2 Mean modularity score of 100 runs for different settings of l for the synthetic data set

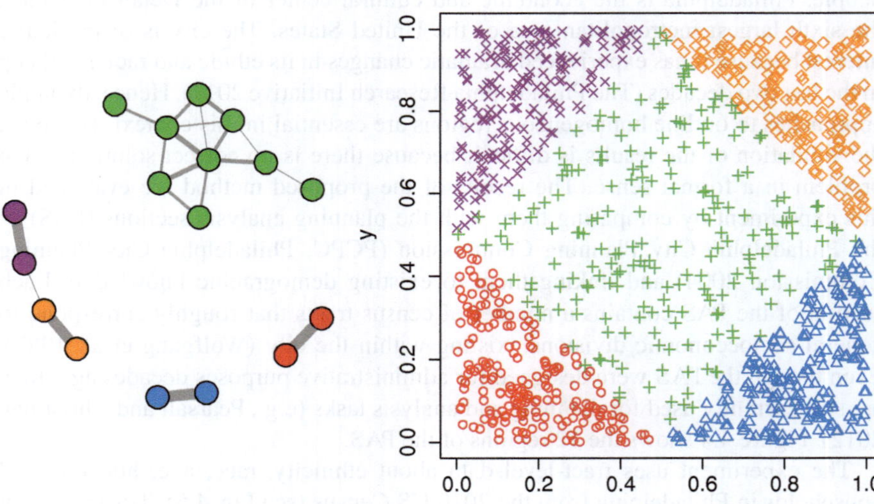

Fig. 4.3 Clustering results for $l = 7$. The data points (*right*) are colored according to the colors of the detected clusters of the graph (*left*). The thickness of the graph's edges corresponds to the weights of the connections

identified the five clusters of the synthetic data set. It is notable that the middle cluster consists of more vertices than the other clusters, even though the middle cluster consists of the same number of data points as the clusters in the corner. The reason for this is that, because of the low value of the l-parameter, the distribution of the CNG's neurons depends heavily on the spatial distribution of observations. Hence, since the spatial extent of the middle cluster is four times that of the clusters in the corners, it is mapped by more neurons.

Finally, it can be seen that three of the graph's clusters in the corners are connected, which is likely due to the small distance between them. Because the MLMO algorithm has taken the weighting of the connections into account, the corner clusters are correctly distinguished.

4.4.2 Practical Application

To evaluate the practical applicability of the proposed method, it is used for delineating homogeneous regions in the city of Philadelphia, Pennsylvania. The city is situated in the northeastern United States along the Delaware and Schuylkill rivers and consists of an area of approximately 369 km^2. Philadelphia is currently the fifth largest city in the United States with an estimated population in 2012 of 1.5 million people. Philadelphia is the economic and cultural center of the Delaware Valley, the sixth largest metropolitan area of the United States. The city is of particular interest because it has experienced dramatic changes in its ethnic and racial makeup in the last two decades (The Philadelphia Research Initiative 2011). Hence, dynamic approaches to outline homogeneous regions are essential in this context. However, the validation of the results is difficult because there is no correct solution to the problem in a formal sense. The results of the proposed method are evaluated in this experiment by comparing them with the planning analysis sections (PAS) of the Philadelphia City Planning Commission (PCPC; Philadelphia City Planning Commission 2004) and linking them to existing demographic knowledge. Each section of the PAS contains a number of census tracts that roughly correspond to general socioeconomic divisions existing within the city (Wolfgang et al. 1987). Even though the PAS were designed for administrative purposes decades ago, they are still currently used for planning and analysis tasks (e.g., Pearsall and Christman 2012). Figure 4.4 shows the 12 regions of the PAS.

The experiment uses tract-level data about ethnicity, race, age, housing, and households in Philadelphia from the 2010 US Census (see Fig. 4.5). Tracts without a significant population are removed from the data set and all attributes are standardized to zero mean and unit variance to make them comparable. Overall, the study site consists of 380 census tracts.

Similar to the previous experiment, it is unclear how much spatial dependence should be incorporated into the CNG's learning process to obtain reasonable results. Figure 4.6 shows the mean modularity scores of 100 runs for different settings of l.

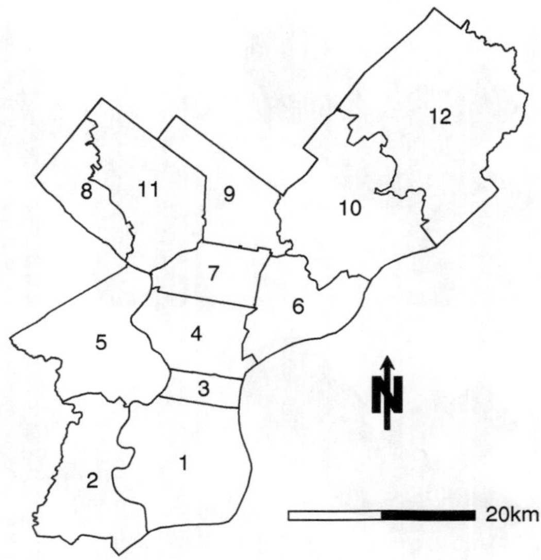

Fig. 4.4 Philadelphia analysis sections (PAS)

The highest mean modularity score (0.646) is achieved with $l = 2$ and $l = 3$. Notable local maxima can be observed for $l = 10$ (0.598) and $l = 13$ (0.600). For $l > 19$, the modularity score is basically at its minimum value.

In contrast to the previous experiment, no prior knowledge about clusters in the data set is available; any parameter l might be as reasonable as any other one. However, based on the objective of this experiment, three demands on l should be met: First, parameter l should be chosen so that the modularity score is high, because a high modularity score is a strong indicator of a clear-cut clustering structure. Second, parameter l should be high enough so that a fair portion of the tracts' social and demographic characteristics is taken into account in the process of clustering. Third, parameter l should be low enough so that the resulting clusters tend to be spatially contiguous. Spatial contiguity is in particular a useful property for spatial planning and policy making, because spatially contiguous clusters can typically be described by a single spatial outline, which eases perception and understanding of the clusters.

Figure 4.3 exemplarily shows the graph resulting from a CNG that has been trained with $l = 2$, $l = 10$, and $l = 13$ alongside the regions that result from clusters of the graphs. While the clusters for $l = 2$ are the most spatially contiguous, there is little difference between $l = 10$ and $l = 13$ observable. Additionally, the graph for $l = 2$ seems much more clearly structured than the graphs for $l = 10$ and $l = 13$.

Furthermore, comparing Fig. 4.7 with Fig. 4.4, it can be seen that the PAS, which were designed for planning purposes, do not correspond well to the obtained clusters. The PAS consist of 12 different regions, whereas the proposed method determined only 5–6 clusters.

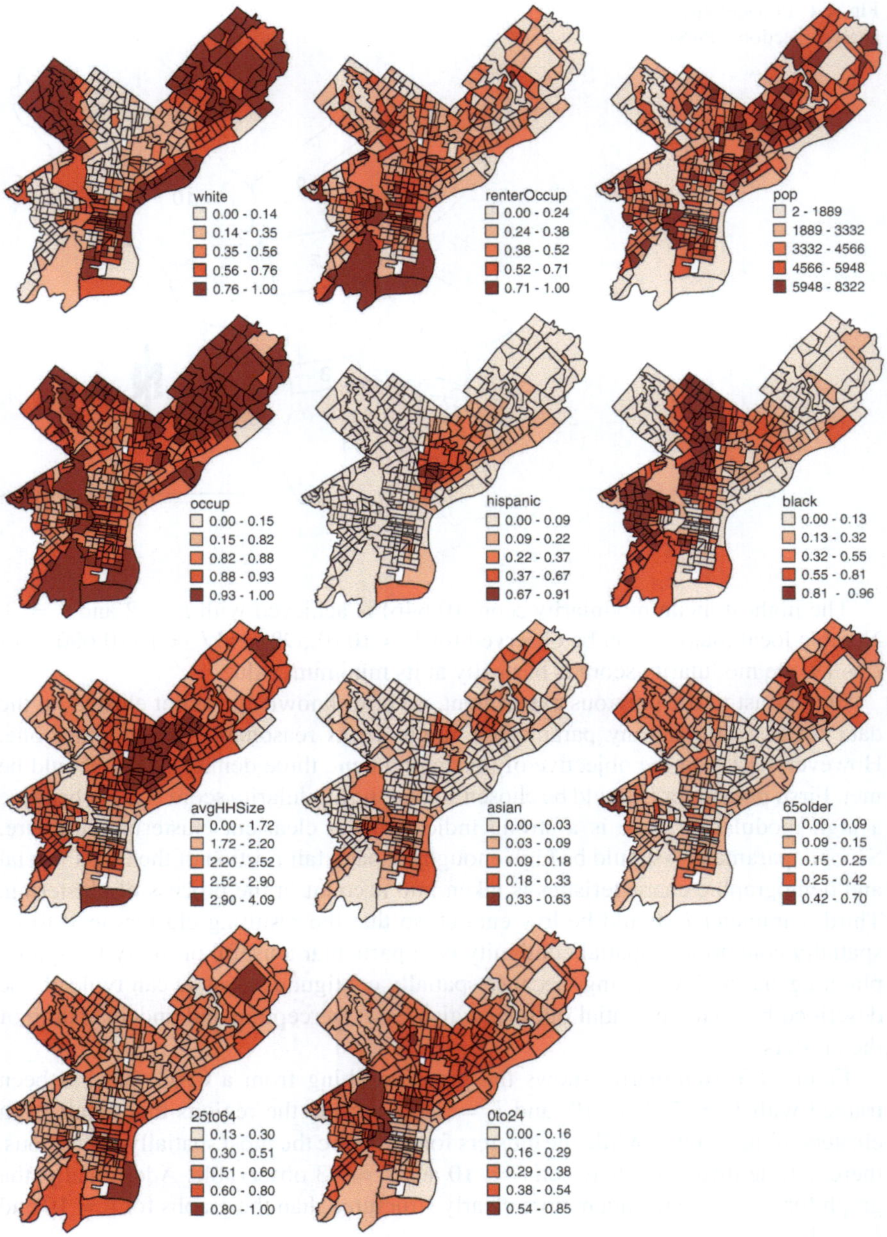

Fig. 4.5 Variables used for the experiment: Rate of Whites (white), Blacks (black), Asians (asian), Hispanics (hispanics), renter-occupied houses (renterOccup), occupied houses (occup), population younger than 25 years (0to24), population between 25 and 64 years (25to64), population older than 64 years (65older), average size of households (avgHHSize), total population (pop)

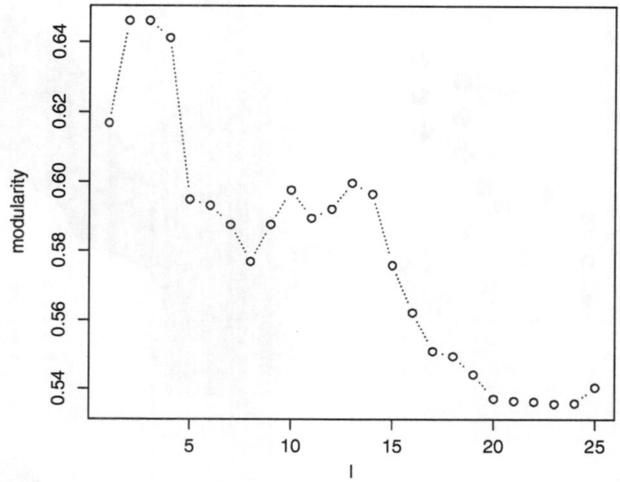

Fig. 4.6 Mean modularity scores of 100 runs for different settings of l for the real-world data set

In order to compare the nonspatial characteristics of the clustering results and of the PAS, their mean homogeneity with respect to the different attributes is compared. The homogeneity of a cluster is calculated as the sample standard deviation of the differences between the cluster's center and the data that is assigned to it. Table 4.1 shows the mean homogeneity values of the attributes for the different clusterings. Notably, the homogeneity for all attributes decreases with increasing l. Furthermore, even though the PAS consists of double as much clusters as the clustering for $l = 13$, its mean homogeneity is mostly equal or worse. However, in contrast to the other clusterings, the PAS is perfectly spatially contiguous. The clustering for $l = 2$ is nearly as spatially contiguous as the PAS, but it is less homogeneous than the PAS with respect to the attributes 65older, black, hispanic, and occup. However, for the majority of the attributes, the clustering for $l = 2$ is still more homogeneous than the PAS.

Philadelphia is one of the most segregated cities in the United States; even the most affluent Blacks live in neighborhoods that are close to majority black (Logan 2011). Hence, it can be expected that these neighborhoods emerge as distinct clusters in the clustering results. Comparing Fig. 4.4 with Fig. 4.5 reveals that the predominantly Black neighborhoods, especially in North Philadelphia, are mixed with non-black neighborhoods or are separated by the outlines of the PAS (e.g., Sects. 7, 9, and 11 in Fig. 4.4). Also the clustering for $l = 2$ (compare Fig. 4.7 with Fig. 4.5) does not clearly identify the predominantly Black neighborhoods. However, these neighborhoods are clearly outlined by cluster 3 of the clustering for $l = 10$ and cluster 5 of the clustering for $l = 13$.

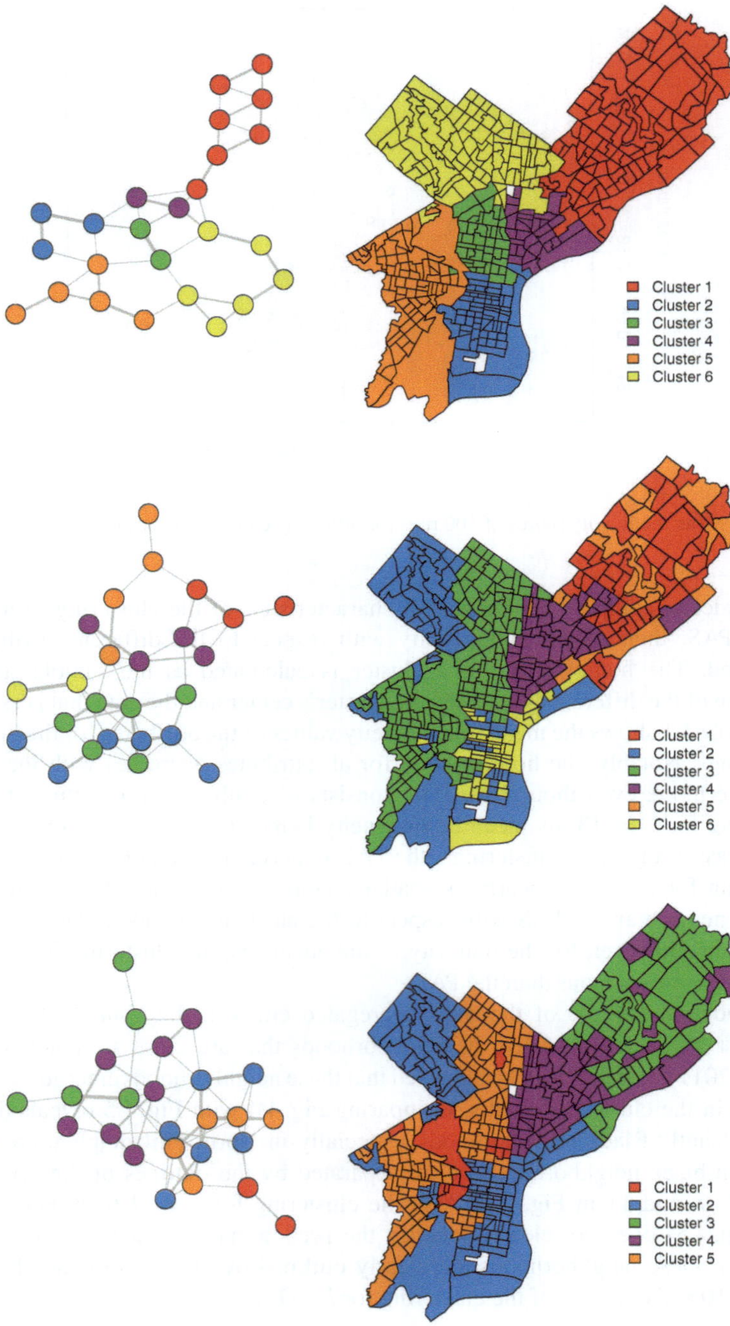

Fig. 4.7 Clustering results for $l = 2$ (*top*), $l = 10$ (*middle*), and $l = 13$ (*bottom*). The census tracts (*right*) are colored according to the colors of the detected clusters of the graph (*left*). The thickness of the graph's edges corresponds to the weights of the connections

Table 4.1 Mean homogeneity of the attributes for different clusterings

	$l = 2$	$l = 10$	$l = 13$	PAS
pop	1,157.834	1,044.499	1,006.730	1,257.946
0–24	0.055	0.037	0.035	0.071
25–64	0.046	0.033	0.027	0.058
65older	0.041	0.032	0.026	0.026
white	0.122	0.103	0.092	0.184
black	0.119	0.101	0.090	0.090
asian	0.039	0.038	0.038	0.043
hispanic	0.040	0.037	0.033	0.033
avgHHSize	0.241	0.242	0.228	0.295
occup	0.043	0.043	0.040	0.400
renterOccup	0.112	0.113	0.106	0.106

4.5 Conclusion and Further Work

This study presented a new method which combines CNG, topology learning, and graph clustering to outline homogeneous regions, taking into account spatial dependence. The proposed method does not require prior knowledge about the actual number of clusters in the data, because it utilizes the modularity score when clustering the learned topology. Two experiments, one using a synthetic data set and another one using a demographic data set of Philadelphia, PA confirmed the usefulness of the method for delineating homogeneous clusters. Because of this property, the proposed method is in particular well suited for spatial analysis and planning tasks.

There are some considerations that must be taken into account when applying the proposed method. The CNG algorithm uses a nonlocal update scheme, which prevents it from being easily stuck in local optima. However, repeated runs of the experiments have shown that the final positions of the neurons and consequently the learned topology can differ slightly with each run. This difference can possibly affect the clustering of the topology.

The results of the proposed method depend also on its parametrization. The method combines multiple algorithms, and each one's parameter setting can critically affect the final results. It is unclear how to choose the parameters so that the final results meet the analyst's requirements. In particular, the choice of the parameter l, which controls the degree of spatial dependence incorporated into the clustering, has a significant impact on the homogeneity and spatial contiguity of the clustering. However, although the chance that the resulting clusters are contiguous increases with high values of l, there is no guarantee that the clusters will ever be spatially contiguous.

In this study, the proposed method utilizes a modified version of the CHL algorithm to learn a topology from CNG, but other approaches might also be reasonable. For example, instead of connecting the first and second neuron of the rank ordering to form a topology, it is also possible to remove the first neuron temporarily from the set of neurons, determine a new rank order using the CNG's ordering scheme, and then connect the first neuron of the resulting rank order with the previously found first neuron. How this strategy performs in comparison to the one used in this study is unclear and deserves further research. Additionally, CHL is sensitive to noisy data and outliers (Aupetit 2005). Using alternative algorithms for topology learning bears potential to improve the results.

This study uses the MLMO algorithm to cluster the CNG's topology. The MLMO algorithm uses a greedy heuristic to optimize the modularity score of the graph. Although the algorithm has been shown to generally perform very well, it lacks accuracy, like any greedy clustering method (Fortunato 2010). In principle, any other graph clustering algorithm can be applied within the graph clustering step of the method.

The proposed method combines different methods from different but related disciplines for clustering spatial data. As scientific research for each of these disciplines is going to continue, it can be expected that more powerful methods will be developed. Utilizing these methods has the potential to further increase the value of the proposed method. In particular, improving the CNG algorithm with regard to convergence and parametrization seems worth pursuing.

Finally, the presented method is rather technical and difficult to understand and apply by nonexperts. In order to be of real practical value for spatial planners and policy makers, it is necessary to integrate the method into a combined software toolkit which provides powerful analytical and visual means in order to validate the results and which is also easy to use.

References

Amedeo D (1969) An optimization approach to the identification of a system of regions. Pap Reg Sci 23(1):25–44

Arribas-Bel D, Schmidt CR (2013) Self-organizing maps and the US urban spatial structure. Environ Plan B Plan Des 40(2):362–371

Aupetit M (2005) Learning topology with the generative gaussian graph and the EM algorithm. In: Advances in neural information processing systems, Vancouver, pp 83–90

Bação F, Lobo V, Painho M (2005) The self-organizing map, the geo-som, and relevant variants for geosciences. Comput Geosci 31:155–163

Bailey T, Gatrell A (1995) Interactive spatial data analysis. Longman Scientific & Technical, Harlow Essex

Blondel VD, Guillaume JL, Lambiotte R, Lefebvre E (2008) Fast unfolding of communities in large networks. J Stat Mech Theory Exp 2008(10):P10008

Brandes U, Delling D, Gaertler M, Gorke R, Hoefer M, Nikoloski Z, Wagner D (2008) On modularity clustering. IEEE Trans Knowl Data Eng 20(2):172–188

Costa J, De Andrade Netto M (1999) Cluster analysis using self-organizing maps and image processing techniques. In: IEEE international conference on systems, man, and cybernetics 1999. IEEE SMC '99 conference proceedings, Tokyo, vol 5, pp 367–372

Costa J, Oliveira R (2007) Cluster analysis using growing neural gas and graph partitioning. In: International joint conference on neural networks, IJCNN 2007, Orlando, pp 3051–3056

De Silva V, Carlsson G (2004) Topological estimation using witness complexes. In: Proceedings of the first eurographics conference on point-based graphics. Eurographics Association, Aire-la-Ville, pp 157–166

Edelsbrunner H, Shah NR (1997) Triangulating topological spaces. Int J Comput Geom Appl 7(04):365–378

Flexer A (2001) On the use of self-organizing maps for clustering and visualization. Intell Data Anal 5(5):373–384

Fortunato S (2010) Community detection in graphs. Phys Rep 486(3–5):75–174

Fortunato S, Barthelemy M (2006) Resolution limit in community detection. Proc Natl Acad Sci 104(1):36–41

Fritzke B (1995) A growing neural gas network learns topologies. Adv Neural Inf Process Syst 7:625–632

Goodchild MF (1986) Spatial autocorrelation. Concepts and techniques in modern geography. Geo Books, Norwich

Guo D, Peuquet D, Gahegan M (2003) ICEAGE: interactive clustering and exploration of large and high-dimensional geodata. GeoInformatica 7(3):229–253

Hagenauer J, Helbich M (2013) Contextual neural gas for spatial clustering and analysis. Int J Geogr Inf Sci 27(2):251–266

Hagenauer J, Helbich M, Leitner M (2011) Visualization of crime trajectories with self-organizing maps: a case study on evaluating the impact of hurricanes on spatio-temporal crime hotspots. In: Proceedings of the 25th international cartographic conference, International Cartographic Association, Paris

Han J, Kamber M, Tung AKH (2001) Spatial clustering methods in data mining: a survey. In: Miller HJ, Han J (eds) Geographic data mining and knowledge discovery. Taylor and Francis, London, pp 188–217

Helbich M, Brunauer W, Hagenauer J, Leitner M (2013) Data-driven regionalization of housing markets. Ann Assoc Am Geogr 103(4):871–889

Jain AK, Murty MN, Flynn PJ (1999) Data clustering: a review. ACM Comput Surv 31(3):264–323

Kohonen T (1982) Self-organized formation of topologically correct feature maps. Biol Cybern 43:59–69. doi:10.1007/BF00337288

Kohonen T (2001) Self-organizing maps, 3rd edn. Springer, New York/Secaucus

Lancichinetti A, Fortunato S (2009) Community detection algorithms: a comparative analysis. Phys Rev E 80:056117

Logan JR (2011) Separate and unequal: the neighborhood gap for Blacks, Hispanics and Asians in metropolitan America. http://www.s4.brown.edu/us2010/Data/Report/report0727.pdf. Accessed on 28 Mar 2013

Martinetz T (1993) Competitive Hebbian learning rule forms perfectly topology preserving maps. In: Gielen S, Kappen B (eds) ICANN 93, Amsterdam. Springer, London, pp 427–434

Martinetz T, Schulten K (1991) A "neural-gas" network learns topologies. Artif Neural Netw 1:397–402

Martinetz T, Berkovich S, Schulten K (1993) "Neural-gas" network for vector quantization and its application to time-series prediction. IEEE Trans Neural Netw 4(4):558–569

Miller HJ (2004) Tobler's first law and spatial analysis. Ann Assoc Am Geogr 94(2):284–289

Murtagh F (1995) Interpreting the kohonen self-organizing feature map using contiguity-constrained clustering. Pattern Recognit Lett 16(4):399–408

Newman MEJ, Girvan M (2004) Finding and evaluating community structure in networks. Phys Rev E 69:026113

Openshaw S (1984) The modifiable areal problem. Concepts and techniques in modern geography. Geo Books, Norwich

Pearsall H, Christman Z (2012) Tree-lined lanes or vacant lots? Evaluating non-stationarity between urban greenness and socio-economic conditions in Philadelphia, Pennsylvania, USA at multiple scales. Appl Geogr 35(1):257–264

Philadelphia City Planning Commission (2004) The political and community service boundaries of Philadelphia. http://www.phila.gov/CityPlanning/resources/Publications/Political_boundaries.pdf. Accessed on 26 Mar 2013

Schaeffer SE (2007) Graph clustering. Comput Sci Rev 1(1):27–64

Skupin A, Esperbé A (2011) An alternative map of the United States based on an n-dimensional model of geographic space. J Vis Lang Comput 22(4):290–304

The Philadelphia Research Initiative (2011) A city transformed – the racial and ethnic changes in Philadelphia over the last 20 years. http://www.pewtrusts.org/uploadedFiles/wwwpewtrustsorg/Reports/Philadelphia_Research_Initiative/Philadelphia-Population-Ethnic-Changes.pdf. Accessed on 26 Mar 2013

Ultsch A, Siemon HP (1990) Kohonen's self organizing feature maps for exploratory data analysis. In: Proceedings of international neural networks conference, Paris. Kluwer Academic, pp 305–308

Van Der Laan L, Schalke R (2001) Reality versus policy: the delineation and testing of local labour market and spatial policy areas. Eur Plan Stud 9(2):201–221

Wolfgang ME, Figlio R, Sellin T (1987) Delinquency in a birth cohort. Studies in crime and justice. University of Chicago Press, Chicago

Part II
Housing and Real Estate

Chapter 5
Hedonic House Price Modeling Based on Multilevel Structured Additive Regression

Alexander Razen, Wolfgang Brunauer, Nadja Klein, Stefan Lang, and Nikolaus Umlauf

Abstract This chapter reviews recent developments in hedonic modeling of house prices based on structured additive regression (STAR) models. In STAR models, continuous covariates are modeled as P(enalized)-splines. Furthermore, random effects for spatial indexes, smooth functions of two-dimensional surfaces, and (spatially) varying coefficient terms may also be estimated using this methodology. Based on hierarchical STAR models, we discuss a number of useful extensions. With respect to value-at-risk concepts, financial institutions are often not only interested in the expected value but also in different quantiles of the distribution of real estate prices. To meet these requirements, we apply multilevel STAR models for location scale and shape (GAMLSS type regression) and a Bayesian version of quantile regression. As another extension, we sketch multiplicative region-specific scaling factors for nonlinear covariates in order to permit spatial variation in the nonlinear price gradients.

Keywords Bayesian hierarchical models • Hedonic pricing models • GAMLSS • Bayesian quantile regression • MCMC

Opinions expressed by the authors do not necessarily reflect the official viewpoint of UniCredit Bank Austria AG.

A. Razen (✉) • S. Lang • N. Umlauf
University of Innsbruck, Universitätsstraße 15, A-6020 Innsbruck, Austria
e-mail: alexander.razen@uibk.ac.at; nikolaus.umlauf@uibk.ac.at; stefan.lang@uibk.ac.at

W. Brunauer
UniCredit Bank Austria AG, Julius Tandler-Platz 3, A-1090 Wien, Austria
e-mail: wolfgang.brunauer@realevalue.at

N. Klein
University of Göttingen, Platz der Göttinger Sieben 5, D-37073 Göttingen, Germany
e-mail: nklein@gwdg.de

M. Helbich et al. (eds.), *Computational Approaches for Urban Environments*, Geotechnologies and the Environment 13, DOI 10.1007/978-3-319-11469-9_5

5.1 Introduction

The Basel II and III frameworks strictly define the conditions under which financial institutions are authorized to accept real estate as collateral in order to decrease their credit risk, including the evaluation of the properties on a regular basis by means of statistical methods. A widely used concept here is the hedonic pricing model (Rosen 1974). It assumes that the price of a property can be decomposed into implicit prices of its attributes, which are estimated in a regression analysis of price against attributes. Reviews of hedonic price theory in a real estate context can be found, for example, in Follain and Jimenez (1985), Sheppard (1999) or Malpezzi (2003).

The bundle of attributes characterizing a property involves not only individual attributes of the building itself but also locational attributes of the region where the building is located in. Thus, the real estate market intrinsically is spatial; why there is a vast literature on spatial house price modeling, see, for example, Banerjee et al. (2004), Cohen and Coughlin (2008) or Helbich et al. (2014). Typically, residential properties belong to several levels of spatial (administrative) units, which turns the hedonic model into a *multilevel* or *hierarchical* regression problem (Gelman and Hill 2006). For instance, in our case study, house selling prices with associated individual attributes (the elementary level-1) are grouped in municipalities (level-2), which form districts (level-3), which are themselves nested in counties (level-4). Available neighborhood covariates on either of these spatial resolutions that might be important for predicting house prices should be accounted for, and it is furthermore reasonable to assume that unmeasured neighborhood characteristics such as local policy and infrastructure affect individual house prices. Another major problem in hedonic price modeling is that economic theory does not provide clear guidance concerning the functional form of the dependence of price on characteristics, which suggests that hedonic pricing models should allow for nonlinearity in the price functions.

A particularly broad and rich framework for nonlinear and spatial modeling is provided by generalized structured additive regression (STAR) models, described, for example, in Fahrmeir et al. (2013). In STAR models, continuous covariates are modeled as P(enalized)-splines. Furthermore, random effects for spatial indexes, smooth functions of two-dimensional surfaces, and (spatially) varying coefficient terms may also be estimated using this methodology.

The purpose of this chapter is to review recent developments in hedonic modeling of house prices based on STAR models. More specifically, we describe *multilevel versions of STAR models* (Lang et al. 2014; Brunauer et al. 2013) as our basic modeling framework and develop a number of extensions. With respect to value-at-risk concepts, financial institutions are often not only interested in the expected value but also in different quantiles of the distribution of real estate prices. To meet these requirements, we apply multilevel STAR models for location scale and shape (GAMLSS type regression, Klein et al. 2013; Rigby and Stasinopoulos 2005) and a Bayesian version of quantile regression (Waldmann et al. 2013) and compare the results (Razen et al. 2014). As a second extension, we introduce multiplicative region-specific scaling factors for nonlinear covariates in order to permit spatial variation in the nonlinear price gradients. This allows highly nonlinear implicit price functions to vary within a regularized framework, accounting for district-specific spatial heterogeneity, which can lead to a considerable improvement of model quality and predictive power (Brunauer et al. 2010). We finally describe publicly available software to estimate the described complex modeling framework (Umlauf et al. 2012).

5.2 Data Description and Model Specification

We have a dataset of owner-occupied single-family homes in Austria at our disposal which exhibits a quite typical structure for real estate data:

- The set of explanatory variables consists of covariates characterizing the house, namely, the size, age, year of sale, quality, and equipment of the building, which we call *structural attributes/covariates*.
- Individual observations are linked to municipality codes, which allows association with covariates accounting for sociodemographic, economic, and neighborhood attributes. Following, for example, Can (1998), we will call these *neighborhood attributes/covariates*.

5.2.1 Structural Attributes

The dataset containing dated house prices together with the housing attributes has been collected in order to estimate the value of the collateral for mortgages by the UniCredit Bank Austria AG from October 1997 to September 2009. Two slightly different instructions for data collection have been employed, which is why the structural covariates affected thereof are encoded accordingly (see Brunauer et al. (2013) for a detailed description). We use continuous variables measuring the size and the age as well as the time of sale and categorical variables that describe the quality of the house. Guided by economic theory on hedonic house prices, we expect the following directions of the effects:

- Continuous covariates/attributes: As we regress the structural covariates on logged prices per square meter (sq. m.), a decreasing effect of the floor area of the building due to decreasing marginal returns of additional floor area (*area*) and an increasing effect of the size of the plot it is built on (*area_plot*) can be assumed. The age of the building (*age*), which is calculated as the difference between the year of valuation and the year of construction (i.e., the age at the time of sale), reflects depreciation over time and should therefore have a decreasing effect. The time index (*time_index*, the year of purchase of the house) can be considered as the remaining unexplained temporal heterogeneity and is a measure for the quality adjusted development of house prices over time.

- Categorical covariates/attributes: A high quality of the heating system (*heat*) as well as of the bathroom and toilets (*bath*) should have an increasing effect on house prices. Furthermore, the existence of an attic (*attic_dum*), a terrace (*terr_dum*), and a garage (*garage*, further separated into good and bad quality) should raise house prices.

5.2.2 Spatial Resolutions and Neighborhood Attributes

House prices with structural attributes are nested within three spatial resolutions and hence associated with the respective neighborhood attributes, which we use on the most detailed level available. We use various socioeconomic and demographic attributes as well as measures of proximity to work and metropolitan areas, obtained from the sources described in Brunauer et al. (2013), to explain spatial variation in house prices per sq. m.

Level-1 is the individual level, on which house prices and housing attributes are measured (see Sect. 5.2.1). In total, 3,231 observations are available on the individual level after validation.

Level-2 is the municipal level. Observations are available in 946 of the 2,379 Austrian municipalities. On level-2, we employ the following covariates:

- Socioeconomic/demographic characteristics of the neighborhood: On the one hand, we use the purchase power index (*pp_ind*), the average level of education, indicated by the share of academics (*educ*), which both reflect disposable income and should therefore affect prices positively. On the other hand, we use an age index (*age_ind*), constructed as a population-weighted mean of 20 age cohorts, which measures the average age of inhabitants. A high population age index, reflecting excess of age, serves as a proxy for structural weakness and should have a negative effect on house prices.
- Measures of proximity to work and metropolitan areas: Urban economic theory states that commuting to centers of economic activity gives rise to a location rent, which is why a high commuter index (*comm*), that is, many employees commuting from the municipality, should tend to affect prices negatively. However, close proximity to these centers also provides certain disamenities, as the local infrastructure tends to match the needs of residential use worse. Therefore, the effect of a low commuter index is unclear. Furthermore, as a measure of centrality, we employ population density (*dens*). In densely populated areas, land becomes more valuable, which is why we expect a positive effect of this covariate.

Level-3 is the district level. Individual observations are available on 109 of 121 districts, only the inner districts of Vienna are missing. As each of these districts has neighboring units, spatial effects can be regularized using the neighborhood structure. On this level, an externally provided home price index indicating the neighboring house price level, *wko_ind*, is available.

Level-4 is the county level (9 counties); we do not employ any further explanatory covariates on this level.

5.3 Structured Additive Distributional Regression

5.3.1 Hierarchical STAR Models

Suppose that observations $(y_i, \mathbf{z}_i, \mathbf{x}_i)$, $i = 1, \ldots, n$, are given, where y_i is a continuous response variable, and $\mathbf{z}_i = (z_{i1}, \ldots, z_{iq})'$, and $\mathbf{x}_i = (x_{i1}, \ldots, x_{ip})'$ are vectors of covariates. For the variables in $\mathbf{z}$, possibly nonlinear effects are assumed, whereas the variables in $\mathbf{x}$ are modeled in the usual linear way. The components of $\mathbf{z}$ are not necessarily continuous covariates. A component may also indicate a time scale, a cluster, or a spatial index (e.g., municipality, district, or county) a certain observation pertains to. We assume an additive decomposition of the effects of z_{ij} (and x_{ij}) and obtain the model

$$y_i = f_1(z_{i1}) + \ldots + f_q(z_{iq}) + \mathbf{x}_i'\boldsymbol{\gamma} + \varepsilon_i. \tag{5.1}$$

Here, $f_1, \ldots, f_q$ are nonlinear functions of the covariates $\mathbf{z}_i$ and $\mathbf{x}_i'\boldsymbol{\gamma}$ is the usual linear part of the model. The errors ε_i are assumed to be mutually independent Gaussian with mean 0 and variance σ^2, that is, $\varepsilon_i \sim N(0, \sigma^2)$.

The nonlinear effects in (5.1) are modeled by a basis functions approach, that is, a particular function f of covariate z is approximated by a linear combination of basis or indicator functions

$$f(z) = \sum_{k=1}^{K} \beta_k B_k(z). \tag{5.2}$$

The B_k's are known basis functions and $\boldsymbol{\beta} = (\beta_1, \ldots, \beta_K)'$ is a vector of unknown regression coefficients to be estimated. Defining the $n \times K$ design matrix $\mathbf{Z}$ with elements $\mathbf{Z}[i,k] = B_k(z_i)$, the vector $\mathbf{f} = (f(z_1), \ldots, f(z_n))'$ of function evaluations can be written in matrix notation as $\mathbf{f} = \mathbf{Z}\boldsymbol{\beta}$. Accordingly, we obtain

$$\mathbf{y} = \boldsymbol{\eta} + \boldsymbol{\varepsilon} = \mathbf{Z}_1\boldsymbol{\beta}_1 + \ldots + \mathbf{Z}_q\boldsymbol{\beta}_q + \mathbf{X}\boldsymbol{\gamma} + \boldsymbol{\varepsilon}, \tag{5.3}$$

where $\mathbf{y} = (y_1, \ldots, y_n)'$, $\boldsymbol{\eta} = (\eta_1, \ldots, \eta_n)'$ and $\boldsymbol{\varepsilon} \sim N(\mathbf{0}, \sigma^2\mathbf{I})$.

Recently Lang et al. (2014) have proposed a multilevel version of STAR models to cope with the hierarchical nature of the data on house prices. Suppose that covariate $z_j \in \{1, \ldots, K\}$ is a unit or cluster index and z_{ij} indicates the cluster observation i pertains to. In our case, z_{ij} denotes the municipality where the ith house is located. Then the design matrix $\mathbf{Z}_j$ is a $n \times K$ incidence matrix with $\mathbf{Z}_j[i,k] = 1$ if the i-th observation belongs to cluster k and zero else. The $K \times 1$ parameter vector $\boldsymbol{\beta}_j$ is the vector of regression parameters, that is, the k-th element in $\boldsymbol{\beta}$ corresponds to the regression coefficient of the k-th cluster. We now define the second-level equation

$$\boldsymbol{\beta}_j = \boldsymbol{\eta}_j + \boldsymbol{\varepsilon}_j = \mathbf{Z}_{j1}\boldsymbol{\beta}_{j1} + \ldots + \mathbf{Z}_{jq_j}\boldsymbol{\beta}_{jq_j} + \mathbf{X}_j\boldsymbol{\gamma}_j + \boldsymbol{\varepsilon}_j, \tag{5.4}$$

where the terms $\mathbf{Z}_{j1}\boldsymbol{\beta}_{j1}, \ldots, \mathbf{Z}_{jq_j}\boldsymbol{\beta}_{jq_j}$ correspond to additional nonlinear functions $f_{j1}, \ldots, f_{jq_j}$ and $\mathbf{X}_j\boldsymbol{\gamma}_j$ comprises additional linear effects of cluster level covariates. For the house price data, these are the covariates on municipality level, that is, the purchase power index, share of academics, etc. The "errors" $\boldsymbol{\varepsilon}_j \sim N(\mathbf{0}, \tau_j^2\mathbf{I})$ comprise a vector of i.i.d. Gaussian random effects. Using the compound prior (5.4), we obtain an additive decomposition of the cluster-specific effect. By allowing a full STAR predictor (as in the level-1 equation), a rather complex decomposition of the cluster effect $\boldsymbol{\beta}_j$ including interactions is possible. A special case arises if cluster-specific covariates are not available. Then the prior for $\boldsymbol{\beta}_j$ collapses to $\boldsymbol{\beta}_j = \boldsymbol{\varepsilon}_j \sim N(\mathbf{0}, \tau_j^2\mathbf{I})$, and we obtain a simple i.i.d. Gaussian cluster-specific random effect with variance parameter τ_j^2.

A third or fourth level in the hierarchy is possible by assuming that the second or third level regressions contain additional cluster-specific random effects whose parameters are again modeled through STAR predictors of cluster-level covariates.

In our model, we distinguish four levels: Single-family homes (level-1) belong to municipalities (level-2), which are nested in districts (level-3), which are themselves nested in counties (level-4). Then our model can be written as the following four-level hierarchical STAR model:

$$\begin{aligned}
\text{level-1: } \mathbf{lnp_{qm}} &= \mathbf{f}_1(\mathit{area}) + \mathbf{f}_2(\mathit{areaplot}) + \mathbf{f}_3(\mathit{age}) + \mathbf{f}_4(\mathit{timeindex}) + \\
&\quad \mathbf{f}_5(\mathit{muni}) + \mathbf{X}\boldsymbol{\gamma} + \boldsymbol{\varepsilon} \\
&= \mathbf{Z}_1\boldsymbol{\beta}_1 + \mathbf{Z}_2\boldsymbol{\beta}_2 + \mathbf{Z}_3\boldsymbol{\beta}_3 + \mathbf{Z}_4\boldsymbol{\beta}_4 + \mathbf{Z}_5\boldsymbol{\beta}_5 + \mathbf{X}\boldsymbol{\gamma} + \boldsymbol{\varepsilon} \\
\text{level-2: } \boldsymbol{\beta}_5 &= \mathbf{f}_{5,1}(\mathit{ppind}) + \mathbf{f}_{5,2}(\mathit{lneduc}) + \mathbf{f}_{5,3}(\mathit{ageind}) + \mathbf{f}_{5,4}(\mathit{comm}) + \\
&\quad \mathbf{f}_{5,5}(\mathit{lnden}) + \mathbf{f}_{5,6}(\mathit{dist}) + \boldsymbol{\varepsilon}_5 \\
&= \mathbf{Z}_{5,1}\boldsymbol{\beta}_{5,1} + \mathbf{Z}_{5,2}\boldsymbol{\beta}_{5,2} + \mathbf{Z}_{5,3}\boldsymbol{\beta}_{5,3} + \mathbf{Z}_{5,4}\boldsymbol{\beta}_{5,4} + \\
&\quad \mathbf{Z}_{5,5}\boldsymbol{\beta}_{5,5} + \mathbf{Z}_{5,6}\boldsymbol{\beta}_{5,6} + \boldsymbol{\varepsilon}_5 \\
\text{level-3: } \boldsymbol{\beta}_{5,6} &= \mathbf{f}_{5,6,1}(\mathit{wkoind}) + \mathbf{f}^{mrf}_{5,6,2}(\mathit{dist}) + \mathbf{f}_{5,6,3}(\mathit{county}) + \boldsymbol{\varepsilon}_{5,6} \\
&= \mathbf{Z}_{5,6,1}\boldsymbol{\beta}_{5,6,1} + \mathbf{Z}_{5,6,2}\boldsymbol{\beta}_{5,6,2} + \mathbf{Z}_{5,6,3}\boldsymbol{\beta}_{5,6,3} + \boldsymbol{\varepsilon}_{5,6}, \\
\text{level-4: } \boldsymbol{\beta}_{5,6,3} &= \mathbf{1}\gamma_0 + \boldsymbol{\varepsilon}_{5,6,3}.
\end{aligned} \tag{5.5}$$

The level-1 equation contains the main predictor. Apart from usual linear effects, the predictor is composed of possibly nonlinear effects of the continuous covariates *area*, *areaplot*, *age*, and *time_index* as well as an uncorrelated municipality effect controlling for spatial heterogeneity. The municipality effect is a rather complex spatial effect and is further decomposed through the remaining levels in the hierarchy of the model. The level-2 equation contains nonlinear effects of continuous municipality-specific covariates and a spatial district effect, which is decomposed in the level-3 equation. Two of the covariates on level-2 enter the equation logarithmically (denoted by the prefix "*ln_*"), namely, the share of academics and the population density. The reason for this is that the distributions of these covariates are strongly positively skewed, which results in volatile estimation results on the natural scale. District-specific spatial heterogeneity is modeled through the correlated spatial effect *dist* in the level-3 equation by Markov random fields (see Sect. 5.4.3). We denote this by the superscript "*mrf*". The level-3 equation is additionally composed of a nonlinear effect of the district-specific covariate *wko_ind* and a spatial county effect. The level-4 equation constitutes a usual county specific i.i.d. random effect and for technical reasons (improved mixing) the intercept of the model.

On all levels, for continuous covariates, possibly nonlinear functions $\mathbf{f}_1, \mathbf{f}_2, \ldots$ modeled by P-splines (see Sect. 5.4.2) are assumed. The categorical covariates on level-1, describing the quality and condition of the house, are encoded as dummy variables and subsumed in the design matrix $\mathbf{X}$ with estimated parameters $\boldsymbol{\gamma}$.

5.3.2 *Beyond Mean Modeling: Distributional Regression*

The model implied by Eq. (5.3) can equivalently be written as

$$\mathbf{y} \sim N(\boldsymbol{\eta}, \sigma^2\mathbf{I}) = N(\mathbf{Z}_1\boldsymbol{\beta}_1 + \ldots + \mathbf{Z}_q\boldsymbol{\beta}_q + \mathbf{X}\boldsymbol{\gamma}, \sigma^2\mathbf{I}), \tag{5.6}$$

that is, given the covariates the response vector is multivariate normal with mean $\boldsymbol{\eta}$ and homoscedastic covariance matrix $\sigma^2\mathbf{I}$. Hence, so far only the mean of the response distribution is modeled in dependence of covariates. As already pointed out in Fahrmeir et al. (2004), not only the mean but also the variances of the response may depend on covariates when modeling real estate data. For instance, the analysis of data on the monthly rent of apartments in Munich in Fahrmeir et al. (2004) revealed that apartments built in the 1950s–1970s generally exhibit lower variability than modern apartments that have been built recently. The reason is that the postwar period in Germany is characterized by a quite homogeneous construction style based on low-quality construction material.

To model the variance of the responses, we may assume

$$\sigma^2 = \exp(\tilde{\boldsymbol{\eta}}) = \exp(\tilde{\mathbf{Z}}_1\tilde{\boldsymbol{\beta}}_1 + \ldots + \tilde{\mathbf{Z}}_q\tilde{\boldsymbol{\beta}}_q + \tilde{\mathbf{X}}\boldsymbol{\gamma}),$$

where the $\tilde{\mathbf{Z}}_j$ and $\tilde{\mathbf{X}}$ are design matrices of further covariates $\tilde{z}_1, \ldots, \tilde{z}_q$ and $\tilde{x}_1, \ldots, \tilde{x}_p$ modeling the variance. Of course, some (or all) of these covariates may be identical to the covariates that enter the mean equation. In general, $\tilde{\eta}$ may be another STAR predictor including further hierarchical levels.

The heteroscedastic multilevel STAR model is a special case of the much more general class of multilevel STAR models for location scale and shape (GAMLSS type regression, recently introduced in Klein et al. (2013) as structured additive distributional regression). Here, we assume that observations on a scalar response variable $y_1, \ldots, y_n$ as well as covariate information $\boldsymbol{\nu}_i$, $i = 1, \ldots, n$, have been collected. The conditional distribution of observation y_i given the covariate information $\boldsymbol{\nu}_i$ is assumed to be from a pre-specified class of K-parametric distributions $f_i(y_i|\vartheta_{i1}, \ldots, \vartheta_{iK})$ indexed by the (in general covariate-dependent) parameters $\vartheta_{i1}, \ldots, \vartheta_{iK}$. Each parameter ϑ_{ik} is linked to a semiparametric regression predictor η_{ik} formed of the covariates via a suitable (one-to-one) response function such that $\vartheta_{ik} = h_k(\eta_{ik})$ and $\eta_{ik} = h_k^{-1}(\vartheta_{ik})$. The response function is usually chosen to ensure appropriate restrictions on the parameter space such as the exponential function $\vartheta_{ik} = \exp(\eta_{ik})$, to ensure positivity.

As an example, we consider the two-parametric gamma distribution with mean parameter $\vartheta_1 = \mu > 0$, shape parameter $\vartheta_2 = \sigma > 0$, and probability density function

$$f(y_i|\mu_i, \sigma_i) = \left(\frac{\sigma_i}{\mu_i}\right)^{\sigma_i} \times \frac{y_i^{\sigma_i - 1}}{\Gamma(\sigma_i)} \times \exp\left(-\frac{\sigma_i}{\mu_i} \times y_i\right),$$

where $\Gamma(x) = \int_0^\infty u^{x-1} \exp(-u) \mathrm{d}u$ for $x > 0$ is the gamma function. Due to the positivity constraints, both parameters are linked to a semiparametric regression predictor via the exponential function:

$$\mu_i = \exp(\eta_i^\mu) > 0,$$
$$\sigma_i = \exp(\eta_i^\sigma) > 0.$$

For each predictor, we set up a four level hierarchical STAR model similar to model (5.5).

5.3.3 *Bayesian Quantile Regression*

The GAMLSS framework allows to model the most important characteristics of the response distribution as a function of covariates. However, we still rely on a specific parametric probability distribution like the normal or gamma distribution. In contrast, quantile regression aims at directly modeling the *quantiles* of the response distribution in dependence of covariates without resorting to a specific parametric distribution family. For $0 < \tau < 1$, let q_τ be the τ-quantile of the response distribution, for example, $q_{0.75}$ is the 75 % quantile. Then in linear quantile regression, we assume

$$q_{\tau,i} = \beta_{\tau,0} + \beta_{\tau,1}x_{i1} + \ldots + \beta_{\tau,p}x_{ip},$$

that is, the quantile q_τ of the response distribution is a linear combination of the covariates as in the multiple linear regression model. Generalizations to structured additive predictors are conceptually straightforward (although estimation is truly a challenge). The response distribution is implicitly determined by the estimated quantiles q_τ provided that quantiles for a reasonable dense grid of τ-values are estimated. In contrast to the GAMLSS framework, a specific *parametric* distribution is not specified a priori which makes quantile regression a distribution-free approach.

Estimation of the quantile-specific regression coefficients $\boldsymbol{\beta}_\tau$ is achieved by minimizing the asymmetrically weighted absolute error criterion

$$\hat{\boldsymbol{\beta}}_\tau = \text{argmin}_{\boldsymbol{\beta}} \sum_{i=1}^{n} w_\tau(y_i, \eta_{i\tau})|y_i - \eta_{i\tau}| \tag{5.7}$$

where $\eta_{i\tau} = \mathbf{x}_i'\boldsymbol{\beta}_\tau$ and

$$w_\tau(y_i, \eta_{i\tau}) = \begin{cases} 1-\tau & y_i < \eta_{i\tau} \\ 0 & y_i = \eta_{i\tau} \\ \tau & y_i > \eta_{i\tau}. \end{cases}$$

Frequentist quantile regression as outlined above is extensively treated in Koenker (2005), see also Fahrmeir et al. (2013).

Bayesian quantile regression has been developed utilizing the equivalence between posterior mode and maximum likelihood estimation under noninformative priors $\boldsymbol{\beta}_\tau \propto$ const; see Yu and Moyeed (2001) and Yue and Rue (2011). Therefore, we have to define a specific distributional assumption for the error terms (or equivalently the responses) to make the Bayesian standard machinery work. If we start with the model,

$$y_i = \mathbf{x}_i'\boldsymbol{\beta}_\tau + \varepsilon_{i\tau}, \quad i = 1, \ldots, n,$$

we will assume independent and identically distributed errors following an asymmetric Laplace distribution, that is, $\varepsilon_{i\tau} \mid \sigma^2$ i.i.d. ALD$(0, \sigma^2, \tau)$ with density

$$p(\varepsilon_{i\tau} \mid \sigma^2) = \frac{\tau(1-\tau)}{\sigma^2} \exp\left(-w_\tau(\varepsilon_{i\tau}, 0)\frac{|\varepsilon_{i\tau}|}{\sigma^2}\right).$$

For the responses, the error distribution induces $y_i \mid \boldsymbol{\beta}_\tau, \sigma^2 \sim$ ALD$(\mathbf{x}_i'\boldsymbol{\beta}_\tau, \sigma^2, \tau)$, such that the density of the responses is given by

$$p(y_i \mid \boldsymbol{\beta}_\tau, \sigma^2) = \frac{\tau(1-\tau)}{\sigma^2} \exp\left(-w_\tau(y_i, \mathbf{x}_i'\boldsymbol{\beta}_\tau)\frac{|y_i - \mathbf{x}_i'\boldsymbol{\beta}_\tau|}{\sigma^2}\right).$$

It then turns out that maximizing the corresponding posterior (for fixed σ^2)

$$p(\boldsymbol{\beta}_\tau \mid \mathbf{y}, \sigma^2) \propto \prod_{i=1}^{n} p(y_i \mid \boldsymbol{\beta}_\tau, \sigma^2)$$

$$\propto \exp\left(-\sum_{i=1}^{n} w_\tau(y_i, \mathbf{x}_i'\boldsymbol{\beta}_\tau)\frac{|y_i - \mathbf{x}_i'\boldsymbol{\beta}_\tau|}{\sigma^2}\right),$$

with respect to $\boldsymbol{\beta}_\tau$, is equivalent to minimizing the optimization criterion (5.7).

While the asymmetric Laplace distribution allows to conveniently express quantile regression in a Bayesian framework, it complicates inference based on Markov chain Monte Carlo (MCMC) simulations due to the absolute value contained in its definition. It is therefore advantageous to represent the asymmetric Laplace distribution as a scale mixture of normal distributions as suggested in Yue and Rue (2011): Let $u_i \mid \sigma^2 \sim \text{Expo}(1/\sigma^2)$, $i = 1, \ldots, n$, be i.i.d. exponentially distributed with rate parameter σ^2 and

$$y_i \mid u_i, \boldsymbol{\beta}_\tau, \sigma^2 \sim N(\mathbf{x}_i'\boldsymbol{\beta}_\tau + \xi u_i, \sigma^2/w_i)$$

with

$$\xi = \frac{1-2\tau}{\tau(1-\tau)}, \qquad w_i = \frac{1}{\delta^2 u_i}, \qquad \delta^2 = \frac{2}{\tau(1-\tau)}.$$

Then the marginal distribution $y_i \mid \boldsymbol{\beta}_\tau, \sigma^2$ is obtained by integrating out u_i and is indeed an asymmetric Laplace distribution, that is,

$$y_i \mid \boldsymbol{\beta}_\tau, \sigma^2 \sim \text{ALD}(\mathbf{x}_i'\boldsymbol{\beta}_\tau, \sigma^2, \tau).$$

Bayesian inference can now efficiently be implemented after imputing the scale variables u_i as additional unknowns. Basically, the resulting model is a conditionally Gaussian regression model with offsets ξu_i and weights w_i; see Waldmann et al. (2013) and Fahrmeir et al. (2013) for details.

Finally note that the linear predictor $\eta_\tau = \mathbf{x}_i'\boldsymbol{\beta}_\tau$ in Bayesian quantile regression can be replaced by a (hierarchical) structured additive predictor as in (5.1) without any further difficulties. This is in contrast to frequentist quantile regression as, for example, in Koenker and Mizera (2004). Here, statistical inference is considerably complicated by replacing linear by additive predictors.

5.4 Effect Modeling and Priors

Effect modeling and priors depend on the covariate or term type. We first describe the general form of priors. Sections 5.4.2 and 5.4.3 give specific examples for effect modeling using specific design matrices and forms of the basic prior.

5.4.1 General Form of Basic Priors

In a frequentist setting, overfitting of a particular function $\mathbf{f} = \mathbf{Z}\boldsymbol{\beta}$ is avoided by defining a roughness penalty on the regression coefficients; see, for instance, Fahrmeir et al. (2013) in the context of structured additive regression. In a Bayesian framework, a standard smoothness prior is a (possibly improper) Gaussian prior of the form

$$p(\boldsymbol{\beta}|\tau^2) \propto \left(\frac{1}{\tau^2}\right)^{\mathrm{rk}(\mathbf{K})/2} \exp\left(-\frac{1}{2\tau^2}\boldsymbol{\beta}'\mathbf{K}\boldsymbol{\beta}\right) \times I(\mathbf{A}\boldsymbol{\beta} = \mathbf{0}), \tag{5.8}$$

where $I(\cdot)$ is the indicator function. The key components of the prior are the penalty matrix $\mathbf{K}$, the variance parameter τ_j^2, and the constraint $\mathbf{A}\boldsymbol{\beta} = \mathbf{0}$. Usually the penalty matrix is rank deficient, that is, $\mathrm{rk}(\mathbf{K}) < K$, resulting in a partially improper prior.

The amount of smoothness is governed by the variance parameter τ^2. A conjugate inverse Gamma prior is employed for τ^2 (as well as for the error variance parameter σ^2 in models with Gaussian responses), that is, $\tau^2 \sim IG(a, b)$ with small values such as $a = b = 0.001$ for the hyperparameters a and b resulting in an uninformative prior on the log scale.

The term $I(\mathbf{A}\boldsymbol{\beta} = \mathbf{0})$ imposes required identifiability constraints on the parameter vector. A straightforward choice is $\mathbf{A} = (1, \ldots, 1)$, that is, the regression coefficients are centered around zero.

5.4.2 Continuous Covariate Effects

For a continuous covariate z, our basic approach for modeling, a smooth function f is using P-splines introduced in a frequentist setting by Eilers and Marx (1996) and in a Bayesian version by Lang and Brezger (2004). P-splines assume that the unknown functions can be approximated by a polynomial spline which can be written in terms of a linear combination of B-spline basis functions. Hence, the columns of the design matrix $\mathbf{Z}$ are given by the B-spline basis functions evaluated at the observations z_i. Lang and Brezger (2004) propose to use first- or second-order random walks as smoothness priors for the regression coefficients, that is,

$$\beta_k = \beta_{k-1} + u_k, \quad \text{or} \quad \beta_k = 2\beta_{k-1} - \beta_{k-2} + u_k, \tag{5.9}$$

with Gaussian errors $u_k \sim N(0, \tau^2)$ and diffuse priors $p(\beta_1) \propto$ const, or $p(\beta_1)$ and $p(\beta_2) \propto$ const, for initial values. This prior is of the form (5.8) with penalty matrix given by $\mathbf{K} = \mathbf{D}'\mathbf{D}$, where $\mathbf{D}$ is a first- or second-order difference matrix.

5.4.3 *Spatial Effects*

Assume now that $\mathbf{z}$ represents the location a particular observation pertains to. If exact locations are available, $\mathbf{z} = \left(z^{(1)}, z^{(2)}\right)'$ is two-dimensional, and the components $z^{(1)}$ and $z^{(2)}$ correspond to the coordinates of the location. In this case, the spatial effect $f\left(z^{(1)}, z^{(2)}\right)$ could be modeled by two-dimensional extensions of P-splines as described in Lang and Brezger (2004). An alternative approach widely used in the geostatistics literature is to model the spatial effect by stationary Gaussian random fields; see Kamman and Wand (2003).

If exact locations are not available as in our application, the correlated district-specific heterogeneity effect $f_{5,6,2}^{mrf}(dist)$ in Eq. (5.5) can be modeled by Markov random fields (MRF). Suppose that $z \in \{1, \ldots, K\}$ is the indicator for the district in which a house is located. MRFs define one parameter for every discrete geographical unit (districts in our case), that is, $f(z) = \beta_z$, and are defined via the conditional distributions of β_z given the parameters β_s of neighboring sites s. We denote the set of neighbors of site z by $N(z)$. Typically sites are assumed to be neighbors if they share a common boundary. MRFs assume that the conditional distribution of β_z given neighboring sites $s \in N(z)$ is Gaussian with

$$\beta_z \mid \beta_s, s \neq z \sim N\left(\frac{1}{|N(z)|} \sum_{s \in N(z)} \beta_s, \frac{\tau^2}{|N(z)|}\right),$$

where $|N(z)|$ denotes the number of neighbors of site z.

The joint (prior) distribution of $\boldsymbol{\beta}$ is of the form (5.8) with penalty matrix $\mathbf{K}$ given by

$$\mathbf{K}[z, s] = \begin{cases} -1 & z \neq s, s \in N(z), \\ 0 & z \neq s, s \notin N(z) \\ |N(z)| & z = s. \end{cases} \tag{5.10}$$

If a Markov random field is used in the level-1 equation, the design matrix $\mathbf{Z}$ is a 0/1 incidence matrix whose entry in the i-th row, and k-th column is 1 if the i-th observed house is located in district k and 0 else. In our application, the MRF is specified in the level-3 equation to model smooth district-specific heterogeneity. In this case, the design matrix is the identity matrix, that is, $\mathbf{Z}_{5,6,2} = \mathbf{I}$.

5.4.4 *Bayesian Inference Based on Markov Chain Monte Carlo Simulations*

It is beyond the scope of this chapter to present detailed algorithms for Bayesian inference. Instead, we refer to the recent literature. An overview about Bayesian inference in additive models based on MCMC simulations is given in Fahrmeir et al. (2013). Algorithms for hierarchical structured additive regression are given in Lang et al. (2014). Bayesian distributional regression in the spirit of GAMLSS is treated in Klein et al. (2013). Bayesian quantile regression is discussed in detail in Waldmann et al. (2013) and Fahrmeir et al. (2013).

5.5 Generalized Random Slope Modeling

A common phenomenon observed for real estate data is spatial heterogeneity in the sense that (possibly nonlinear) effects of covariates vary in size (and possibly also shape) from one spatial unit to another. This is primarily the case if the spatial units, for example, districts, can be regarded as submarkets of one larger market.

In order to model spatially heterogeneous effects, we discuss in this section nonlinear generalizations of random slopes. Suppose that for a continuous covariate z a nonlinear effect $f(z)$ is assumed. Moreover, suppose that there might be heterogeneity with respect to a cluster variable $c \in \{1, \dots, C\}$ in the sense that the nonlinear function is not homogeneous from cluster to cluster. However, completely different functional forms in each cluster are not likely a priori. Instead, one might think that only a particular feature of the function is subject to cluster-specific heterogeneity. Here, we assume homogeneity for the functional form but heterogeneity for the scaling of the function. This leads to a term of the form

$$(1 + \alpha_{c_i}) f(z_i) = f(z_i) + \alpha_{c_i} f(z_i),$$

where the possibly nonlinear function f of z is scaled by the factor $(1 + \alpha_c)$. In matrix notation, we obtain

$$\text{diag}(1 + \alpha_{c_1}, \dots, 1 + \alpha_{c_n}) \times \mathbf{Z}\boldsymbol{\beta}, \tag{5.11}$$

where $\mathbf{Z}$ is the design matrix corresponding to the nonlinear function f. An equivalent formulation in terms of the cluster-specific scaling parameter vector $\boldsymbol{\alpha} = (\alpha_1, \dots, \alpha_C)'$ and a 0/1 incidence (design) matrix $\mathbf{C}$ for the cluster-specific scaling effect is given by

$$\mathbf{f} + \text{diag}(f(z_1), \dots, f(z_n)) \times \mathbf{C}\boldsymbol{\alpha}. \tag{5.12}$$

Similar to pure additive models, the prior for the scaling parameter vector $\boldsymbol{\alpha} = (\alpha_1, \dots, \alpha_C)'$ may obey another structured additive model, that is,

$$\boldsymbol{\alpha} = \mathbf{Z}_{\alpha,1}\boldsymbol{\beta}_{\alpha,1} + \dots + \mathbf{Z}_{\alpha,q_\alpha}\boldsymbol{\beta}_{\alpha,q_\alpha} + \mathbf{X}_\alpha \boldsymbol{\gamma}_\alpha + \boldsymbol{\varepsilon}_\alpha.$$

Some care has to be taken regarding identifiability of the parameters. In particular, there is an arbitrary multiplicative constant for the nonlinear function f.

A possible way to assure identifiability is to assume monotonicity for f (either monotonically increasing or decreasing) and to restrict the spread of f, for example, by assuming

$$\sum_{k=1}^{K} \beta_k^2 = c.$$

The constant c can, for example, be chosen such that the squared sum of the coefficients is identical to that of the model without scaling factors. Imposing monotonicity constraints is easily done using the methodology of Brezger and Steiner (2008).

An application of generalized random slope modeling in the context of real estate data is given in Brunauer et al. (2010) when modeling rents of apartments in dependence of covariates. The nonlinear price gradients are assumed to be district specific and modeled in the form of generalized random slopes as proposed above.

5.6 Software

The multilevel STAR models described above can be estimated with the open source software package **BayesX** (Brezger et al. 2005). To facilitate exploration and visualization of fitted models, the R (R Development Core Team 2013) package **R2BayesX** (Umlauf et al. 2012) has been developed, which provides a fully interactive R interface to **BayesX** with the usual R modeling "*look & feel*". In the following, we exemplify the usage of the software estimating the four-level hierarchical STAR model (5.5).

We first load the required packages and data sets.

```
R> library("R2BayesX")
R> library("spdep")
R> load("AustriaHouse.rda")
R> load("DistrictsBnd.rda")
```

The file `AustriaHouse.rda` contains four data sets, one for each spatial resolution as described in Sect. 5.2. The data set with the highest resolution including the house prices and housing attributes is called `HousePrice`, data on the municipal level is provided in the data frame `Municipal`, on the district and county level in objects `District` and `County`, respectively. The file `DistrictsBnd.rda` contains a boundary map object `DistrictsBnd` that is used to compute the necessary neighborhood structure for estimating the level-3 correlated spatial effect of the districts in Austria. After transforming the class "`bnd`" object to an object of class "`SpatialPolygons`" with

```
R> DistrictsSp <- bnd2sp(DistrictsBnd)
```

the final neighborhood object `DistrictsNb`, which is used for fitting the model, can be generated by

```
R> DistrictsNb <- poly2nb(DistrictsSp)
```

Here, districts are identified as neighbors if they share a common border, but different neighborhood structures can be employed; see, for example, function `dnearneigh()` or `tri2nb()` in package **spdep** (Bivand 2014). The four-level hierarchical STAR model is then estimated with

```
R> b <- bayesx(lnp_qm ~ -1 + heat_o2 + heat_o3 + heat_neu1 +
+    heat_neu2 + bath_o1 + bath_o3 + bath_neu1 + bath_neu2 +
+    garage_1 + garage_2 + marker + attic_dum + cellar_dum +
+    terr_dum + sx(c_area) + sx(c_area_plot) + sx(c_age) +
+    sx(c_time_ind) +

+      ## Level-2
+      sx(municipal ~ -1 + sx(c_pp_ind) + sx(c_ln_educ) +
+      sx(c_age_ind) + sx(c_comm) + sx(c_ln_dens) +

+        ## Level-3
+        sx(district ~ -1 + sx(c_wko_ind) +
+        sx(district, bs="mrf", map=DistrictsNb) +

+          ## Level-4
+          sx(county ~ 1, bs="re", data=County),

+        bs="re", data=District),

+      bs="re", data=Municipal),

+    data=HousePrice, method="HMCMC", iterations=120000,
+    step=100, burnin=20000)
```

where the (possibly) nonlinear smooth terms are per default set up using P-splines within the smooth term constructor function `sx()`. The spatially correlated effect is specified by changing the basis-type argument of `sx()` to `bs = "mrf"` and providing the neighborhood object to argument `map`. The random effects of the municipals, districts, and counties are specified with `bs = "re"`. Here, the first argument of `sx()` is a `formula` that specifies the terms of the random effect equation. This means that higher-level formulas can be defined within the formula of the previous levels, representing the multilevel structure of the data. In addition, the corresponding level-specific data set needs to be supplied to argument `data` within `sx()`.

By setting the number of iterations of the MCMC sampler to 120,000, estimation takes approximately 2 1/2 min on a Windows system with an Intel i7-3740QM 2.70 GHz processor. A model summary can then be printed by typing

```
R> summary(b)
```

which returns the estimation results for all levels. The estimated smooth and random effects, for example, level 1, can be plotted with

```
R> plot(b, model=1, term="sx")
```

For spatial and random effects the plotting method per default shows the kernel densities of the mean posterior coefficients. In addition, to further analyze the correlated or uncorrelated spatial effects, map effect plots can be drawn by supplying the corresponding boundary map object, for example, for the level-3 correlated district effect, this is done by

```
R> plot(b, model=3, term="sx(district)", map=DistrictsBnd)
```

One way to inspect convergence of the MCMC chains is to look at the resulting sampling paths, for example, for term `c_area`, the sampling paths of the coefficients are plotted with

```
R> plot(b, model=1, term="sx(c_area)", which="coef-samples")
```

The estimated effects can also be extracted, for example, the effect of `c_age_ind` of level-2, using function `fitted()`

```
R> fit <- fitted(b, model=2, term="sx(c_age_ind)")
```

where the object `fit` is a data frame containing the estimated mean as well as the 2.5, 50, and 97.5 % quantiles of the effect, among others.

For a detailed description of the package **R2BayesX**, including a number of examples, see also Umlauf et al. (2012).

5.7 Results

We now present the estimation results for the mean regression, the GAMLSS regression based on the gamma distribution, and the quantile regression. The results are based on a final MCMC run with 120,000 iterations and a burn in period of 20,000 iterations. We stored every 100th iteration resulting in a sample of 1,000 practically independent draws from the posterior. Computing times for the MCMC sampler were approximately 2 1/2 min (mean regression), 55 min (GAMLSS regression), and 4 min (quantile regression) on a modern desktop computer (Intel quad-core processor 2.7 GHz). Note that no more than 32,000 iterations are typically enough in preliminary MCMC runs to obtain sufficiently exact estimation results. However, we used the comparably large number of iterations in the final run to be absolutely sure about the precision of estimates.

We first show in Sect. 5.7.1 the effects of the continuous covariates on the expected house price per sq. m. received from the mean regression and the GAMLSS regression based on the gamma distribution. Next, we will focus on different quantiles of the house price per sq. m. and compare the results of these two models with those of the quantile regression (5.7.2). The last Sect. 5.7.3 is devoted to the spatial effects.

5.7.1 Continuous Covariate Effects

5.7.1.1 Structural Covariates

Figure 5.1 shows the effects of the structural continuous covariates. In order to get an impression of the magnitude of effects and make the results comparable, we hold the other structural covariates constant at mean level of attributes and the

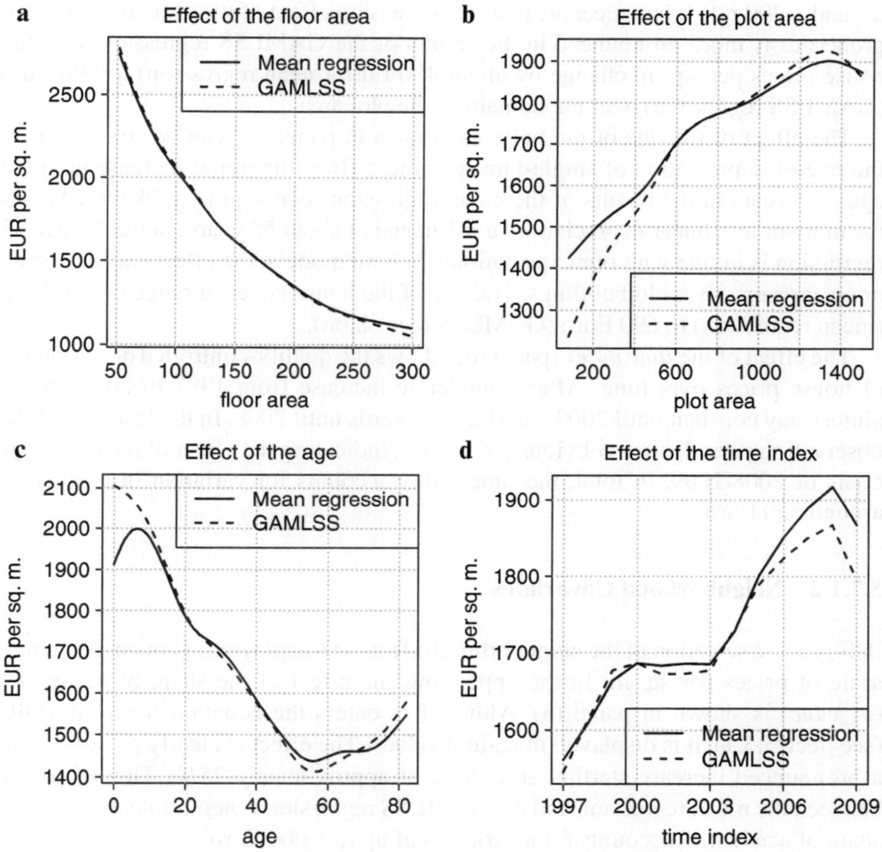

Fig. 5.1 Effects of the continuous structural covariates of level-1. (**a**) Effect of the floor area (variable area); (**b**) effect of the plot area (area_plot). (**c**) Effect of the age of the building (age); (**d**) effect of the time index (time_index). Shown are the posterior mean estimates of the mean regression and the GAMLSS regression based on the gamma distribution

categorical variables at their mode level and evaluate all neighborhood covariates and spatial effects at the mode of the municipals (which we call the *average effect*) and transform the functions from the mean regression to natural units (prices in Euro per sq. m.). Since the effects are quite different in magnitude, we do not show them on the same scale.

In panel (a), the effect of the floor area (variable *area*) is shown. We find a monotonically decreasing and very pronounced effect of additional floor area on prices per sq. m., accounting for a variation of more than 1,650 Euro. However, the decreasing effect weakens as the floor area becomes larger. The results of the mean regression and the GAMLSS regression are virtually the same.

Additional plot area (*area_plot*, panel (b)) yields higher prices per sq. m. of floor area, although this effect becomes weaker as plot area increases and levels off at around 1,200 m^2. It is noticeable that the increasing effect of the plot area for small properties is more pronounced in the results of the GAMLSS regression. In total, house prices per sq. m. change by about 480 Euro (mean regression) to 690 Euro (GAMLSS regression) over the domain of the plot area.

The effect of the *age* of the building, shown in panel (c), can be considered as the rate of depreciation of single-family homes. Thus, the initial increase up to an age of 7 years in the results of the mean regression seems quite unlikely, whereas the more or less linear depreciation (until an age of about 55 years) in the GAMLSS regression is in line with our expectations. In both models, the effect stays constant or even reverses for old buildings. The age of the house covers a range of 560 Euro (mean regression) to 690 Euro (GAMLSS regression).

The effect of the *time index* (panel (d)) shows the quality-controlled development of house prices over time. After a moderate increase from 1997 to 2000, prices almost stay constant until 2003 and rise afterwards until 2008. In the last year of the observation period, prices obviously decrease, indicating the effect of the economic crisis of 2008/2009. In total, the time index accounts for variation in a range of around 350 Euro.

5.7.1.2 Neighborhood Covariates

In Fig. 5.2, a selection of the neighborhood effects are displayed, again on the natural scale of prices per sq. m. In the upper row, the effect of the share of academics (*ln_educ*) is shown in panel (a). Although it enters the equation logarithmically (see Sect. 5.3.1), it is displayed in natural values. The effect is clearly positive, with a pronounced increase starting at a share of approximately 25 %. The difference between the mean regression and the GAMLSS regression is negligible. In total, the share of academics accounts for a variation of up to 1,000 Euro.

The effect of the age index (*age_ind*, displayed in panel (b)) is more or less linear for both the mean regression and the GAMLSS regression. The negative direction of this effect could be interpreted as a decreasing attractiveness of municipalities that exhibit an excess of age, which could be expected from our considerations in Sect. 5.2.2. The effect of the age index has a bandwidth of up to 535 Euro.

The effect of the population density *ln_dens*, displayed in natural values in panel (c), shows a tendency toward higher house prices in more densely populated areas. Here, we can find a considerable difference of up to 260 Euro between the results of the mean regression and those of the GAMLSS regression for highly densely populated areas. In all, this effect accounts for a variation between 630 Euro (mean regression) and 930 (GAMLSS regression) per sq. m.

Finally, the effect of the house price index *wko_ind* (the only covariate on the district level) is shown in panel (d). As expected, this effect is increasing, although for index values of more than 140, it becomes lightly weaker in the GAMLSS regression and clearly weaker in the mean regression. Prices per sq. m. increase by 470 Euro (mean regression) to 560 Euro (GAMLSS regression) from the lowest to the highest house price index.

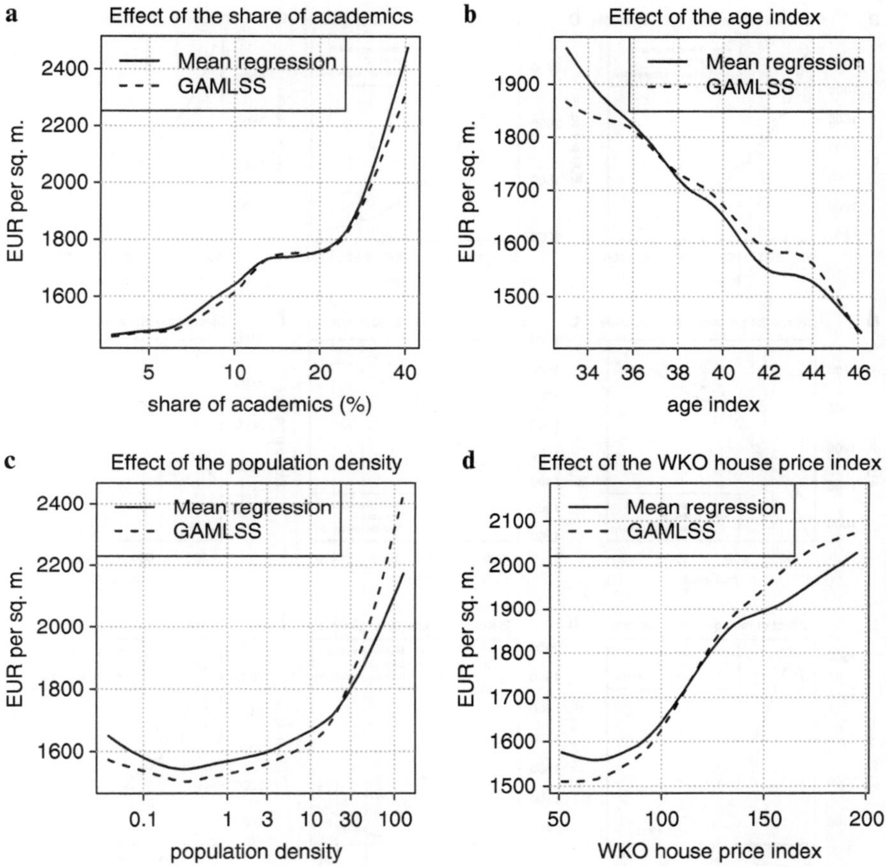

Fig. 5.2 Effects of the neighborhood covariates. *First row*: effect of the share of academics (educ) (**a**) and the age index (age_ind) (**b**). *Second row*: effect of the log of population density (ln_dens) (**c**) and the house price index (wko_ind) (**d**). Shown are the posterior mean estimates of the mean regression and the GAMLSS regression based on the gamma distribution

5.7.2 *Quantiles*

Figure 5.3 shows the effects for the structural covariates for the 20-, 50-, and 80 %-quantiles. Beside the results of the mean regression and the GAMLSS regression, we now also compare the effects of the quantile regression. Again, we hold all other covariates constant at mean level of attributes and – if necessary – transform the functions to natural units.

In general, the effects for the different quantiles are similar to those for the expected house prices per sq. m., displayed in Fig. 5.1. However, we can find some interesting differences between the three models: Quantile regression estimates

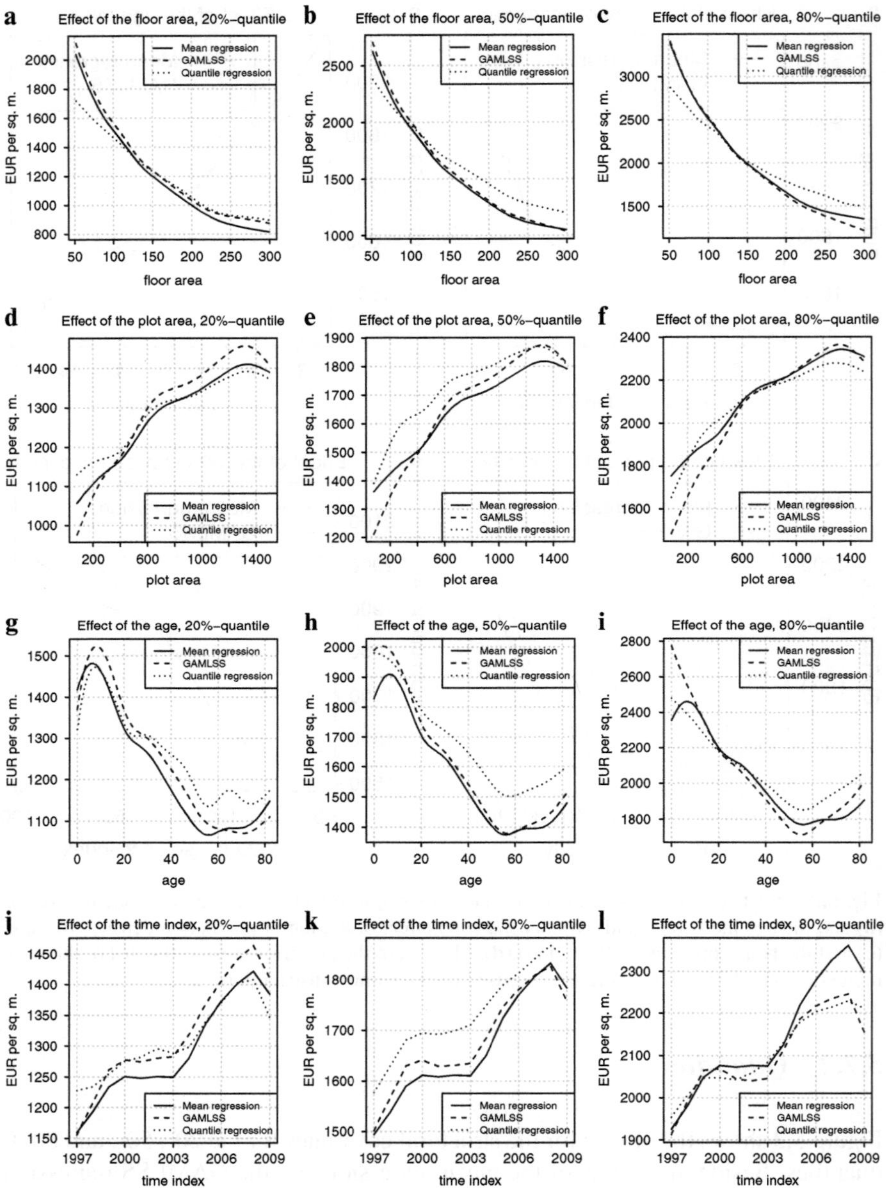

Fig. 5.3 Effects of the continuous structural covariates of level-1. (**a**), (**d**), (**g**), (**j**) effects of the 20 %-quantile; (**b**), (**e**), (**h**), (**k**) effects of the 50 %-quantile; (**c**), (**f**), (**i**), (**l**) effects of the 80 %-quantile

individual effects for each and every quantile. GAMLSS regression also allows for different marginal effects due to the two estimated parameters of the gamma distribution. Mean regression, in contrast, only reveals one marginal effect for each covariate, shifts it according to the estimated variance and converts it to natural units using the transformation property of the lognormal distribution.

While the differences in the estimated effects between the mean regression and the GAMLSS regression were rather small for the expected house price per sq. m., there are now substantial differences for the quantiles. Compared to the results of the quantile regression and the GAMLSS regression, the mean regression seems to slightly underestimate the prices for the 20- and 50 %-quantile, while the results for the 80 %-quantile are more similar. This indicates a more skewed distribution of house prices than can be captured by the mean regression.

Figure 5.4 illustrates this distribution by showing the posterior 10-, 20-, 50-, 80-, 90 %-quantile estimates of the three different methods for each structural covariate. Additionally, the posterior mean estimates are displayed for the mean regression and the GAMLSS regression. Particularly the effects of the age of the building considerably differ between the individual quantiles both in the GAMLSS and the quantile regression, revealing the limits of the mean regression where the marginal effects are almost the same for all quantiles.

5.7.3 *Spatial Effects*

The total amount of spatial heterogeneity is composed of spatial effects on municipal (level-2), district (level-3), and county level (level-4). Continuous neighborhood effects (see Sect. 5.7.1) explain spatial heterogeneity explicitly to a certain extent on two of these levels (municipal and district). We call this the *explained* spatial heterogeneity. The remaining i.i.d. spatial random effects ε_5, $\varepsilon_{5,6}$ and $\varepsilon_{5,6,3}$ as well as the correlated district-specific effect $f_{5,6,2}(dist)$ in (5.5) account for *unexplained* spatial heterogeneity. In the following, we analyze the *distribution of spatial heterogeneity* over Austria. For the sake of clarity, we only show the estimation results for the 50 %-quantiles.

5.7.3.1 Total Spatial Heterogeneity

Figure 5.5 visualizes the posterior 50 %-quantile estimates of the total spatial effects (explained plus unexplained heterogeneity) for the mean regression, the GAMLSS regression and the quantile regression evaluated at the average effect. We can see a very pronounced spatial heterogeneity showing single-family homes in the western counties to be considerably more expensive than in the eastern and southern counties. Furthermore, we find a clearly positive effect in urban areas with a strong peak in Vienna.

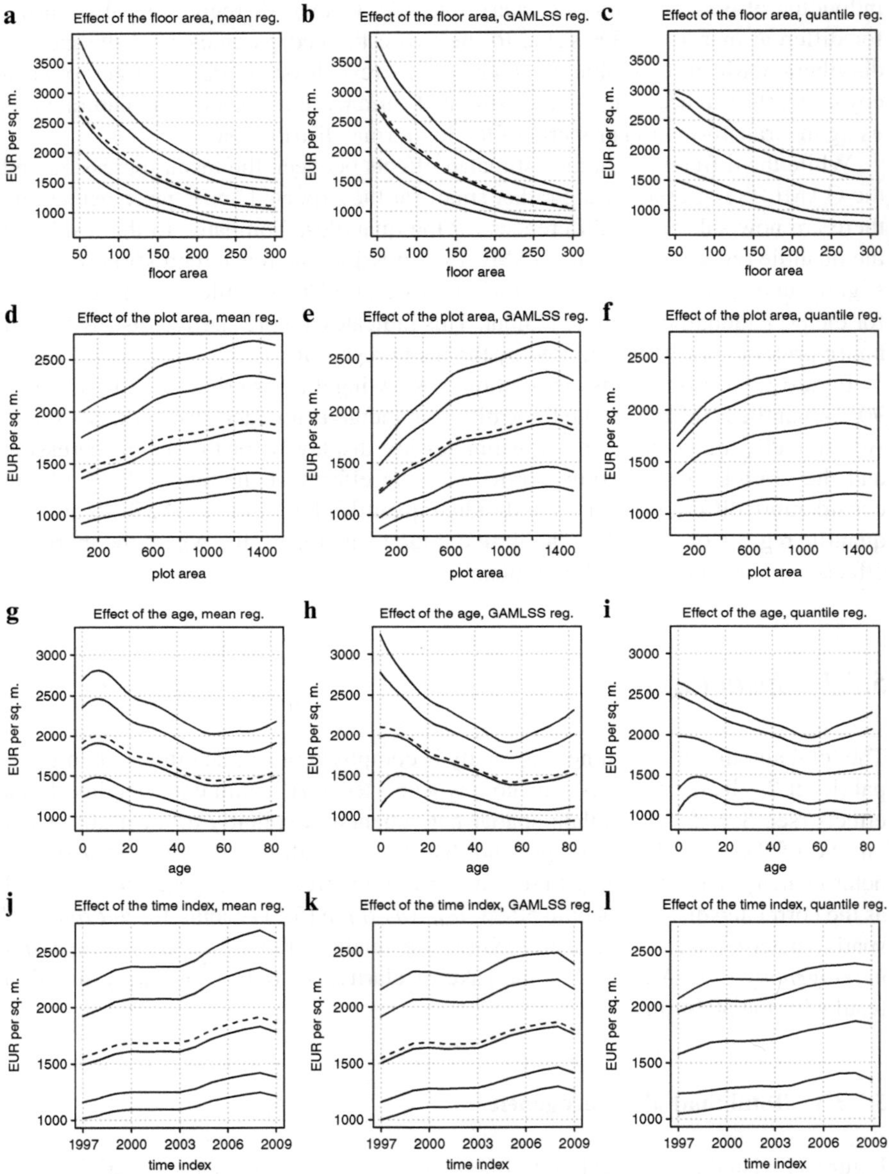

Fig. 5.4 Effects of the continuous structural covariates of level-1. (**a**), (**d**), (**g**), (**j**) effects of the mean regression; (**b**), (**e**), (**h**), (**k**) effects of the GAMLSS regression; (**c**), (**f**), (**i**), (**l**) effects of the quantile regression. Shown are the posterior 10-, 20-, 50-, 80-, 90 %-quantile estimates (*solid lines*) as well as the posterior mean estimates of the mean regression and the GAMLSS regression (*dashed lines*)

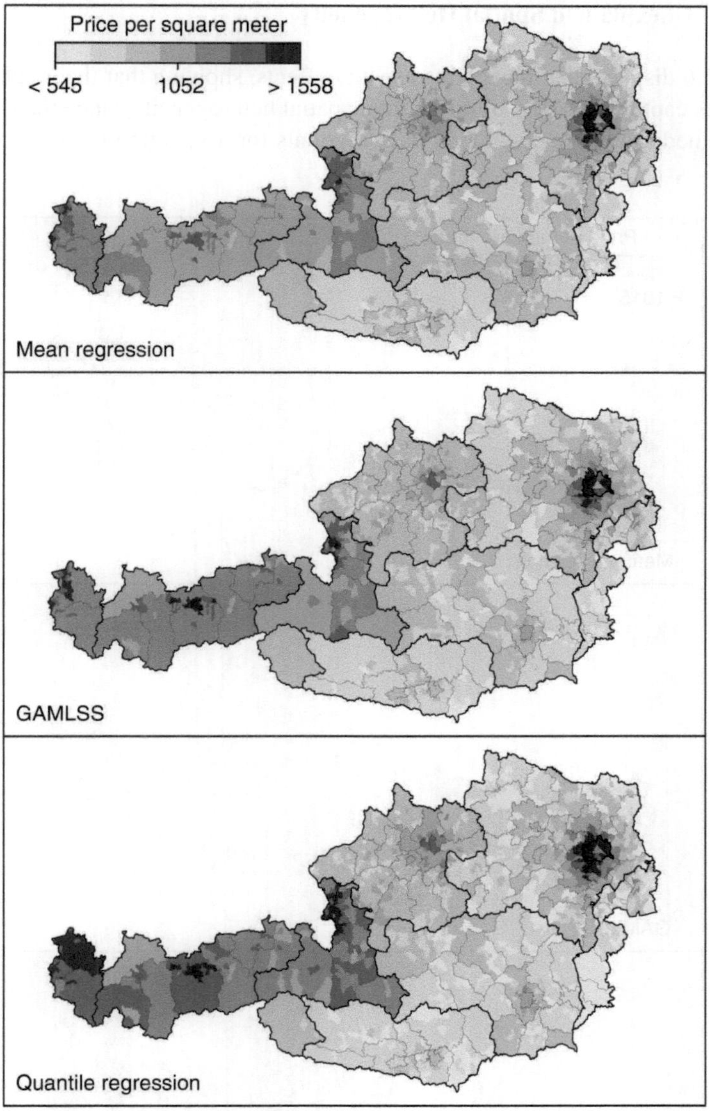

Fig. 5.5 Distribution of the total spatial heterogeneity (evaluated at the average effect). Shown are the posterior 50 %-quantile estimates for the mean regression, the GAMLSS regression, and the quantile regression

Comparing the total spatial effects according to the different estimation methods, we only can find minor differences. In total, the spatial heterogeneity has a bandwidth of 1,950 Euro (mean regression) or 2,175 Euro (GAMLSS and quantile regression).

5.7.3.2 Unexplained Spatial Heterogeneity

Figure 5.6 displays the unexplained spatial effects, showing that the neighborhood attributes capture a large part of the total spatial heterogeneity, since the remaining unexplained spatial heterogeneity only accounts for a variation of about 520 Euro

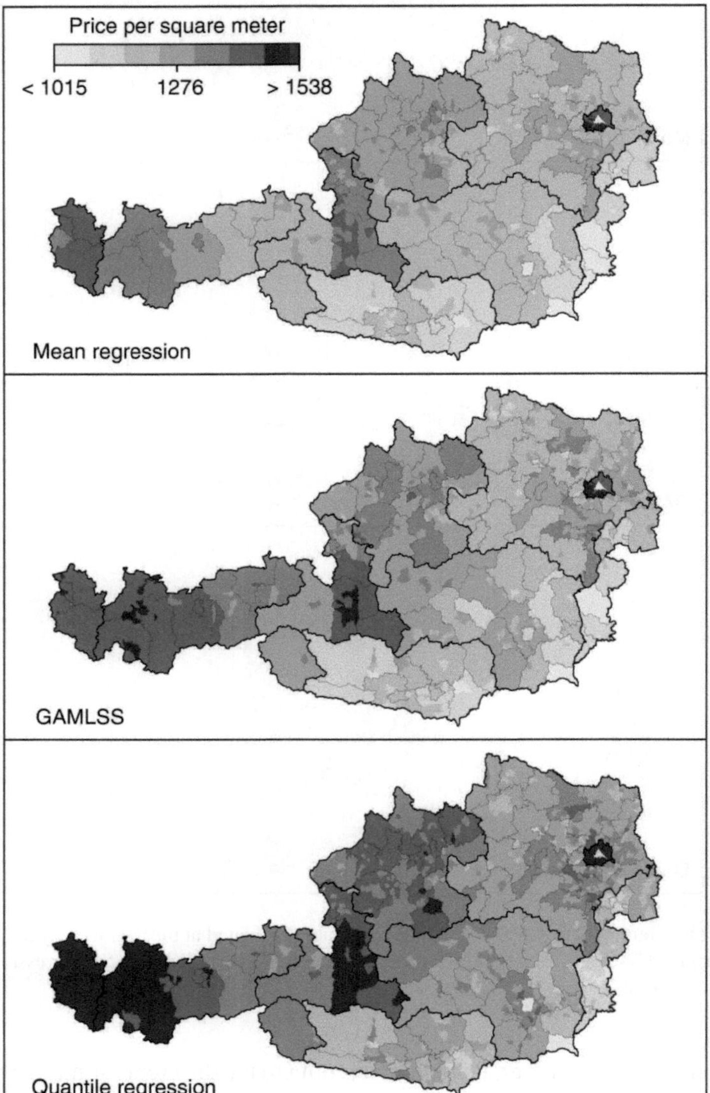

Fig. 5.6 Distribution of the unexplained spatial heterogeneity (evaluated at the average effect). Shown are the posterior 50 %-quantile estimates for the mean regression, the GAMLSS regression and the quantile regression

(mean regression) to 640 Euro (quantile regression). Again, the highest values of the estimated effects can be found in the western counties and in Vienna.

Acknowledgements This work was supported by funds of the Oesterreichische Nationalbank (Oesterreichische Nationalbank, Anniversary Fund, project number: 15309).

References

Banerjee S, Gelfand AE, Knight JR, Sirmans CF (2004) Spatial modeling of house prices using normalized distance-weighted sums of stationary processes. J Bus Econ Stat 22:206–213

Bivand R (2014) **spdep**: spatial dependence: weighting schemes, statistics and models. R package version 0.5–71

Brezger A, Steiner W (2008) Monotonic regression based on Bayesian P-splines: an application to estimating price response functions from store-level scanner data. J Bus Econ Stat 26:90–104

Brezger A, Kneib T, Lang S (2005) **BayesX**: analyzing Bayesian structured additive regression models. J Stat Softw 14(11):1–22

Brunauer W, Lang S, Wechselberger P, Bienert S (2010) Additive hedonic regression models with spatial scaling factors: an application for rents in Vienna. J Real Estate Finance Econ 40:390–410

Brunauer W, Lang S, Umlauf N (2013) Modeling house prices using multilevel structured additive regression. Stat Model 13:95–123

Can A (1998) GIS and spatial analysis of housing and mortgage markets. J Hous Res 9(1):61–86

Cohen JP, Coughlin CC (2008) Spatial hedonic models of airport noise, proximity, and housing prices. J Reg Sci 48:859–878

Eilers PHC, Marx BD (1996) Flexible smoothing using B-splines and penalized likelihood. Stat Sci 11:89–121

Fahrmeir L, Kneib T, Lang S (2004) Penalized structured additive regression for space-time data: a Bayesian perspective. Stat Sin 14:731–761

Fahrmeir L, Kneib T, Lang S, Marx B (2013) Regression: models, methods and applications. Springer, Berlin/New York

Follain J, Jimenez E (1985) Estimating the demand for housing characteristics: a survey and critique. Reg Sci Urban Econ 15:77–107

Gelman A, Hill J (2006) Data analysis using regression and multilevel/hierarchical models. Cambridge University Press, Leiden

Helbich M, Brunauer W, Vaz E, Nijkamp P (2014) Spatial heterogeneity in hedonic house price models: the case of Austria. Urban Stud 51:390–411

Kamman EE, Wand MP (2003) Geoadditive models. J R Stat Soc C 52:1–18

Klein N, Kneib T, Lang S (2013) Bayesian structured additive distributional regression. Working paper 2013–23, working papers in economics and statistics, research platform empirical and experimental economics, University of Innsbruck

Koenker R (2005) Quantile regression. Cambridge University Press, Cambridge/New York

Koenker R, Mizera I (2004) Penalized triograms: total variation regularization for bivariate smoothing. J R Stat Soc B 66:145–163

Lang S, Brezger A (2004) Bayesian P-splines. J Comput Graph Stat 13:183–212

Lang S, Umlauf N, Wechselberger P, Harttgen K, Kneib T (2014) Multilevel structured additive regression. Stat Comput 24:223–238

Malpezzi S (2003) Hedonic pricing models: a selective and applied review. In: O'Sullivan T, Gibb K (eds) Housing economics and public policy. Blackwell Science Ltd., Oxford/Malden, pp 67–89

Razen A, Brunauer W, Klein N, Kneib T, Lang S, Umlauf N (2014) Modeling the volatility of real estate market values using bayesian distributional and quantile regression. Working paper 2014-12, working papers in economics and statistics, research platform empirical and experimental economics, University of Innsbruck

R Development Core Team (2013) R: a language and environment for statistical computing. Vienna, Austria: R Foundation for Statistical Computing. ISBN 3-900051-07-0

Rigby RA, Stasinopoulos D (2005) Generalized additive models for location, scale and shape. Appl Stat 54:507–554

Rosen S (1974) Hedonic prices and implicit markets: product differentiation in pure competition. J Pol Econ 82:34–55

Sheppard S (1999) Hedonic analysis of housing markets. In: Cheshire PC, Mills ES (eds) Handbook of regional and urban economics, vol 3. Elsevier Science, Amsterdam, pp 1595–1635

Umlauf N, Adler D, Kneib T, Lang S, Zeileis A (2012) Structured additive regression models: an R interface to BayesX. Working paper 2012-10, working papers in economics and statistics, research platform empirical and experimental economics, University of Innsbruck

Waldmann E, Kneib T, Lang S, Yue Y (2013) Bayesian semiparametric additive quantile regression. Stat Model 13:223–252

Yu K, Moyeed RA (2001) Bayesian quantile regression. Stat Probab Lett 54:437–447

Yue Y, Rue H (2011) Bayesian inference for additive mixed quantile regression models. Comput Stat Data Anal 55:84–96

Chapter 6
Simple Agents, Complex Emergent City: Agent-Based Modeling of Intraurban Migration

Shipeng Sun and Steven M. Manson

Abstract Intraurban migration—residential movement within a metropolitan area—defines the nature of urbanization. Housing location decision making is a complex process driven by the interactions between the housing market and home searchers. Researchers have paid much attention to the environmental, socioeconomic, cultural, and policy features of housing markets. In contrast, housing search has been relatively neglected due to challenges of theory, methodology, and data. This article addresses these challenges by presenting an agent-based model of intraurban migration featuring straightforward and empirically specified rules for housing search. This model is calibrated and validated against real-world housing vacancies and relocation origin–destination pairs extracted from parcel records for the Twin Cities of Minnesota, USA, for 2005–2007. Drawing on these unique data sidesteps a long-standing issue, the prohibitive costs of identifying, recording, and quantifying housing search activities for an entire metropolitan region. Conceptually, this model updates geographic theories of intraurban migration that focus on intervening opportunities and spatial bias. It also methodologically advances the agent-based modeling of urbanization with a high-resolution, empirically specified model that demonstrates how urban pattern emerges from simple rules and interactions. Overall, the model demonstrates that relatively straightforward housing search rules can simulate realistic patterns of intraurban migration.

Keywords Intraurban migration • Agent-based modeling • Housing search • Housing locational decisions

S. Sun (✉)
Department of Environmental Studies, University of Illinois Springfield, One University Plaza, Springfield, IL 62703, USA
e-mail: sunsp.gis@gmail.com; ssun32@uis.edu

S.M. Manson
Department of Geography, Environment and Society, University of Minnesota, 267 19th Avenue South, Minneapolis, MN 55455, USA
e-mail: manson@umn.edu

M. Helbich et al. (eds.), *Computational Approaches for Urban Environments*, Geotechnologies and the Environment 13, DOI 10.1007/978-3-319-11469-9_6

6.1 Introduction

Intraurban migration, or residential moves within a metropolitan area, is a complex process involving the interaction of housing market characteristics with the perceptions of home searchers. Intraurban migration research has a long and rich history in geography and other social sciences, engineering fields, and policy disciplines (Dorigo and Tobler 1983; Clark 1986; Brown and Moore 1970; Roseman 1971; Simmons 1968; Simpson et al. 2008; Clark 2008). There remain significant challenges in terms of data, method, and theory in understanding this form of migration. Data on specific individuals who drive intraurban migration is difficult to obtain and use. Methodologically, there is a need to combine the most common approach, statistical methods, with fast-emerging simulation modeling methods. In terms of theory, the large number of competing explanations for intraurban migration points to the need for continuing work on foundational research on individual behavior. When these challenges of data, method, and theory are taken together, they indicate the need for empirically based approaches that combine statistical and simulation models to develop and test straightforward frameworks for understanding how individual behavior gives rise to aggregate patterns and processes of intraurban migration.

We developed an agent-based model that draws on novel data derived from land parcels to develop and test an updated form of the intervening opportunity theory of intraurban migration. This work is significant in several respects. This conceptual model brings together underexamined geographical and sociological findings to develop and test straightforward spatial behavioral rules that capture key features of intraurban migration. To specify and test this model we developed a new data source, individual migration chains extracted from tax parcel data that allowed us to track the movements of actual households in space and over time. We bring these data and the conceptual model together in an agent-based model that is calibrated and validated with mathematical and statistical approaches, leveraging the relative strengths of these different methods. More broadly, this work addresses the need for simple and generalizable to complement the large and growing body of work that focuses on representing complicated dynamics with extensive and detailed datasets (Brown et al. 2008; Torrens 2012). It also contributes to the fast expanding body of work seeking to simplify complex urban dynamics by using new data sources to develop relatively straightforward and generalized models that capture significant features of urban form and processes (Batty 2008, 2012).

The rest of this paper examines this confluence of data, method, and theory. The next section reviews locational decision-making theories of intraurban migration and proposes a model for housing location decisions with two different strategies. Section 6.3 applies this model to intraurban migration of homeowners in the Twin Cities Metropolitan Area of the USA (TCMA), along the way introducing the use of land parcel data to calibrate and validate an agent-based model of individual migration. Section 6.4 presents the model results, including model validation. The paper concludes with discussion of our findings and their implications for urban agent-based modeling and our understanding of intraurban migration generally.

6.2 Conceptual Model of Intraurban Migration

A core conceptual challenge in understanding intraurban migration is developing theories of how individual behavior leads to complex urban patterns and processes. Intraurban migration has three interrelated components, as introduced by Wolpert (1965) and expanded over the years: (1) housing conditions, or the broad social, demographic, economic, and environmental conditions that trigger household migration; (2) housing utilities, expressed as the balance between utilities of the current housing and expected utilities of other housing opportunities; and (3) housing search, or the search process and perceptions of housing by potential buyers. Intraurban migration research focuses primarily on the first and second components, while a smaller body of work centers on the third component of housing search as a sociospatial process that guides the first and second.

We advance this third component by testing a modified intervening opportunity theory, drawing on sociospatial conceptions of housing perception to examine simple decision-making rules that lead to realistic complex migration patterns in aggregate. This third component helps guide the first two and can be examined separately, which does not minimize the fact that the residential choices of households are driven in part by a host of demographic and socioeconomic characteristics of migrants combined with housing utilities, including housing structure, the biophysical environment, neighborhood quality, as well as accessibility to services (Adams 1984; Quigley and Weinberg 1977; Clark 2008). Additional considerations include household factors ranging from income and race to environmental preferences (Choldin 1973; Pellegrini and Fotheringham 2002; Jones et al. 2004). These factors in turn modify the effect of housing conditions and their attendant perceived difference in utilities (De Jong and Roempke Graefe 2008; Geist and McManus 2008; Mulder 2007; Cooke 2008), interactions with commuting and transportation infrastructure (Clark, Huang, and Withers 2003; Rouwendal and Rietveld 1994), government policies and developer decisions (Brown and Chung 2008), and the lending practices of financial institutions (Brown and Longbrake 1970; Brown and Moore 1970; Clark 1982). In sum, a wide variety of factors influence housing conditions and housing utilities and, by extension, intraurban migration and urbanization more generally.

Despite the importance of housing conditions and housing utilities, the third component of housing search and perception has much to do with the nature of intraurban migration because it guides the effects of the first two components. This component has a distinguished research history, but overall it has received far less attention than the first two components. Theories of housing search complement our understanding of these other components because they posit that there are fundamental regularities in how households perceive housing opportunities. Much of this research emphasizes the distinct spatial and temporal limits of homebuyers' search strategies in local and regional housing markets (Clark 1982; Clark and Flowerdew 1982; Smith et al. 1979). Related work examines how home search and job search interact, particularly in how people think about where they want to live as a function of where they want to work (Waddell 1993; van Ommeren et al. 1997;

cf. Clark and Withers 1999). This research points to the importance of incomplete information and bounded rationality of decision making in intraurban migration, as much of the work on the first two components of housing condition and utility assumes that households have complete information during the housing search and will go to any lengths to find the optimal home. Instead, research on search and perception highlights how housing search is often bounded in space and time, whether by homebuyers' greater knowledge of and comfort with local neighborhoods, bounds on how much time households (or their real estate agents) can devote to the housing search, or a willingness to settle for a new house that is good enough instead of being perfect.

The importance of spatial distance between current and potential housing is a unifying theme in much work on housing search, along with direction to a lesser extent. Most of this research relies on the theoretical antecedent of intervening opportunity theory, developed by Stouffer (1940) to describe the relationship between housing opportunities and moving distances within a metropolitan area. Assuming the quantity of vacant housing units is proportional to the distance from a household's current dwelling, Stouffer posited that the number of households that move a given distance has a logarithmic relationship with the housing opportunities located within that distance because people are likely to choose a vacancy near their current dwelling. The related exponential distribution of moving distances has been validated by empirical research on various metropolitan areas (Clark and Burt 1980; Clark et al. 2003; Quigley and Weinberg 1977). The basic form of this negative exponential distribution of move distances is

$$f(d) = \lambda e^{-\lambda d}, \quad \lambda > 0 \text{ and } d > 0 \tag{6.1}$$

where $f(d)$ is the probability of a household relocating by distance of d and λ is a shape parameter. Mathematically, λ is the reciprocal of average d, or in other words $1/\lambda$ is the average move distance. The intervening opportunities model was extended to the case of interurban migration and evolved into the influential gravity model and related family of spatial interaction models (Guldmann 1999; Jayet 1990; Fotheringham 1983; Cochrane 1975; Ruiter 1967; Erlander 2010). Regardless of variant or degree of sophistication, these models retain at their heart a focus on logarithmic or exponential distance decay in space.

The direction people move is an important consideration alongside distance. Adams (1969) argued that spatial search and residential locational behavior are based on a limited mental map or image of the city. Importantly, this image is sectoral in that it comprises a wedge-shaped region centered on the work-home axis. While Adams took the central city is a proxy for work location, the theory was validated using specific workplace data as well (Clark and Burt 1980; Clark et al. 2003). Move directions can be modeled as a von Mises distribution (Gaile and Burt 1980), which is the counterpart of the normal distribution for directional data spanning 0–360°, with a density function of

$$f\left(x \middle| \mu, \kappa\right) = \frac{e^{-\kappa \cos(x-\mu)}}{2\pi I_0(\kappa)} \tag{6.2}$$

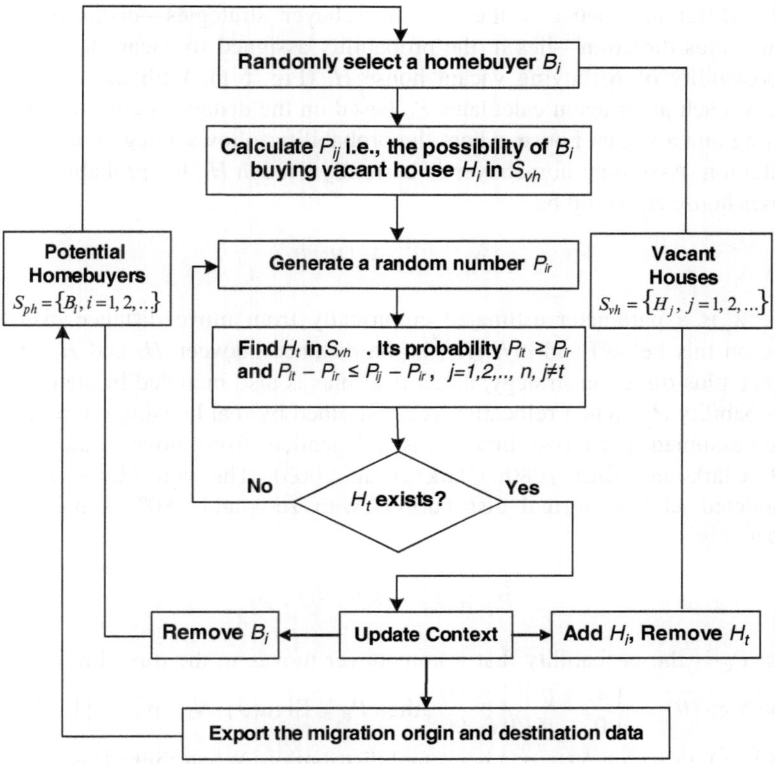

Fig. 6.1 Modeled intraurban migration process

where μ is the mean direction, κ is a measure of variance of directions around μ, and $I_0(\kappa)$ is a modified Bessel function with order zero. When κ is zero, the distribution is uniformly circular (i.e., with equal probability in any direction); when it is larger, the distribution will concentrate around μ in a similar fashion to the normal distribution.

We bring together these sociospatial findings on housing perception via a conceptual agent-based model of the distance and directional relationships among housing vacancies and current dwellings of potential migrants (Fig. 6.1). This model adopts two housing search and relocation strategies—distance-only and distance-plus-direction—that condition intervening opportunity theory with the statistical distribution of moving distances and directions as specified by the negative exponential and von Mises distributions (after Clark et al. 2003). The model is based on two lists for the regional housing market: one of potential homebuyers and another of vacant houses. Each model year, the model iterates over homebuyers randomly. Each homebuyer generates a random number and the vacant house with the closest greater probability is chosen as the destination. When a homebuyer moves, he or she is removed from the buyer list and the vacated house is added into the vacant house list.

The difference between the two homebuyer strategies—distance-only and distance-plus-direction—lies in the probability assigned to vacant houses, P_{ij}, or the possibility of B_i buying vacant house H_j (Fig. 6.1). With the distance-only strategy, each actor agent calculates P_{ij} based on the distance between her current dwelling and a vacant house, where the probability follows a negative exponential distribution. Assuming homebuyer B_i currently lives in H_i, the probability that she chooses house H_j would be

$$P_{ij} = \lambda e^{-\lambda d(H_i,\, H_j)} \tag{6.3}$$

where λ is a parameter estimated empirically from move distance distribution (more on this below) and $d(H_i, H_j)$ is the distance between H_i and H_j. With the distance-plus-direction strategy, directional bias is also included in the calculation of probability P_{ij}. When relocation is constrained by real housing opportunities, it can be assumed that move direction is independent from move distance (Adams 1969; Clark and Burt 1980; Clark et al. 2003). The von Mises distribution is modeled as two normal distributions with zero and 180° as mean values, respectively:

$$P_{ij} = \lambda e^{-\lambda d(H_i,\, H_j)} \cdot P_\theta \tag{6.4}$$

where P_θ is the probability that a homebuyer moves in the direction of θ. If we define $\text{Sign}\,(\theta) = \begin{cases} 1, & \textit{if}\ |\theta| \le 90 \\ 0, & \textit{if}\ |\theta| > 90 \end{cases}$, then $P_\theta = \text{Sign}(\theta) \cdot N(0, \sigma_1^2) + [1 - \text{Sign}(\theta)] \cdot N(180, \sigma_2^2)$, in which $N(\mu, \sigma^2)$ is normal distribution. When Sign(θ) is one, homebuyers move toward the suburbs; when it is zero, they move toward downtown. The standard deviation σ_1 and σ_2 control the extent to which migrant household moves concentrate along the home–downtown corridor. When these deviations are small, houses near the corridor are more likely to be chosen, but when they are large, more houses have greater odds of being chosen. When σ_1 is greater than σ_2, households are more likely to move to suburbs; when smaller, households move toward downtown.

6.3 Methods

A key methodological challenge in understanding intraurban migration is developing straightforward, empirically specified approaches to model how the choices of individuals generate aggregate migration patterns and processes. The primary form of intraurban migration modeling is mathematical and statistical, ranging from gravity modeling to hedonic specifications to various flavors of (new) economic geography of urban areas. Less common but long-standing is simulation modeling, which has enjoyed renewed interest in the form of agent-based modeling (ABM). We develop an ABM of urban intramigration and use a new form of data to calibrate this model, namely, empirically specified migration chains from land parcel data. While they have some drawbacks, these data offer several advantages over many other forms of data used to understand the migration choices of individuals.

6.3.1 Data

The paucity of data on the migration choices of individuals remains a critical challenge in understanding intraurban migration. While migration evinces clear patterns such as suburbanization, gentrification, or decline when examined at gross temporal and spatial scales, our understanding of migration at the scales of individuals is limited by the dearth of public data available on movements of individual households in space at the scale of specific housing units and in time at the scale of a year (Adams 1969; Clark 1976, 1986). There are several different ways to garner these data, although we focus on the advantages of parcel data below.

A common approach to measuring intraurban migration is surveying individuals and then reporting on them over large enumeration units. These surveys ask questions about recent moves, such as time since last move or change in commuting time, and range from travel surveys to general instruments such as the American Community Survey (ACS), the American Housing Survey (AHS), the Current Population Survey (CPS), and the Public Use Microdata Samples (PUMS). These surveys are taken of individuals but when reported are aggregated to regions such as census tracts or traffic analysis zones. As a result, these sources offer good information about intraurban migration in general but lack the spatial resolution necessary to analyze individual moves at subregional scales. These data may be downscaled to create statistically plausible individuals (e.g., giving agents an income from a statistical distribution and giving them a random location within a census tract), but this does not link to actual individuals and places (Berger and Schreinemachers 2006). In sum, census-like surveys offer good attribute detail over broad extents at the cost of spatial specificity.

Another common approach to measuring intraurban migration is to gather data on specific households or houses in an area. Directly surveying migrants is a good way to understand their home-seeking behavior, but this approach is expensive and typically reaches only a small subset of migrants. Other sources include city or telephone directories and utility records that can be used to track the moves of individuals from one address to another, although these data are often incomplete and, in cases such as utility records, subject to confidentiality provisions. A related approach is using home sales data to capture attributes of specific houses, but these data usually say little about the search and migration behavior of specific individuals. Overall, data on specific households and houses offer spatial specificity not found in aggregate data noted above, but their use is not without challenges.

We developed a novel form of information on household intraurban migration to address key data challenges, namely, migration chains from land parcel data for an entire region. A migration chain establishes linked pairs of moves, each defined by a household that leaves a property and one that moves into the just-vacated property. Parcel data are suited to this task when they encompass all home ownership for a specific area; in the Twin Cities, for example, these data describe over one million lots. This research utilizes the annual regional parcel dataset in the TCMA compiled and managed by the regional government, the Metropolitan Council, spanning the seven counties of Anoka, Carver, Dakota, Hennepin, Ramsey, Scott, and Washington. Relevant information includes owner's name and date last sold; other data vary by jurisdiction, such as square footage of houses and their lots or dwelling type (for a review of these data and those from other locations, see Manson et al. 2009). We identified about 4,800 origin–destination pairs for the years 2005 through 2007, which contain the most complete information for the region and pertain to the period before the US housing market collapsed in 2008.

While developing migration chains from land parcel data is laborious, it can be semiautomated. We developed migration chains for the Twin Cities by comparing the owners of a parcel across years, detecting valid owner changes and matching owners across years. We weeded out transactions, such as speculation and bank sales, that represent ownership change without a household move. We also left out condominiums and apartments given that many are not owner occupied (so renters are not included). We developed software that embodied a multipart strategy to deal with variations and errors in names. All names were uniformly formatted into the order of first name, middle name, and last name. Then an intelligent name comparison routine determined if two different names actually refer to the same person, family, or organization. It employed a dictionary of abbreviations, which records various forms of names for a single institution such as the city of Minneapolis and MPLS and the Minnesota Department of Transportation and MNDOT. It also scanned all parts and letters in two names, and if the percentage of matched parts or letters is beyond a predefined criterion, the two names are defined as the same. For instance, George Washington and G. Washington would be judged as the same person, and George Washington and George and Martha Washington are the same household. We then reviewed all matches manually to minimize dataset errors.

Semiautomated extraction of migration chains from parcel data is not a panacea for migration research, but it offers significant advantages over other approaches. While it can identify housing attributes, such as square feet of number of rooms, it does not provide characteristic of movers, such as age or size of household. It is far more extensive in coverage than most sales databases, much less expensive than surveys of individuals, and provides a level of specificity not found in higher-level data such as the census. As a result, this novel approach to migration data provides critical spatial and temporal information at resolutions sufficient to test theories of individual migration.

6.3.2 Agent-Based Modeling of Migration

We develop an agent-based model of the modified intervening opportunity theory presented above that is calibrated and validated against migration chains derived from parcel data. The ABM treats residential choice as primarily influenced by distance and direction between movers and vacancies, updating the classic intervening opportunity model with individual agents acting on real-world evidence. Agent-based modeling has garnered a lot of attention for spatially explicit modeling of urbanization and land use more broadly (Gimblett 2002; Parker et al. 2003a; Irwin et al. 2009; see Batty 2008; O'Sullivan 2008). An agent-based model is a computational system composed of semiautonomous software programs (termed agents) that can represent entities ranging from atoms through households to cities. Each agent in the system has its own resources, local context, knowledge, behavioral rules, and goals. Importantly, agents interact with each other and their larger environment. ABMs are increasingly used to understand urban issues such as growth and sprawl, land use and transportation, and racial segregation and residential structure because they explain how simple microbehavior leads to complex macro patterns and processes (Torrens 2006; Fossett 2006; Salvini and Miller 2005; Miller et al. 2004). Using an ABM is important given the intractability of deriving analytical solutions to a system of equations defined by real-world spatial data on thousands of individuals outside of a simplifying mathematical approach or use of a statistical model (Krzanowski and Raper 2001; Kwasnicki 1999). These approaches are commonly used in part because they are powerful, but an ABM, by instantiating in agents the underlying mathematical formulation of intervening opportunity theory, allows exploration of the theory in a real-world context. Marrying mathematical and statistical formalism with agent-based modeling is increasingly seen as a way forward for theoretically derived and empirically tested models of human behavior (Irwin et al. 2009).

The model developed here joins other related efforts that use ABM to understand urban processes. There is a fast-growing body of research that applies this approach to construct models centered on representing the decision-making processes of individuals and their resultant mobility (Haase and Schwarz 2009; Torrens 2012; Kennedy 2012; Parker et al. 2003b; Macy and Willer 2002; An 2012; Matthews et al. 2007; O'Sullivan et al. 2012). These models vary broadly in their degree of specificity and extent to which they are conceptually stylized models. Some attempt to simulate classical urban residential processes and patterns, such as monocentric cities and residential segregation (Benenson and Torrens 2004; Crooks et al. 2008), with highly generalized and stylized models. Others build on these simpler models via greater empirical specification, seeking to simulate urban residential processes including gentrification (Jackson et al. 2008; Diappi and Bolchi 2008; O'Sullivan 2002; Torrens and Nara 2007) and urban sprawl (Brown et al. 2008; Fernandez et al. 2005; Loibl and Toetzer 2003). Other models go even further by offering intricately detailed and data-rich explorations of urban processes underlying complex residential choices within the urban sphere (Birkin and Wu 2012; Zaidi and Rake 2001).

This model is implemented in a spatially explicit agent-based model of land change (Manson and Evans 2007). Agents are software objects, or semiautonomous programs that have their own properties and routines, that exist in an environment composed of raster and vector format layers. Importantly, agents update the environment by virtue of changing spatial layers after taking actions such as building new houses or moving between houses. The process of modeling TCMA intraurban migration in the model has four steps: (1) establishing the spatiotemporal context, (2) populating agents and environment, (3) running the model to create vacancies and simulate migration, and (4) validating model output (Fig. 6.2).

6.3.2.1 Step 1: Spatiotemporal Context

We model intraurban migration in the seven-county TCMA from 2005 to 2007. As an organizing framework, we adopt standard housing submarkets that map onto well-established neighborhoods as defined by the regional real estate board (see Fig. 6.3). We interpolate the population of each submarket as given by regional government surveys and land-use zoning to fit in these submarkets as a series of raster data layers with a resolution of 100 m.

6.3.2.2 Step 2: Agent Specification

The chief actors are households in the owner-occupied housing sector, housing developers, and governmental institutions. Populating actor agents involves significant simplification, because the core strength of agent-based modeling is illustrating how complex results can arise from simple actions. We focus on three types of agent.

Institutional Agents They shape housing development and migration destination options. The model incorporates the policy effects of the regional planning agency, the Metropolitan Council, and local governments through a set of areas that are off limits to new housing (land reserved for agricultural use or wetland offsets) and areas that are designated for new development (defined by growth zones and sewerage availability). These effects are coded as rules that denote locations where development can and cannot occur.

Developers They expand the housing stock and add new vacancies into the housing market to join the existing vacancies given by the parcel dataset. Developers build new houses that are added to the vacancy lists. Their key characteristic is the rate at which they build houses, which is given empirically by the parcel dataset as 5,392 per year. They build houses at random locations in areas designated by institutional agents. Using developers is a straightforward way of ensuring the growth in housing mirrors that in reality while maintaining an analog to the real world, but their decision making is far simpler than that of real developers.

Households These are the primary agent of interest. Agents are placed in the study area via a polygon file where the number of households within each spatial unit is determined by the actual household population in a given neighborhood, listed in the population data noted above. The migration rate is around 7 % per year for owner-occupied housing and this number of households is placed. Households are assigned

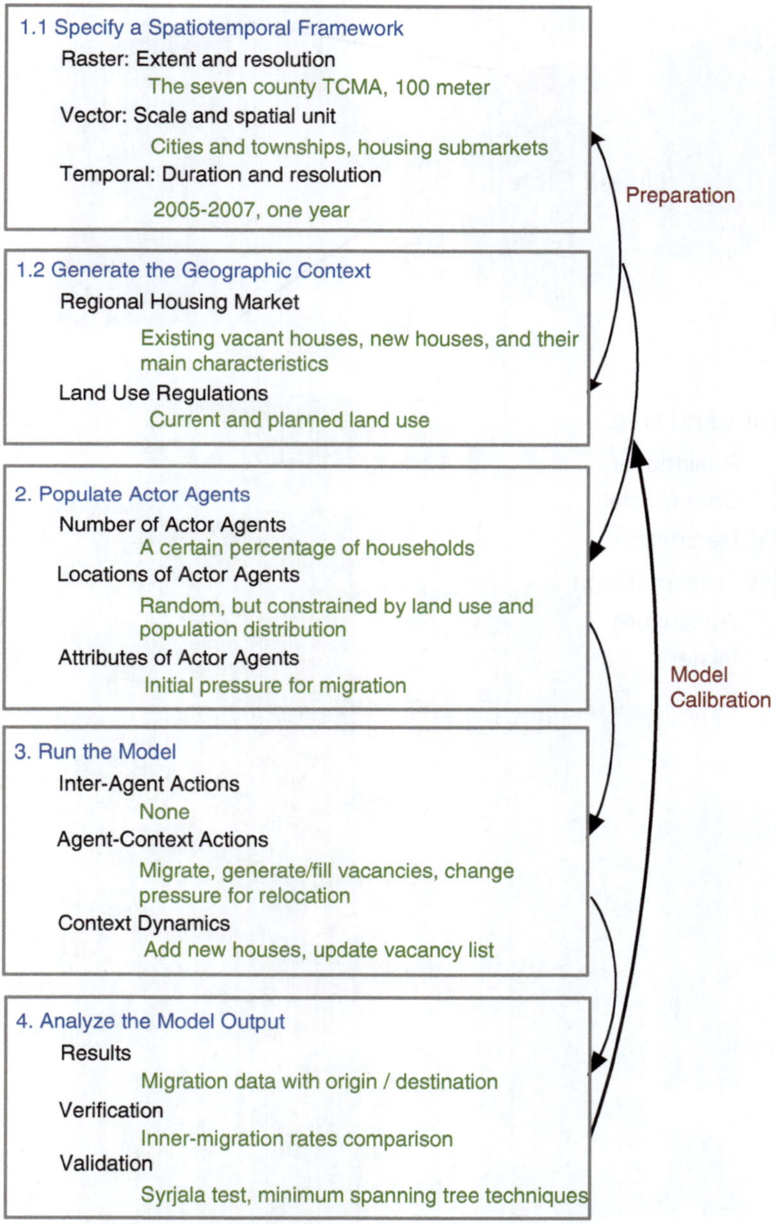

Fig. 6.2 Main modeling steps

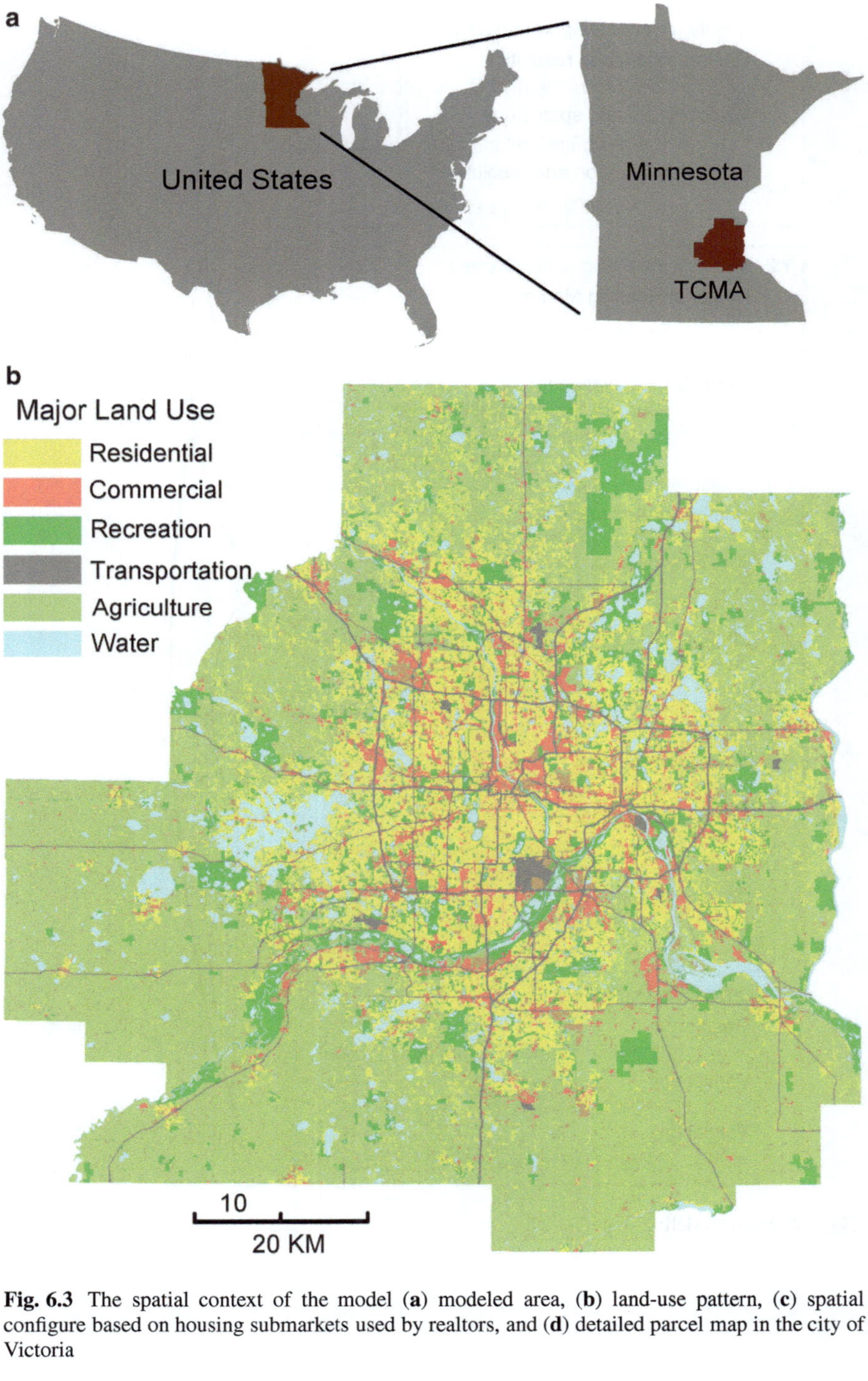

Fig. 6.3 The spatial context of the model (**a**) modeled area, (**b**) land-use pattern, (**c**) spatial configure based on housing submarkets used by realtors, and (**d**) detailed parcel map in the city of Victoria

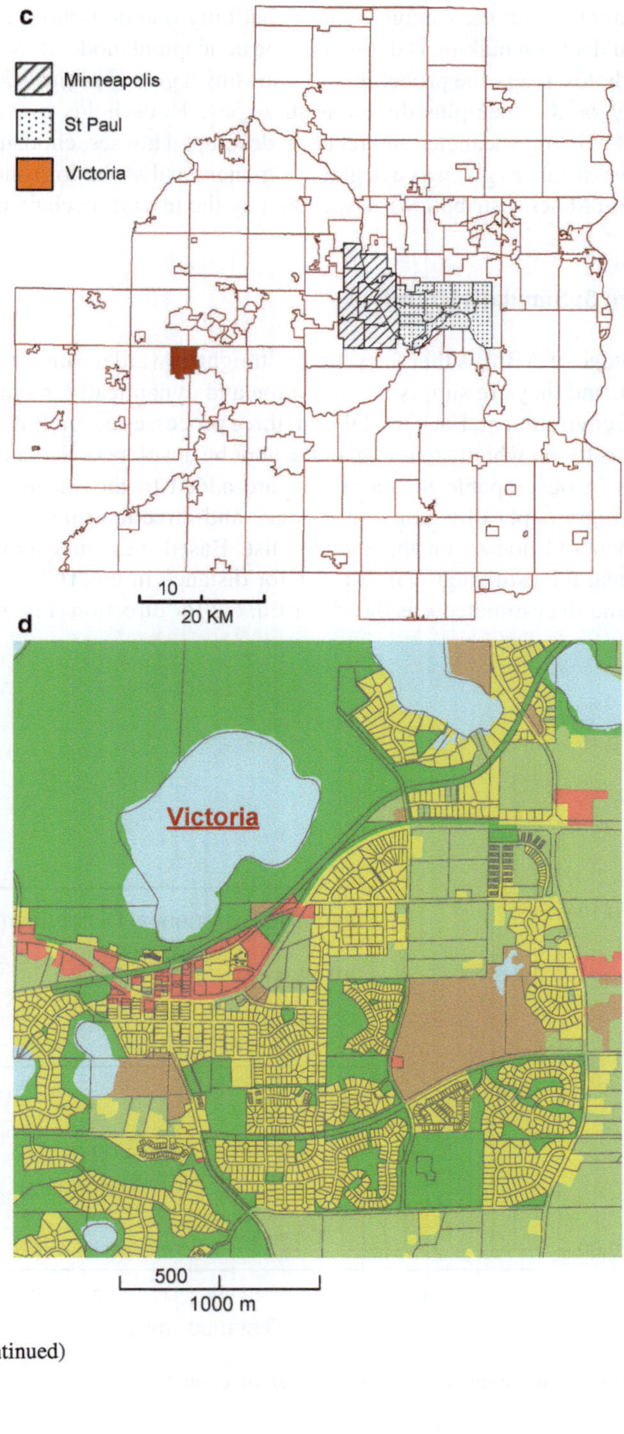

Fig. 6.3 (continued)

a random parcel within their neighborhood, but only one household can occupy a parcel. Agent decision making is defined by the conceptual model developed above, where households assess the probability of choosing a given vacancy (P_{ij}) via either distance-only or distance-plus-direction strategies. Households assess vacancies comprised of existing vacancies and recently developed houses, choosing vacancies per their move distance (given as a negative exponential distribution) and directional bias (per a von Mises distribution) as specified by the migration chain data.

6.3.2.3 Step 3: Simulate Migration

A key advantage of ABM is that they can be straightforward to run; for once agents are specified, and they are simply set in motion and dynamically interact with each other and the environment. Each model year, three processes occur. First, institutions apply policy rules on which areas can and cannot be developed. Second, developers build houses in developable locations that are added to the vacancy list. Third, households migrate, per direction and distance-and-direction rules, to new parcels and place their old houses on the vacancy list. Based on actual moves given by the parcel data, the estimated λ in Eq. 6.1 for distance in the TCMA is 0.160 for 2005–2007 and the estimated κ is 0.085 for Eq. 6.2 for direction (Fig. 6.4).

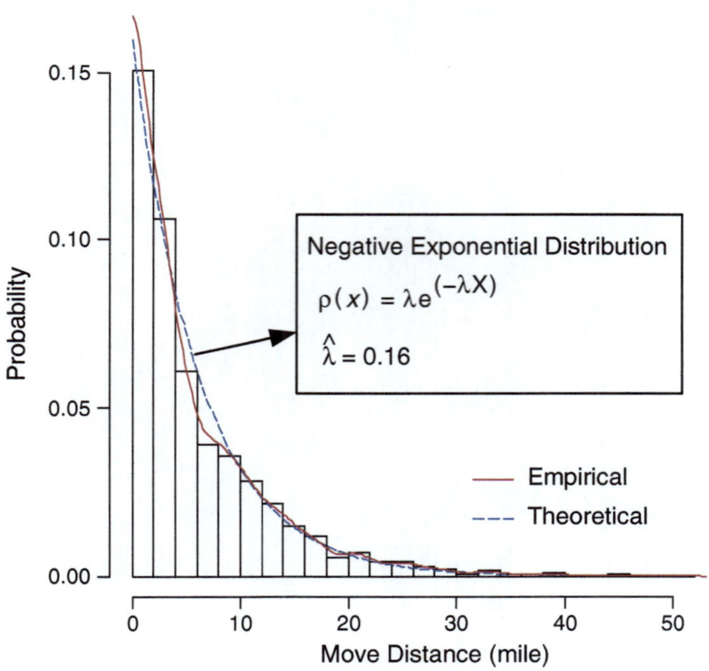

Fig. 6.4 Empirical distribution of move distance-and-direction

6.3.2.4 Step 4: Model Validation

Model validation involves measuring how well the model duplicates real-world phenomena. Model validation in the absolute or predictive sense is theoretically infeasible as no single model can reproduce every aspect of a complex open system (Oreskes 1998). That said, statistical measures can provide a useful benchmark for assessing how well different complex model configurations perform (Windrum et al. 2007; Manson 2007). Validation requires comparing model results to empirical data, to which end we used three different metrics: inner-migration rates, Syrjala tests, and minimum spanning trees. As model migration rates are calculated from actual relocation data, the number of modeled movers equals the actual number of migrations and values of λ in Eq. 6.1 and κ for Eq. 6.2 for direction. The main difference is therefore the spatial distribution of these migrant households. The three approaches employed to assess spatial fit are well suited to the problem at hand given that point pattern methods vary in their sensitivity and accuracy, as determined by their capacity to discriminate between point patterns, remain stable over different samples, and deal a range of underlying distributions (Wallet and Dussert 1998).

The three model validation approaches offer specific advantages while complementing one another. First, inner-migration rates compare the percentage of households that move within housing submarkets (i.e., those that stay within a given area or neighborhood). We use a multiscalar model specification across several submarket specifications to develop a strong measure of comparison between modeled and actual migration. Second, Syrjala tests compare the spatial distribution patterns of simulated and actual destinations of migrant households, offering the advantage over many standard point pattern analyses in assessing not just locations but also quantities across several scales of aggregation via a modified procedure that apportions simulated and actual destination points. Third, use of minimum spanning trees (MST) offers an optimized nearest-neighbor distance analysis that, instead focusing on local nearest neighbors, describes the shortest, noncircular path connecting all points. Each of these three approaches offers distinct advantages as well as overlaps in validating how well simulated and actual migration match.

6.4 Results

In order to compare the simulated results with the actual distribution of migration destinations, we employ inner-migration rate comparison alongside Syrjala tests and MST to compare the similarity between the spatial distributions of actual and simulated migration destinations. Both distance and distance-and-direction yield realistic moves across scales of aggregation. Inner-migration rates at various scales indicate that the model recreates realistic aggregate spatial patterns of intraurban migration. Inner-migration rates measure the percentage of migrants who remain in the originating spatial unit and indicate the extent to which simulated moves match real relationships among vacant housing supply, move distance distribution, and residential locations. In calculating inner-migration rates, it is necessary to correct for the fact that inner-migration rates are defined by arbitrary spatial units (Turner et al. 1989), in that small-area data can be combined at different resolutions. We measured the inner-migration rates across a series of 29 regular grids ranging from a coarse one 2 × 2 of grid cells to a fine-scaled 3 × 3grid for the entire TCMA. At finer scales, up to a third of the grid cells fall outside of the seven-county region given its irregular boundary and are not included in the count because they would inflate the number of seemingly correct moves.

The simulated inner-migration rates mirror actual rates given by the parcel data, which indicates that the model captures key relationships between vacant housing opportunities, move distance distribution, and land-use patterns (Fig. 6.5a). Both decision-making strategies—distance-only and distance-plus-direction—produce inner-migration rates that are close to the actual values. Distance-and-direction outperforms just distance, as illustrated by the total root mean squared errors, which compares how well the simulation does against actual moves measured by inner-migration rates (Fig. 6.5b).

Syrjala tests offer an advantage over inner-migration rates in that they demonstrate how well a simulated distribution resembles an actual one (Syrjala 1996). The Syrjala test compares the values of two sets of samples or, in the case of intraurban migration, destinations tessellated onto a regular grid. The test produces two measures, a Syrjala statistic and a *p* value. The Syrjala statistic measures the differences between the cumulative distribution functions of the two samples. The smaller the statistic, the closer the two sample distributions, while the *p* value indicates the probability that the two samples are from the same population and spatial distribution. There is no simple analytical solution for *p*; instead, it is calculated through sample permutation and denotes the percentage of randomized permutations that have bigger Syrjala statistic than the sample. If *p* is 0.03, for example, only 3 % of these random distributions are more similar to the distribution of one sample A than the other sample B, implying the spatial distribution of A is statistically different from that of B.

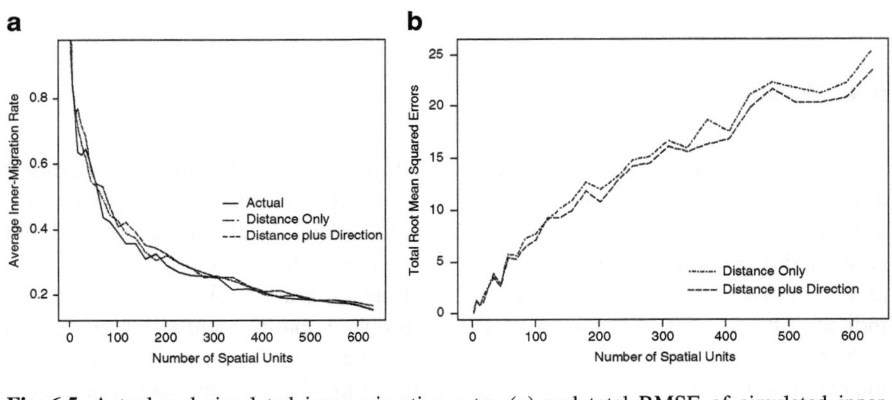

Fig. 6.5 Actual and simulated inner-migration rates (**a**) and total RMSE of simulated inner-migration rates against actual rates (**b**)

The Syrjala test of intraurban migration in the Twin Cities sheds light on complex patterns. Key measures are the percentage of subdivisions (not the number of households) that are not statistically different from the actual distribution (H_1), the mean Syrjala statistic $\overline{S}$, and the mean p value $\overline{p(S)}$. First, both decision-making models score well on H_1, where 71 % of subdivisions for the distance-only strategy and 69 % for the distance-plus-direction strategy match reality. Second, the distance-only strategy fares slightly better than distance-plus-direction strategy in recreating real migration patterns given lower average Syrjala statistics (0.696 vs. 0.771) and higher average p value (0.212 vs. 0.182). Overall, these two strategies are similarly successful in how they replicate real-world migration destinations.

The minimum spanning tree (MST) method focuses more on the relative position among intraurban migration destinations than the other two methods. Besides providing trees for visual inspection, the approach generates simple mean path length $\overline{d}$ and variance $\sigma(d)$, where a short path length indicates that points are close to each other and a small variance means the points are evenly distributed. An MST is network structure that connects all nodes with a minimum total distance (Zahn 1971; West 2001). MSTs treat individual locations as nodes of a network in which each is connected to neighboring locations, which preserves information about the adjacency of nodes (Fig. 6.6). Importantly, an MST minimizes the length of the path connecting location while guaranteeing that every location linked to another one (Guo 2008). This approach preserves both absolute and topological spatial characteristics in a way that heightens sensitivity and accuracy (Wallet and Dussert 1998), as well as offering the benefit of identifying spatial hierarchies of migration.

Two specific examples illustrate how MST analysis compares the spatial distribution of simulated and actual migration destinations. In the exurban city of Norwood, migration extends toward the Minneapolis downtown (Fig. 6.7), which mirrors the simulated results. However, both decision-making models also produce two extra spurs on the MST that trend south and north, which is not consistent with the true situation.

For the inner-ring suburban city of Robbinsdale, simulated results have a more concentrated pattern than the real situation, implying that the average path length of simulated move destinations is shorter than the reality (Fig. 6.8).

A comprehensive comparison using MST features provides insights into the predictive powers of the two different decision-making strategies (Table 6.1). In terms of the mean shortest path length $\overline{d}$, the distance-only strategy produces the smallest minimum root of mean squared errors (RMSE) compared to actual migration. Both methods generate a smaller average path length than the real migration data, which means more compact patterning of moves. The lower value of the direction-plus-direction method compared with the distance-only method is expected because the directional bias compresses the migration destinations into a smaller region. The significantly shorter average path length of the distance-only method, together with the lower variance, implies that the distance-based methods tend to generate a more compact pattern than found in reality. In other words, they will underestimate urban growth and sprawl.

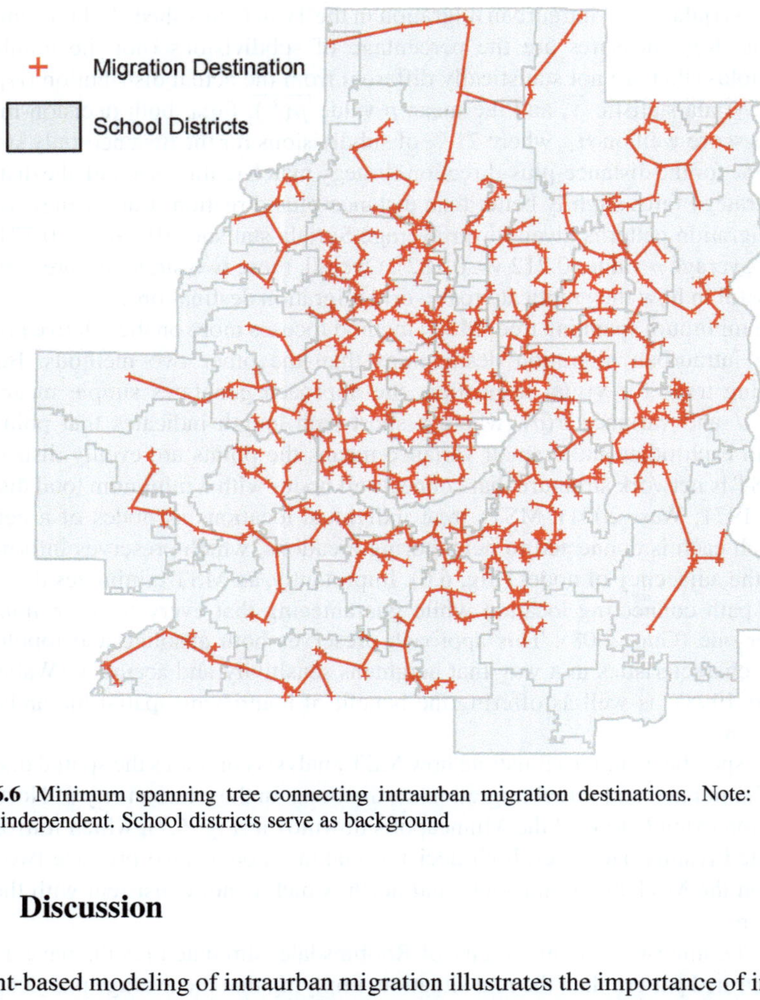

Fig. 6.6 Minimum spanning tree connecting intraurban migration destinations. Note: MST is scale independent. School districts serve as background

6.5 Discussion

Agent-based modeling of intraurban migration illustrates the importance of interaction between vacancy distribution and housing search, particularly in how complex intraurban migration patterns exhibited in the aggregate can arise from simple behavioral rules. The key finding of this work is that while it is a safe assumption that people take into account a range of personal, social, and environmental factors when making momentous housing decisions, distance-and-direction handily captures key facets of intraurban migration.

The prime import of this work is that it demonstrates a straightforward means for modeling the housing search process. The pure distance-based decision-making strategy, when applied to appropriately specified housing vacancies, can generate spatially realistic aggregate migration patterns. The addition of migration direction improves the fit somewhat at the cost of introducing greater complexity, given that it appears to capture the small but significant effect of directional bias even when using just a single downtown center instead of actual working places as the source

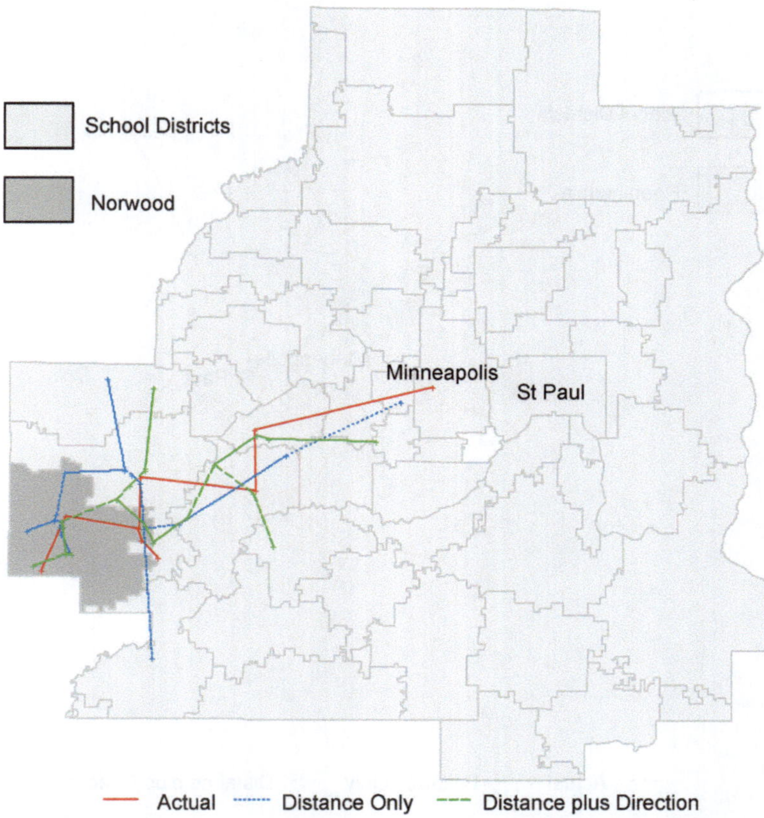

Fig. 6.7 Intraurban migration from Norwood

of this bias. Even then, underestimation of sprawl in suburban and exurban locales points to a prolonged housing search, which implies that people who live in areas with low population density tend to move less frequently and longer distances (see also Van der Vlist et al. 2002).

This work provides a basis for more complicated, utility-comparison-based migration models of housing search strategies. The modified intervening opportunity theory as instantiated in an agent-based model and empirically calibrated with migration chains captures fundamental features of the migration process and can complement deeper investigation of specific factors and locales. Alone, they can help capture key migration dynamics in the absence of many kinds of information usually needed to understand migration, creating a simple and powerful perspective on migration and urbanization. When combined with other data, they provide part of the foundation of a broader and deeper examination of intraurban migration. Directions for future research include more complicated spatial and social landscapes, such as a multi-nodal preference landscape or myriad public and

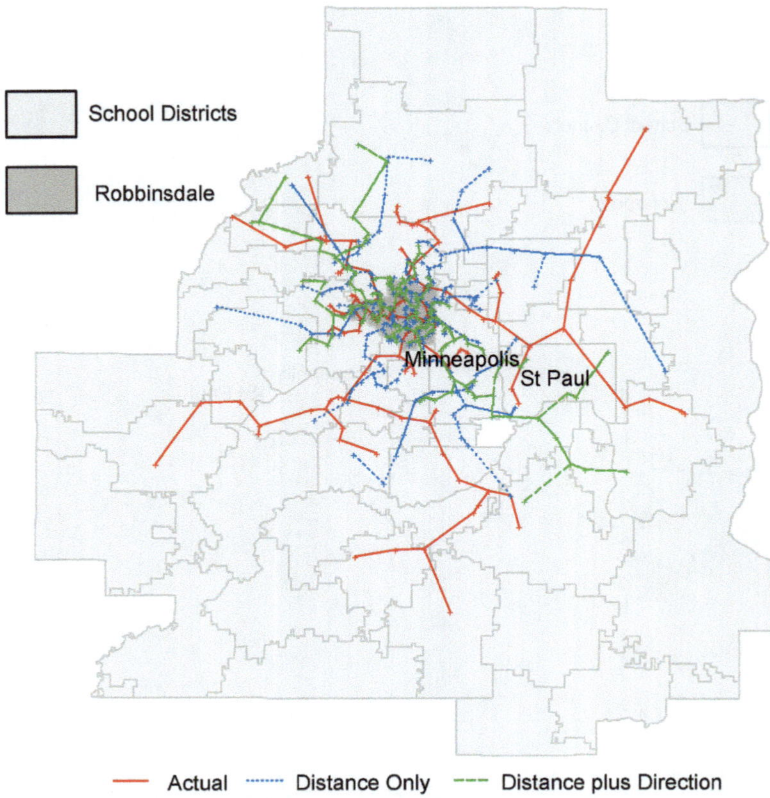

Fig. 6.8 Intraurban migration from Robbinsdale

Table 6.1 Comparison of migration strategy using MST path length distribution

Decision-making strategy	RMSE (m)	$\overline{d}$	$\sigma(d)$
Actual		4,962.08	4,626.66
Distance-only	3,908.477	3,655.031	3,042.45
Distance-plus-direction	5,051.047	3,042.447	2,713.88

private incentives related to housing, and consideration of how these landscapes interact with personal and household attributes.

More broadly, this work addresses a key methodological challenge for many urban modeling approaches, and especially for ABM, resisting the temptation to make models complicated. With data becoming more plentiful and methods growing increasingly sophisticated, models run the risk of committing what Lee (1973) termed the key "sins" of urban models, namely, being hyper-comprehensive and complicated at the cost of parsimony and generalizability (Lee 1973; Klosterman 1994). ABMs are at particular risk because their core strength is demonstrating how complexity arises from actions and interactions of simple agents. There is a fundamental tension between the desire to create realistic models by incorporating many urban processes and the desire to explain how features of a city emerge from the simple interactions among entities such as households and properties (Clarke 2004; Brown et al. 2008). This tension gives rise to the need for simple, empirically based agent-based models of migration that can complement the host of more complicated models. Overall, this approach is deliberately straightforward in that it does not examine the characteristics of movers or the broader organization of housing—the primary foci of migration research—but instead centers on combining long-standing geographical findings to provide a straightforward sociospatial conceptualization of the intraurban migration process. Overall, this work complements existing approaches while breaking new ground in understanding how individual behavior scales up to the urban region.

This model also gives insight into how complexity emerges from simplicity by examining how specific housing opportunities and individual housing search behavior influence the aggregate pattern of intraurban migration. By combining intervening opportunities theory with behavioral evidence on the spatial characteristics of intraurban migration in an agent-based model, we can explore the extent to which real-world migration patterns can result from simple behavioral rules of household search in the context of housing opportunities. When households live in an area with fewer housing opportunities, for example, they are less likely to find a vacant house that meets their needs and thus require more iterations (i.e., more time) to accomplish their housing search. Importantly, while there are many different conceptual frameworks seek to explain migration, and while this diversity signifies healthy inquiry, it highlights the need for simple models of individual actions coupled to broader, generalizable (and admittedly simple) models of urban processes (Batty 2008, 2012). Methodological challenges abound, as evidenced by both the large array of statistical and simulation approaches used in migration analysis and the extent to which they are increasingly combined in hybrid models. Many of these theoretical and methodological issues have at their heart the need for better data, particularly on specific individuals and households who collectively drive intraurban migration. Taken together, these challenges indicate a pressing need for hybrid statistical and simulation models based on data on specific individuals to develop stronger conceptual frameworks of how individual actions give rise to the aggregate patterns and processes of intraurban migration.

Overall, while agent-based modeling can help explain complex systems by integrating many possible interacting components, it is also a valuable way to explore how straightforward behavioral rules of individuals can lead to processes and patterns of complexity. By examining, incorporating, and validating spatial behavioral theories, the modified intervening opportunities model offered here can serve as a sociospatial foundation for more comprehensive urban models as well as contribute to ongoing research on developing and validating theories of human behavior in urbanization.

Acknowledgments This work is supported in part by the National Aeronautics and Space Administration New Investigator Program in Earth-Sun System Science (NNX06AE85G), the National Institutes of Health supported Minnesota Population Center (R24 HD041023), and the Resident Fellowship program of the Institute on the Environment. The authors gratefully acknowledge the assistance of the editor and anonymous reviewers. Responsibility for the opinions expressed herein is solely that of the authors.

References

Adams JS (1969) Directional bias in intra-urban migration. Econ Geogr 45(4):302–323

Adams JS (1984) The meaning of housing in America. Ann Assoc Am Geogr 74(4): 515–526

An L (2012) Modeling human decisions in coupled human and natural systems: review of agent-based models. Ecol Model 229:25–36

Batty M (2008) The size, scale, and shape of cities. Science 319(5864):769–771

Batty M (2012) Building a science of cities. Cities 29:S9–S16

Benenson I, Torrens PM (2004) Geosimulation: automata-based modeling of urban phenomena. Wiley, Chichester

Berger T, Schreinemachers P (2006) Creating agents and landscapes for multiagent systems from random samples. Ecol Soc 11(2):54–71

Birkin M, Wu B (2012) A review of microsimulation and hybrid agent-based approaches. In: Crooks AT, See LM, Batty M, Heppenstall AJ (eds) Agent-based models of geographical systems. Springer, Dordrecht, pp 51–68

Brown LA, Chung S-Y (2008) Market-led pluralism: rethinking our understanding of racial/ethnic spatial patterning in U.S. cities. Ann Assoc Am Geogr 98(1):180–212

Brown LA, Longbrake DB (1970) Migration flows in intraurban space: place utility considerations. Ann Assoc Am Geogr 60(2):368–384

Brown LA, Moore EG (1970) The intra-urban migration process: a perspective. Geogr Ann B 51(1): 1–13

Brown DG, Robinson DT, An L, Nassauer JI, Zellner M, Rand W, Riolo R, Page SE, Low B, Wang Z (2008) Exurbia from the bottom-up: confronting empirical challenges to characterizing a complex system. Geoforum 39(2):805–818

Choldin HM (1973) Kinship networks in the migration process. Int Migr Rev 10:163–175

Clark WAV (1976) Migration in Milwaukee. Econ Geogr 52(1):48–60

Clark WAV (1982) Modeling housing market search. St. Martin's Press, New York

Clark WAV (1986) Human migration. Sage, Beverly Hills

Clark WAV (2008) Geography, space, and science: perspectives from studies of migration and geographical sorting. Geogr Anal 40(3):258–275

Clark WAV, Burt JE (1980) The impact of workplace on residential relocation. Ann Assoc Am Geogr 70(1):59–67

Clark WAV, Flowerdew R (1982) A review of search models and their application to search in the housing market. In: Clark WAV (ed) Modeling housing market search. St. Martin's Press, New York, pp 4–29

Clark WAV, Withers SD (1999) Changing jobs and changing houses: mobility outcomes of employment transitions. J Reg Sci 39:653–673

Clark WAV, Huang Y, Withers S (2003) Does commuting distance matter? Commuting tolerance and residential change. Reg Sci Urban Econ 33(2):199–221

Clarke KC (2004) The limits of simplicity: toward geocomputational honesty in urban modeling. In: Atkinson P, Foody G, Darby S, Wu F (eds) GeoDynamics. CRC Press, Boca Raton, pp 213–232

Cochrane RA (1975) A possible economic basis for the gravity model. J Transp Econ Policy 9(1):34–49
Cooke TJ (2008) Migration in a family way. Popul Space Place 14(4):255–265
Crooks A, Castle C, Batty M (2008) Key challenges in agent-based modelling for geo-spatial simulation. Comput Environ Urban Syst 32(6):417–430
De Jong GF, Roempke Graefe D (2008) Family life course transitions and the economic consequences of internal migration. Popul Space Place 14(4):267–282
Diappi L, Bolchi P (2008) Smith's rent gap theory and local real estate dynamics: a multi-agent model. Comput Environ Urban Syst 32(1):6–18
Dorigo G, Tobler W (1983) Push-pull migration laws. Ann Assoc Am Geogr 73(1):1–17
Erlander S (2010) Behavioral foundations of spatial interaction models. In: Erlander SB (ed) Cost-minimizing choice behavior in transportation planning, Advances in spatial science. Springer, Berlin, pp 15–30. doi:10.1007/978-3-642-11911-8_3
Fernandez LE, Brown DG, Marans RW, Nassauer JI (2005) Characterizing location preferences in an exurban population: implications for agent-based modeling. Environ Plan B: Plan Des 32(6):799–820
Fossett M (2006) Ethnic preferences, social distance dynamics, and residential segregation: theoretical explorations using simulation analysis. J Math Soc 30(3):185–273
Fotheringham AS (1983) A new set of spatial-interaction models: the theory of competing destinations. Environ Plan A 15(1):15–36
Gaile G, Burt J (1980) Directional statistics. Concepts and techniques in modern geography, vol 25. University of East Anglia, Norwich
Geist C, McManus PA (2008) Geographical mobility over the life course: motivations and implications. Popul Space Place 14(4):283–303
Gimblett HR (2002) Integrating geographic information systems and agent-based modeling techniques for simulation of social and ecological processes. Santa Fe Institute Studies on the Sciences of Complexity. Oxford University Press, New York
Guldmann JM (1999) Competing destinations and intervening opportunities interaction models of inter-city telecommunication flows. Pap Reg Sci 78(2):179–194
Guo D (2008) Regionalization with dynamically constrained agglomerative clustering and partitioning (REDCAP). Int J Geogr Inf Sci 22(7):801–823
Haase D, Schwarz N (2009) Simulation models on human–nature interactions in urban landscapes: a review including spatial economics, system dynamics, cellular automata, and agent-based approaches. Living Rev Landsc Res 3(2)
Irwin E, Jayaprakash C, Munroe D (2009) Towards a comprehensive framework for modeling urban spatial dynamics. Landsc Ecol 24(9):1223–1236
Jackson J, Forest B, Sengupta R (2008) Agent-based simulation of urban residential dynamics and land rent change in a gentrifying area of Boston. Trans GIS 12(4):475–491
Jayet H (1990) Spatial search processes and spatial interaction: 1. Sequential search, intervening opportunities, and spatial search equilibrium. Environ Plan A 22(5):583–599
Jones C, Leishman C, Watkins C (2004) Intra-urban migration and housing submarkets: theory and evidence. Hous Stud 19(2):269–283
Kennedy WG (2012) Modelling human behaviour in agent-based models. In: Crooks AT, See LM, Batty M, Heppenstall AJ (eds) Agent-based models of geographical systems. Springer, Dordrecht, pp 167–179
Klosterman RE (1994) Large-scale urban models retrospect and prospect. J Am Plan Assoc 60(1):3–6
Krzanowski R, Raper J (2001) Spatial evolutionary modeling. Spatial information systems. Oxford University Press, Oxford
Kwasnicki W (1999) Evolutionary economics and simulation. In: Brenner T (ed) Computational techniques for modelling learning in economics. Kluwer Academic Publishers, Boston, pp 3–44
Lee DB Jr (1973) Requiem for large scale urban models. J Am Inst Plan 39(3):163–178

Loibl W, Toetzer T (2003) Modeling growth and densification processes in suburban regions – simulation of landscape transition with spatial agents. Environ Model Softw 18(6):553–563

Macy MW, Willer R (2002) From factors to actors: computational sociology and agent-based modeling. Annu Rev Sociol 28:143–166

Manson SM (2007) Challenges in evaluating models of geographic complexity. Environ Plan B Plann Des 34(2):245–260

Manson SM, Evans T (2007) Agent-based modeling of deforestation in southern Yucatán, Mexico, and reforestation in the Midwest United States. Proc Natl Acad Sci 104(52):20678–20683

Manson SM, Sander HA, Ghosh D, Oakes JM, Orfield MW Jr, Craig WJ, Luce TF Jr, Myott E, Sun S (2009) Parcel data for research and policy. Geogr Compass 3(2):698–726

Matthews R, Gilbert N, Roach A, Polhill J, Gotts N (2007) Agent-based land-use models: a review of applications. Landsc Ecol 22(10):1447–1459

Miller EJ, Douglas Hunt J, Abraham JE, Salvini PA (2004) Microsimulating urban systems. Comput Environ Urban Syst 28(1–2):9–44

Mulder CH (2007) The family context and residential choice: a challenge for new research. Popul Space Place 13(4):265–278

O'Sullivan D (2002) Toward micro-scale spatial modeling of gentrification. J Geogr Syst 4(3): 251–274

O'Sullivan D (2008) Geographical information science: agent-based models. Prog Hum Geogr 32(4):541–550

O'Sullivan D, Millington J, Perry G, Wainwright J (2012) Agent-based models – because they're worth It? In: Crooks AT, See LM, Batty M, Heppenstall AJ (eds) Agent-based models of geographical systems. Springer, Dordrecht, pp 109–123

Oreskes N (1998) Evaluation (not validation) of quantitative models. Environ Health Perspect 106(Suppl 6):1453

Parker DC, Manson SM, Janssen M, Hoffmann MJ, Deadman PJ (2003a) Multi-agent systems for the simulation of land use and land cover change: a review. Ann Assoc Am Geogr 93(2): 316–340

Parker DC, Manson SM, Janssen MA, Hoffmann MJ, Deadman P (2003b) Multi-agent systems for the simulation of land-use and land-cover change: a review. Ann Assoc Am Geogr 93(2): 314–337

Pellegrini PA, Fotheringham AS (2002) Modelling spatial choice: a review and synthesis in a migration context. Prog Hum Geogr 26(4):487–510

Quigley JM, Weinberg DH (1977) Intra-urban residential mobility: a review and synthesis. Int Reg Sci Rev 2(1):41

Roseman CC (1971) Migration as a spatial and temporal process. Ann Assoc Am Geogr 61(3): 589–598

Rouwendal J, Rietveld P (1994) Changes in commuting distances of Dutch households. Urban Stud 31(9):1545–1557

Ruiter ER (1967) Toward a better understanding of the intervening opportunities model. Transp Res 1(1):47–56

Salvini P, Miller EJ (2005) ILUTE: an operational prototype of a comprehensive microsimulation model of urban systems. Netw Spat Econ 5(2):217–234

Simmons JW (1968) Changing residence in the city: a review of intraurban mobility. Geogr Rev 58(4):622–651

Simpson L, Gavalas V, Finney N (2008) Population dynamics in ethnically diverse towns: the long-term implications of immigration. Urban Stud 45(1):163–183

Smith TR, Clark WAV, Huff JO, Shapiro P (1979) A decision-making and search model for intraurban migration. Geogr Anal 11(1):1–22

Stouffer SA (1940) Intervening opportunities: a theory relating mobility and distance. Am Sociol Rev 5(6):845–867

Syrjala SE (1996) A statistical test for a difference between the spatial distributions of two populations. Ecology 77(1):75–80

Torrens PM (2006) Simulating sprawl. Ann Assoc Am Geogr 96(2):248–275

Torrens P (2012) Urban geosimulation. In: Heppenstall EAJ, Crooks AT, See LM, Batty M (eds) Agent-based models of geographical systems. Springer, Dordrecht, pp 435–450
Torrens PM, Nara A (2007) Modeling gentrification dynamics: a hybrid approach. Comput Environ Urban Syst 31(3):337–361
Turner MG, Costanza R, Sklar FH (1989) Methods to evaluate the performance of spatial simulation models. Ecol Model 48(1):1–18
Van der Vlist A, Gorter C, Nijkamp P, Rietveld P (2002) Residential mobility and local housing-market differences. Environ Plan A 34(7):1147–1164
van Ommeren J, Rietveld P, Nijkamp P (1997) Commuting: in search of jobs and residences. J Urban Econ 42(3402):421
Waddell P (1993) Exogenous workplace choice in residential location models: is the assumption valid. Geogr Anal 25:65–82
Wallet F, Dussert C (1998) Comparison of spatial point patterns and processes characterization methods. Europhys Lett 42:493–498
West DB (2001) Introduction to graph theory, vol 2, 2nd edn. Prentice Hall, Upper Saddle River
Windrum P, Fagiolo G, Moneta A (2007) Empirical validation of agent-based models: alternatives and prospects. J Artif Soc Soc Simul 10(2):8
Wolpert J (1965) Behavioral aspects of the decision to migrate. Pap Reg Sci 15(1):159–169
Zahn CT (1971) Graph-theoretical methods for detecting and describing gestalt clusters. IEEE Trans Comput 100(20):68–86
Zaidi A, Rake K (2001) Dynamic microsimulation models: a review and some lessons for SAGE. SAGE discussion paper no. 2. Sage, London

Chapter 7
Quantifying Urban Diversity: Multiple Spatial Measures of Physical, Social, and Economic Characteristics

Timothy Rosner and Kevin M. Curtin

Abstract With long-standing trends of rural-to-urban migration, and resultant increasing urban growth, the role the built environment plays in creating a livable urban space will only increase in importance. This research examines Jane Jacobs' four generators of urban diversity, as presented in *The Death and Life of Great American Cities*, and attempts to quantify those concepts in a meaningful way. This chapter presents a methodology for assessing each of the four generators – dwelling density, block length, mix of building age, and mix of uses – as well as a new composite Urban Livability Index that combines all four generators. The resultant values are examined with measures of spatial autocorrelation to determine areas within a city that could benefit from investment in one or more parameters of livability. The methods presented here are intended to create a framework that may be applied to any city in order to assess the built environment and provide useful information to city planners and policy-makers. The District of Columbia is used as a case study for the application and testing of this methodology.

Keywords Urban geography • Diversity • Livability • Jane Jacobs • Geographic information analysis

7.1 Introduction

Urban structure is complex and multifaceted. With increasing urban populations, the effect that the built environment has on the "livability" of urban spaces deserves an increasing amount of consideration. Among some city planners, the concept of

T. Rosner
Rails-to-Trails Conservancy, 2121 Ward Ct. NW; 5th Floor, Washington, DC 20037, USA
e-mail: timothy.rosner@gmail.com

K.M. Curtin, Ph.D. (✉)
Department of Geography and GeoInformation Science, George Mason University, 4400 University Drive, Fairfax, VA 22030, USA
e-mail: curtin@gmu.edu

M. Helbich et al. (eds.), *Computational Approaches for Urban Environments*, Geotechnologies and the Environment 13, DOI 10.1007/978-3-319-11469-9_7

"New Urbanism" has been a motivating force. New Urbanism can trace some of its concepts to the work of Jane Jacobs. In perhaps her most influential work *The Death and Life of Great American Cities*, Jacobs focuses on what she describes as the four generators of urban diversity. These are a fine-grained mix of primary uses; short block lengths; a fine-grained mixing of building age; and sufficient dwelling density to support urban vibrancy (Jacobs 1992, pp. 150–151). Many of these elements have been incorporated into the design principles of New Urbanism, leading to a push for the design of "livable" urban spaces.

In the Charter of the New Urbanism, the Congress for the New Urbanism states, "we stand for the restoration of existing urban centers and towns within coherent metropolitan regions, the reconfiguration of sprawling suburbs into communities of real neighborhoods and diverse districts, the conservation of natural environments, and the preservation of our built legacy" (Talen 1999). The Charter goes on to embrace each of Jacobs' four generators, stating that "neighborhoods should be compact, pedestrian-friendly, and mixed-use"; "interconnected networks of streets should be designed to encourage walking, reduce the number and length of automobile trips, and conserve energy"; "appropriate building densities and land uses should be within walking distance of transit stops, permitting public transit to become a viable alternative to the automobile"; and "preservation and renewal of historic buildings, districts, and landscapes affirm the continuity and evolution of urban society" (Talen 1999). With such clear adoption of Jacobs' ideas, it is easy to see why Jacobs' body of work remains a relevant topic of research.

Increasingly there is a movement among federal, state (provincial), and local governments to invest resources in the interest of creating more "livable" environments. While there is latitude as to what precisely constitutes a livable urban space, most definitions include at least three of Jacobs' generators of urban diversity, namely, dwelling density, short (or frequent) blocks, and mixed primary uses. Often, it is clear to those that are familiar with the many communities within a city which locations need the greatest investment in one or all of these generators. Policymakers must be able to present these decisions in a defensible manner to the public at large. The justification for these decisions is often presented though a qualitative analysis. While the qualitative approach to such work holds great value, and Jacobs' work was almost exclusively qualitative in nature, the key elements that she describes are quantifiable. This chapter seeks to quantify the four generators that Jacobs describes in the form of a single index and four sub-indices in order to permit a quantitative analysis of the livability of different communities within an urban environment; here, Washington D.C. has been chosen as a case study.

While there has been considerable research into one or more of Jacobs' four generators of diversity, as shown in the following section, there does not appear to be research that has considered all four dimensions in combination. Furthermore, nearly all research have been conducted using boundaries that do not necessarily reflect areas that constitute a neighborhood or have been conducted at too coarse a level of detail (such as an entire Metropolitan area). In addition, the existing research has largely been conducted in suburban locations, rather than true urban environments. In her introduction, Jacobs clearly states that "I hope no reader will try to transfer my observations into guides as to what goes on in towns, or little cities, or in suburbs which still are suburban. Towns, suburbs, and even little cities are totally different organisms from great cities... to try to understand towns in terms of big cities will only compound confusion" (Jacobs 1992, p. 16). This chapter seeks to apply Jacobs' four generators of diversity to a truly urban environment at a fine-grained level.

This research does not seek to prove or disprove Jacobs' four generators, but rather to create a methodology that allows for their examination. More importantly, it allows for the investigation and targeted investment by policy-makers toward the success of city neighborhoods. Whether Jacobs' ideas are ultimately validated or not is not the issue here; they have, in fact, been incorporated into the tenets of New Urbanism, Smart Growth, and Transit-Oriented Design. Given the popularity of these ideas among modern planners, it is important to create useful methods for their examination.

The following section presents a review of previous research in the area of livability, followed by a description of the data used to conduct this case study. Next, a detailed description of the methodology is presented, followed by the results of the case study, the conclusions that can be drawn from this case study, as well as possible future research. Please note that the quotations from *The Death and Life of Great American Cities* presented in this paper are from the 1992 edition of the book, although the original was published in 1961.

7.2 Background and Literature Review

When *The Death and Life of Great American Cities* was first published, it was rightfully seen as an unfavorable critique of modern city planning. In the first sentence of the book, Jacobs states explicitly, "This book is an attack on current city planning and rebuilding" (1992, p. 3). Throughout the book, Jacobs names not only the planning concepts to which she objects but also those that she sees as their proponents. Among the parties she identifies on numerous occasions as having had a detrimental effect on the city is Lewis Mumford, a contemporary of hers and a respected planner to this day. Perhaps due in part to inflammatory statements by Jacobs, the level of discord between Jacobs and Mumford has been somewhat exaggerated. In reviewing the relationship between Jacobs and Mumford, Mellon (2009) found that while Jacobs' and Mumford's ideas for what constituted a healthy, diverse, livable urban environment differed, they both strove for the same goal; Mumford is even noted as having encouraged Jacobs to write *Death and Life*.

While the friction between Mumford and Jacobs may have been embellished by history, there is no doubt that she saw futility in the city planning efforts of the time. However, the critique that she presented in Death and Life was not considered by many to be objective (Laurence 2006), especially when compared to the physical sciences being studied at the time. However, in the final chapter of *Death and Life*, Jacobs discusses the scientific theories of Warren Waver and notes that cities are "problems of organized complexity, like the life sciences" (1992, p. 433). This suggests that if we change the way that cities have been studied in the past, it is possible to arrive at methods that produce master plans dramatically different than those that Jacobs railed against. This is the void which the research presented in this chapter attempts to begin filling.

The research presented here is related to other research in the fields of livability and quality of life. Other authors have reviewed the literature relating to these subjects from a broad perspective. These reviews find that the definitions of livability are wide-ranging and note that "concepts such as livability, living quality, living environment, quality of place, residential-perception and satisfaction, the evaluation of residential and living environment, quality of life and sustainability do overlap, and are often used as synonyms – but every so often are contrasted" (van Kamp et al. 2003). It can be noted that some of these concepts, such as quality of life, are usually studied by examining the perceptions of groups of people rather than examining quantitative data that represents the built environment, as is being done in the research presented here. Pacione (2003) also examines a large volume of literature in the field and comes to many of the same conclusions as van Kamp et al. (2003), viewing quality of life and livability studies as having many sub-domains reflecting the researcher's approach to the problem (objective vs. subjective, scale of the study, etc.).

Two examples of the broad ranging nature of quality-of-life and livability studies include the work of Doi et al. (2008) and the work of Wood et al. (2010). The research of Doi et al. (2008) examines quality of life indicators in Takamatsu, Japan, and includes the sweeping statement that "if individuals are enabled to make rational choices about their location over the long term, they can ensure the highest [quality of life] performance all the time." This statement highlights the perceptive nature of this particular study, which does not embrace the notion that individuals are capable of making (and frequently do make) decisions that are irrational. While these types of studies are important (what use is a safe neighborhood if everyone in the community perceives it to be crime-ridden?), it is necessary to include non-perceptive qualitative studies in the literature as well. Wood et al. (2010) attempts to address this by examining the effects of the built environment on the sense of community. However, this particular study used a study area that was largely homogenous in nature, something acknowledged by the authors (Wood et al. 2010). More robust studies are needed to bolster the understanding of the built environment and its role in livability.

As noted earlier, many of the concepts that Jacobs introduced in *Death and Life* have been incorporated into the New Urbanism movement. As such, much of the modern criticism of these ideas comes under the umbrella of critiquing the New Urbanism and Smart Growth planning ideals. Kristen Day (2003) argues that in New Urbanism design, "diversity is not regarded as an existing characteristic of communities . . . the assumption breaks down, however, when New Urbanism is applied to urban neighborhoods in which diversity already exists." This is an important statement, in that it reflects how the ideas of Jacobs differ from how they have been incorporated into New Urbanism. Jacobs actually sees the situation as the reverse of the New Urbanists, stating that, "to be sure, a good city neighborhood can absorb newcomers into itself, both newcomers by choice and immigrants settling by expediency, and it can protect a reasonable amount of transient population too" (1992, pp. 137–138). Here, Jacobs sees the diversity as inherent in the good neighborhood, with the ability to adapt and absorb, as opposed to a diversity that needs to be created. To be fair, she also indicates that diversity is not inherent and thus proposes the four characteristics that she sees as necessary for the creation and sustainability of diversity. In the chapter entitled "Gradual money and cataclysmic money," she warns against sudden infusions of money that "[pour] into an area in concentrated form, producing drastic changes" (Jacobs 1992, p. 293) which may produce challenges because, according to Jacobs, "All city building that retains staying power after its novelty has gone and that preserves the freedom of the streets and upholds citizens' self-management, requires that its locality be able to adapt, keep up to date, keep interesting, keep convenient, and this in turn requires a myriad of gradual, constant, close-grained changes" (1992, p. 294). The distinctions between Jacobs' original work and its incorporation into the New Urbanism movement are important, since much existing research and critique has focused on the principles of the New Urbanists, rather than the original ideas of Jacobs.

Many of the critics of Jacobs and the New Urbanist ideals focus on evaluating them from a single perspective, such as traffic reduction. The research of Filion and Hammond (2003) is an excellent example. The authors question the wisdom of neo-traditional (i.e., New Urbanism) design, noting that they do not "necessarily enhance pedestrian accessibility rates" and "are not as effective at diverting through traffic away from residential streets as those of newer neighborhoods" (Filion and Hammond 2003). Jacobs would likely take this statement as an example of a focus on segregating uses (residential streets versus commercial corridors) and travel modes. Jacobs herself did not see automobiles as an enemy; "we blame automobiles for too much" (1992, p. 338). Instead, she views automobile use as a necessary, though over-used, means of transport – and as especially critical for conducting commerce. "To concentrate on riddance as the primary purpose, negatively to put taboos and penalties on automobiles as children might say, 'Cars, cars go away,' would be a policy not only doomed to defeat but rightly doomed to defeat" (Jacobs 1992, p. 360). The conclusions of Filion and Hammond are likely due to their decision to approach the research from a perspective that differs from Jacobs' with regard to the separation of travel modes and uses.

Similarly, much of the research into Jacobs' four generators of diversity has focused on only one or two of the generators in isolation. A number of studies have been conducted into the effects of mixed-use development. Grant (2002) found that "mixed use districts are becoming more segregated by class, and affordability has not improved. Efforts to mix uses have not stanched the loss of economic vitality for most Canadian cities." However, this research focused on mixed use in the suburban context, a location where Jacobs (as noted earlier) had no intention of her ideas being utilized. Some research has focused on the level of physical deterioration of structures (as a proxy for the success or failure of the community) within a mixed-use context, finding that there are increased levels of deterioration in mixed-use neighborhoods (Taylor et al. 1995). The research of Wansborough and Mageean (2000) examines mixed use in slightly broader terms, focusing on its role in cultural regeneration. Their conclusion differs from Taylor et al. (1995), finding that "the encouragement of ground-floor uses in mixed-use schemes has helped to improve surveillance and soften the boundary between public and private space" (Wansborough and Mageean 2000). Hirt (2007) explores the differences in zoning between the US and German systems finding that, "under the German approach each city block may end up in a different land use category, and this is conducive to a much more fine-grained diversity of uses." The research goes on to note that US zoning techniques "reduce the idea of the mixed-use city, which Jane Jacobs so eloquently advocated, to a small mixed-use part of the city" and "assume that single-family residential areas are inevitable, quite unlike what we find in Germany. This is precisely one of the reasons why Jane Jacobs criticized new urbanism" (Hirt 2007). This is further evidence that while some New Urbanists find inspiration in Jacobs' work, they have not strictly adhered to her philosophy.

There have also been studies that have focused on the density aspect of the four generators of diversity in isolation from the others. Bramley and Power (2009) explored the connection between density and social sustainability within communities and note the trade-offs that occur with increased density. "... Compact forms worsen neighborhood problems and dissatisfaction, while improving access to services" and "policy must therefore think in terms of trade-offs between social objectives" (Bramley and Power 2009). Similarly, Nasar (2003) found that, "the more condensed pattern of development and reduced use of auto did not yield a higher sense of community: residents in [neo-traditional developments] and [standard suburban developments] showed no difference in sense of community." With these two studies, it is not clear just how large of a role self-selection has played in the results. How many of the residents have chosen to live in a particular neighborhood for specific reasons, as opposed to those that live there as a compromise, or through lack of alternative options? A resident that is present in a neighborhood as part of a deliberate locational choice is likely to respond differently than a resident that is there due to a lack of alternatives. This uncertainty strengthens the argument that, as noted earlier, perception is only one aspect that should be evaluated when considering the four generators of diversity.

In contrast to density and mixed use, there has been significantly less research conducted regarding street length and mix of building age in the context of livability. Cozens and Hillier (2008) conducted a review of literature regarding cul-de-sacs and grid street networks (which can be considered a useful proxy for short street segments) and concluded that "the evidence to support New Urbanism's advocacy for permeable street networks is unfounded or largely inconclusive at best." The authors suggest that cul-de-sacs have fallen out of favor due to their association with the Garden City movement; the planning idea against which New Urbanism is sometimes considered a reaction. However, as noted later in this paper, it is not strictly the connectedness of a street layout that is in question with Jacob's work; it is actually the physical length of the network segments. For this reason, Cozen and

Hillier's conclusions must be taken in context and cannot be seen as a refutation of Jacobs' ideas about block length. Along with the limited research into the street length – livability dynamic – there is essentially no current research available that examines the role of building age mixes with regard to livability and diversity.

In addition to the research that has been conducted focusing on a single one of the four generators of diversity, there have been a limited number of studies that have attempted to include elements of at least three of the generators (again, building age mix is absent in these studies). Miles and Song (2009) examined Portland, Oregon, which has utilized many of Jacobs' ideas in its planning and found that the city "has been successful in creating neighborhoods at several economic scales that feature not only the connectivity, accessibly, mixed land use and access to public transit that characterize 'good' neighborhoods from a physical perspective, but also 'good' social environment indicative of strong ties and collective efficacy." This finding echoes the earlier research of Song (2005) who examined three different communities that have utilized "smart growth" policies and found that "only when all these dimensions – connectivity, density, mixed land uses, accessibility, and pedestrian walkability – are combined can they create synergy by having amenities that complement one another." Regarding the four generators of diversity, Jacobs herself writes, "all four in combination are necessary to generate city diversity; the absence of any one of the four frustrates a district's potential" (1992, p. 151). Cervero (2002) also examines multiple parameters, this time strictly from the perspective of travel mode choice. Utilizing Traffic Analysis Zones, he found that "drive-alone and group-ride automobile travel fell relative to transit riding as gross densities increased at both the trip origin and destination. And land-use mixture at both trip ends lowered the probability of driving alone or ride-sharing versus taking a bus or train, ceteris paribus" (Cervero 2002). These papers all underscore the wide-ranging impact that Jacobs' ideas about density, land use, and street networks can have on the livability of the urban environment.

A number of researchers have emphasized the importance of scale when studying the built environment. Tesfazghi et al. (2010) illustrated how having areas of aggregation that are too large can mask the variability that exists at a lower level. This reinforces the work of Openshaw and what he termed the "modifiable areal unit problem," noting that "the definition of these geographical objects is arbitrary and (in theory) modifiable at choice; indeed, different researchers may well use different sets of units" (1983). This highlights the importance of utilizing small, yet standardized units of aggregation. Martinez (2009) chose to study quality of life indicators at a disaggregated level stating that, "when indicators are generated at high levels of aggregation they can give a misleading idea of the problem they address and quantify." Apparicio et al. (2008) who examined a mix of subjective and objective quality of life parameters at an intra-urban level note that, "since individuals' daily lived environment is not on a metropolitan scale, it is important to find the appropriate scale so that the indicators can express the heterogeneity of the conditions faced by urban residents." The research presents a convincing argument to carefully account for the level of detail at which a study is undertaken.

In light of the previous research conducted in the area of livability and the built environment, the research presented here hopes to fill a gap by creating a methodology that effectively quantifies each of Jacobs' four generators of diversity in a manner that may be applied to any large city, as well as creating a single Urban Livability Index which combines all four parameters into a single ranking. By creating a single index as well as four supporting indices, the methodology presented may assist city officials in decision-making during the urban planning process as well as provide a repeatable framework for other researchers in the area of livability.

7.3 Case Study Data

For this study, the data is from two sources. The first source is the U.S. Census Bureau, from which the block group geography is obtained. These serve as the primary units of aggregation and examination for the study. The second source is the Washington, D.C., city government that provided data for street centerlines as well as land ownership data that includes a number of important attributes. All data were projected in the Maryland State Plane Coordinate System (the official coordinate system for the Washington, D.C., city government), which utilizes a specific implementation of the Lambert Conformal Conic projection to minimize the distortion of all measurements within the study area.

The block group file contains 433 block groups and covers the entire city, including areas owned and operated by the federal government, such as the National Mall. The street centerline file includes 34,138 street segments (including freeways, alleyways, driveways, and access ramps) across the entire city, again, including areas under federal jurisdiction, such as Rock Creek Parkway. The ownership file is a point file that contains a single point for each ownership record within Washington, D.C. This file contains a detailed land-use code for each point that corresponds to a list of 109 possible land-use types designated by the city government. This file was appended with data for building construction, renovation, and addition dates for commercial and residential properties, also obtained from the city government. This information was not available for some buildings, such as educational and health-care facilities. The greatest challenge with the ownership data is the different treatment that condominiums and rental units receive. While condominiums in the same building are each represented as a unique point, an apartment building for which a single owner rents all the units only contains one point in this file. This represented a potential difficulty for calculating dwelling densities. However, the same file that contained information on building date also contained information on the number of units in each structure. Thus, apartments units that are represented by a single point for multiple dwellings had the information on the number of units appended to them.

The datasets employed here are largely (and increasingly) available for major cities worldwide. Street centerlines, cadastral data, and population aggregation data are among the most common data layers maintained by city governments that are employing Geographic Information Systems in their planning analyses. This bodes well for the replication of this work, using the proposed methodology in cities across continents.

7.4 Methodology

The method of research for this study consists of the creation of four individual indices for each of Jacobs' four generators of diversity, as well as a single composite index combining all of the four individual indices. The individual indices utilize data that is rescaled to range from 0 to 1. This facilitates the computation of the final index by giving all four parameters equal weights. Other methods, such as the utilization of a z-score, were considered, but this tends to force the parameter values into artificial distributions which may or may not hold true between different cities and require differing transformations, limiting the portability of the methodology. By utilizing rescaled values, this methodology may then be applied to different cities across global regions.

For the first parameter, dwelling density, each of the ownership points is assigned to a block group, and the area for each block group is calculated. Next, all points defined by their use code as residential (see Table 7.1) are selected. For properties where a single dwelling is represented by a single point (single-family homes, condominiums, etc.), a value of "1" is assigned to that point. For properties where multiple dwellings are represented by a single point (such as rental apartments), a value corresponding to the total number of dwellings is assigned. These values

Table 7.1 Truncated list of residential uses for dwelling density

Use Code	Description
003	Residential-transient
011	Residential-row-single family
012	Residential-detached-single family
...	...
022	Residential-apartment-elevator
023	Residential flats-less than 5
024	Residential-conversions-less than 5
025	Residential-conversion-5 units
...	...
127	Coop-vertical-mixed use
216	Condo-investment-horizontal
217	Condo-investment-vertical
316	Condo-duplex

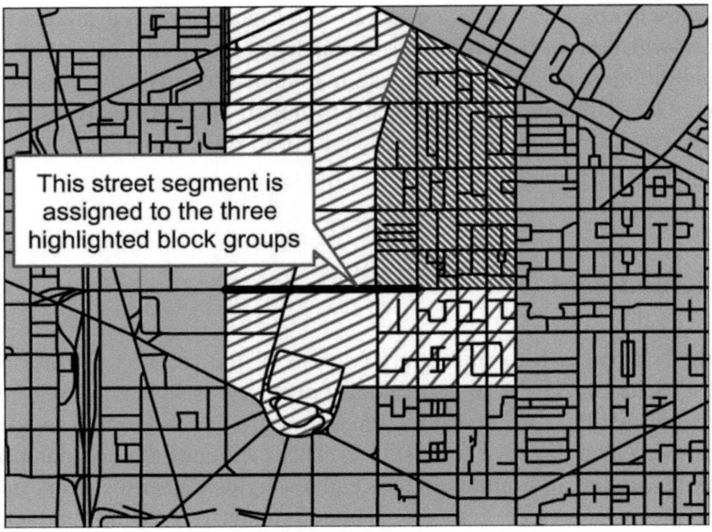

Fig. 7.1 Street segment assigned to multiple block groups

are then totaled for each block group and a density is calculated from these totaled values. Finally, these values are then rescaled from 0 to 1 utilizing the equation:

$$V_r = \frac{(V_i - V_{\min})}{(V_{\max} - V_{\min})} \tag{7.1}$$

where V_i equals the value to be rescaled, $V_{\min}$ equals the lowest density calculated, and the $V_{\max}$ equals the highest density calculated. This results in all values being scaled from 0 to 1, with values closer to 0 having a lower density than values closer to 1. This maintains the natural distribution of the dataset.

For the second parameter, block length, each street segment (excluding alleyways, driveways, and ramps) is assigned to a block group or number of block groups. In some cases, segments are assigned to as many as three different block groups. In Fig. 7.1 a street segment is shown that is assigned to three separate block groups, due to a number of "intersections" actually being over-/underpasses, meaning that this segment runs uninterrupted through multiple block groups. Street segments are assigned to block groups that they are either fully contained by or intersect (including boundary intersection). However, segments that only share one vertex with the block group are not included.

The total number of street segments and their total lengths for each block group are then calculated and a mean street segment length for each block group is determined. Finally, the mean street segment length for each block group is rescaled from 0 to 1 utilizing the same format as Eq. 7.1, where V_i equals the value to be rescaled, $V_{\min}$ equals the lowest mean street segment length, and the $V_{\max}$ equals the highest mean street segment length. In this case, values closer to 0 have a lower mean street segment length than values closer to 1. For this parameter, the lower values are preferred to the higher values.

While there are a number of methods that measure street connectivity, such as the alpha and beta indices (Rodrigue et al. 2009, pp. 29–31), the connectedness of the streets within a block group is not Jacobs' primary concern. A large block group with six long street segments and six intersections would have the same level of connectedness as a small block group with six short street segments and six intersections. However, Jacobs would find the small block group with shorter street segments to be a better generator of diversity than the larger block group with longer street segments. For this reason, the block length parameter examines the mean street segment length, rather than the connectivity within a block group.

For the third parameter – mix of building age – each ownership point and its associated dates are examined. Of the three available dates (construction, renovation, and addition), the newest date is extracted and used for the following calculations. This is based on Jacobs' statement that "a successful city district becomes a kind of ever-normal granary so far as construction is concerned. Some of the old buildings, year by year, are replaced by new ones – or rehabilitated to a degree equivalent to replacement. Over the years there is, therefore, constantly a mixture of building ages and types" (1992, p. 189). First the standard deviation of the dates of the points within a block group is calculated. Then once these standard deviations have been calculated for each block group, they are rescaled from 0 to 1 utilizing the same format from Eq. 7.1, where V_i equals the value to be rescaled, $V_{\min}$ equals the lowest standard deviation, and $V_{\max}$ equals the highest standard deviation. In this case, values closer to 0 are assigned to block groups having a lower building age difference than those with values closer to 1. For this calculation, a higher value is preferred, since by Jacobs' estimation, areas with the greatest variance are better generators of diversity.

For the fourth parameter – mixed uses – two variant methods are used and evaluated. While there are existing methods, such as binary assignments, where geographic units are determined to be mixed use or non-mixed use, based on different criteria (Krizek 2003), the methods employed here seek to create a scale of "mixed-ness" within a block group. The first method consists of calculating the distance between each residential point and the nearest commercial point within the same block group. The mean of these distances is then calculated and rescaled from 0 to 1 utilizing the same format from Eq. 7.1, where V_i equals the value to be rescaled, $V_{\min}$ equals the shortest mean distance, and $V_{\max}$ equals the highest mean distance. The result is values closer to 0 having a shorter mean distance between residential and commercial locations than values closer to 1. Block groups that either do not contain commercial points or do not contain residential points are assigned a value of 1, indicating that they are the least mixed use.

The second method consists of calculating the mean center of all the residential points within a block group and the mean center of all the commercial points within a block group. The distance between the two mean centers is then calculated. This distance is then rescaled from 0 to 1 utilizing the same format from Eq. 7.1, where

Table 7.2 Truncated list of commercial uses for mixed-use calculation

Use Code	Description
001	Residential-single family (commercial use)
031	Hotel-small
032	Hotel-large
...	...
043	Store-department
044	Store-shopping center/mall
045	Store-restaurant
046	Store-barber/beauty shop
...	...
048	Commercial-retail-condo
049	Commercial-retail-misc
051	Commercial-office-small
...	...
067	Commercial-restaurant
068	Commercial-restaurant-fast Food
...	...
465	Vehicle service station-market

values closer to 0 have a shorter distance between residential and commercial mean centers than values closer to 1. As with the first method, block groups that either do not contain commercial points or do not contain residential points are assigned a value of 1, indicating that they are the least mixed use. For both of these methods, "commercial" includes uses described as "retail." See Table 7.2 for a truncated list of the uses included as "commercial" in these calculations.

The idea behind both of these methods is that the calculations reflect the level of intermixing of residential and commercial uses. The closer the shortest distance between a residential location and a commercial location within the same block group, the more intermixed these uses are; the closer the mean centers of these two use types, the more mixed these uses are. For both methods, the Euclidean distance is used for the calculations, as opposed to the Manhattan or network distance. This is done purposefully to separate the measure of mixed-use as much as possible from the actual structure of the street network, which has its own measure in the form of street lengths. This will help to identify areas that may have an adequate street network, but a poor mix of uses, or vice versa. Both methods are evaluated since they may provide slightly different results, though one, the mean center calculation, is far less computationally intensive than the other, making it more repeatable for other cities. As with the block length parameter, lower values (closer to 0) are preferred to higher values (closer to 1), since lower values are indicative of a more fine-grained mix of uses.

It should be noted that in the interest of easing the computational complexity of these calculations, they have both been limited to commercial and residential points within the same block group. This creates the possibility, particularly for the mean shortest distance method, that some residential points may be assigned a distance that is higher than if all commercial points (including those outside of the block group) were used. While a more inclusive computation would be more robust, it has been sacrificed with the intention of reducing the computational burden and with the understanding that the final calculation is derived from the mean which aids in mitigating the impact of these types of occurrences.

The final part of the methodology consists of a composite index, the Urban Livability (UL) Index, which combines all four parameters into a single number. For this composite index, the final score for each of the four parameters is combined. In the case of the block length and mixed-use parameters, the negative of the values is taken, since for these particular indices, lower scores are considered better than higher scores. The final index is as follows:

$$UL_i = d_i + (-a_i) + b_i + (-m_i) \tag{7.2}$$

where UL_i equals the final composite index for the selected block group, d_i equals the rescaled value of the selected block group for the density parameter, b_i equals the rescaled value for the selected block group for the block length parameter, a_i equals the rescaled value of the selected block group for the building age difference parameter, and m_i equals the rescaled value for the selected block group of the mixed-use parameter. While this unweighted additive approach to the composite index may appear simplistic, it corresponds most closely to the work of Jacobs, who did not view one of the parameters to be any more important than the other three. For the final calculation, these composite scores are again rescaled from 0 to 1 utilizing the same format from Eq. 7.1, resulting values closer to 0 having a less "livable" environment across the four parameters than values closer to 1.

For this research, all of the parameters have been examined at the census block group level of aggregation. Jacobs carefully notes that her four generators of diversity operate on a fine-grained level (Jacobs 1992, pp. 150–151). While it would be ideal to examine each of these parameters on a block-by-block basis, there is difficulty in doing so. The most challenging of the parameters to examine at such a small geographic level is that of mixed use. Simple observation in most cities will show us that having a wide variety of uses on a single block is highly uncommon. Figure 7.2 shows an example from Northwest D.C. where uses are clearly delineated by street block. Commercial uses are clearly aligned along U Street, while T Street along with 15th and 16th Streets are all residential in nature.

It is far more common for uses to be mixed by adjacent blocks, such as a residential block adjacent to an office block and a retail block. Thus, it becomes necessary to have a level of aggregation that captures this mixture of uses. Census block groups provide a convenient, yet sufficiently fine-grained level of aggregation for this study, as shown by the number, 433, needed to cover the entire city.

The most apparent downside to utilizing block groups is their irregular size. This creates a challenging situation for comparing certain parameters across block groups – again, mixed uses present a challenge. For areas with particularly high population counts, block groups are smaller. For especially dense populations, the block group may only contain a single residential complex. This creates problems

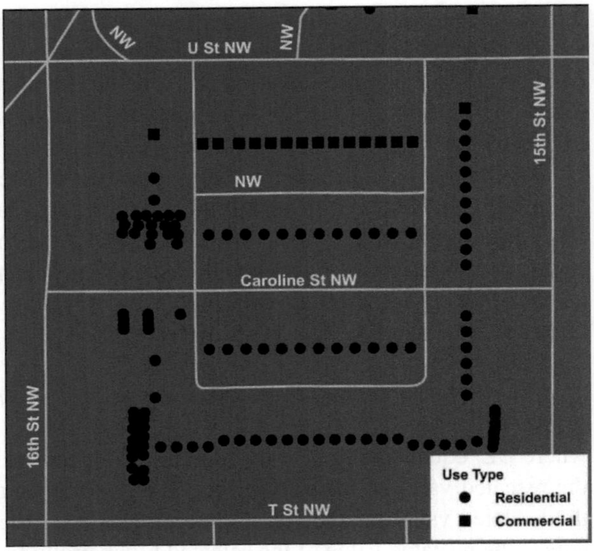

Fig. 7.2 Mix of uses by street block

when calculating the percentage units within a block group that are residential as opposed to commercial. Block groups that are geographically larger inherently have the potential to capture more nonresidential uses. Some authors have addressed this by utilizing a regularly spaced grid overlaid on the study area and aggregating by each of the grid squares (Krizek 2003). However, by utilizing a grid that is unique to that particular study, it becomes very difficult, if not impossible, to verify the study using additional data. Here, by using block groups, there is the possibility of examining the results of the research against certain types of census data (such as poverty rates), used as proxies for livability. The utilization of readily available block groups also facilitates the reproduction of this research within other cities and by other authors. Other researchers have utilized larger areas of aggregation, such as Traffic Analysis Zones (TAZs) or census tracts. By way of comparison, there are 320 TAZs within Washington, D.C. (which are not necessarily contiguous, thus leaving some areas of the city unstudied), and there are 188 census tracts – far fewer than the 433 contiguous block groups used in this study.

7.5 Results

After processing the data according to the methodology outlined above, a number of interesting patterns emerged. Beginning with the first parameter, dwelling density, the number of dwellings per acre ranged from 0 to 65.97, with a mean of 11.22 and a standard deviation of 10.02. Figure 7.3 shows the distribution of the densities, as a histogram, and illustrates the rescaled values as an unclassed choropleth map. Although we do not perform a formal test of similarity to the Poisson distribution

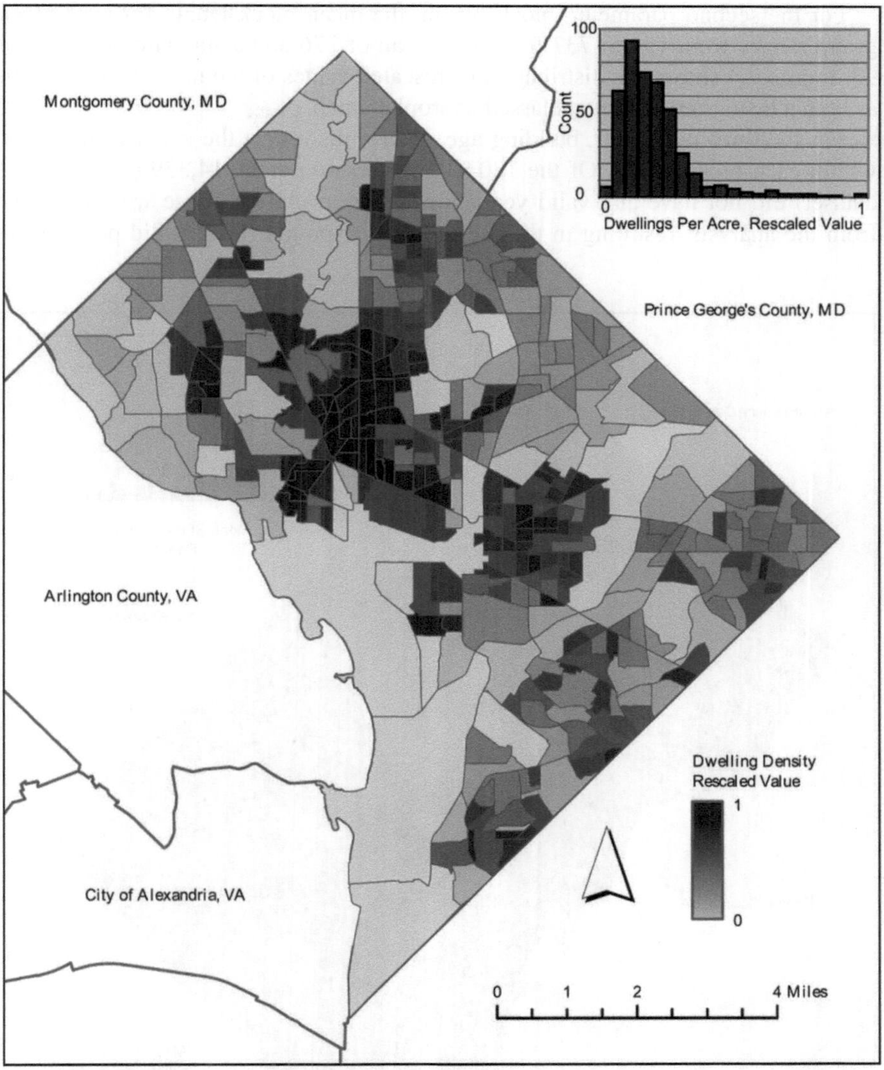

Fig. 7.3 Histogram and map of dwelling density rescaled values

here, this empirical distribution and the descriptive statistics suggest that the Poisson distribution may be useful for future research where a reference distribution is of value. The spatial distribution is as expected, with high densities especially evident in the Foggy Bottom, DuPont Circle, and Columbia Heights/Mount Pleasant neighborhoods, and a wide mix of densities in the Southeast quadrant of the city, east of the Anacostia River. Low densities are clearly present in upper Northwest as well as the Northeast quadrant and along parts of Rock Creek Park and the National Mall.

For the second parameter, block length, the mean block length for each block group ranges from 154 to 737 ft, with a mean of 276 and a standard deviation of 93. Figure 7.4 shows the distribution of rescaled values of the mean block lengths as both a histogram and an unclassed choropleth map.

For the third parameter, building age, the results reflect the standard deviation within each block group. Of the 180,836 ownership points, 14,339 (7.9 % of the dataset) did not have any valid year information available. These were excluded from the analysis, resulting in the use of 166,497 points. The invalid points were

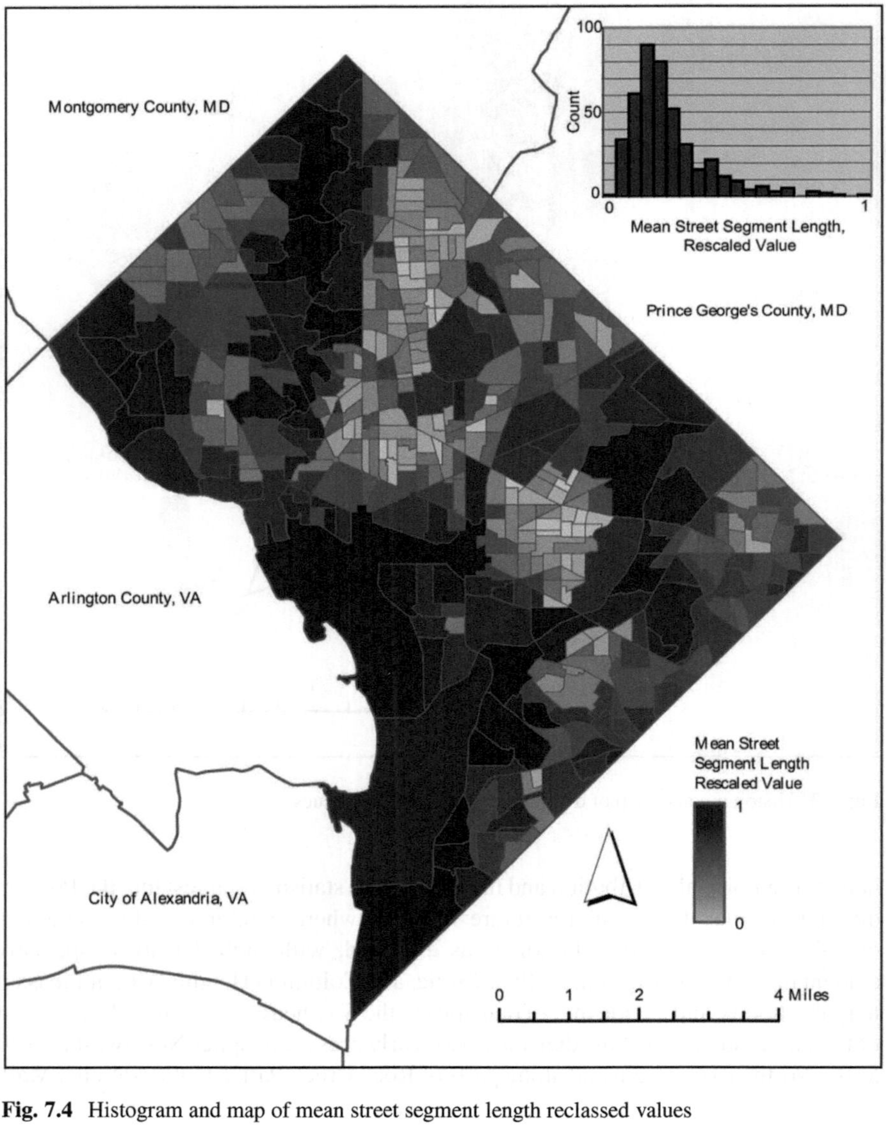

Fig. 7.4 Histogram and map of mean street segment length reclassed values

evenly distributed throughout the city. This was verified by calculating the mean distance to the nearest neighbor for the points with, and the points without, valid year information.

Points with valid year information had a mean nearest neighbor ratio (observed over expected) of 0.48, while points without valid year information had a ratio of 0.46. Using the valid points, the standard deviation for each block group ranged from 0 to 53. These standard deviations had a mean of 21.78 and a standard deviation of 5.67. Figure 7.5 shows the distribution of the standard deviations

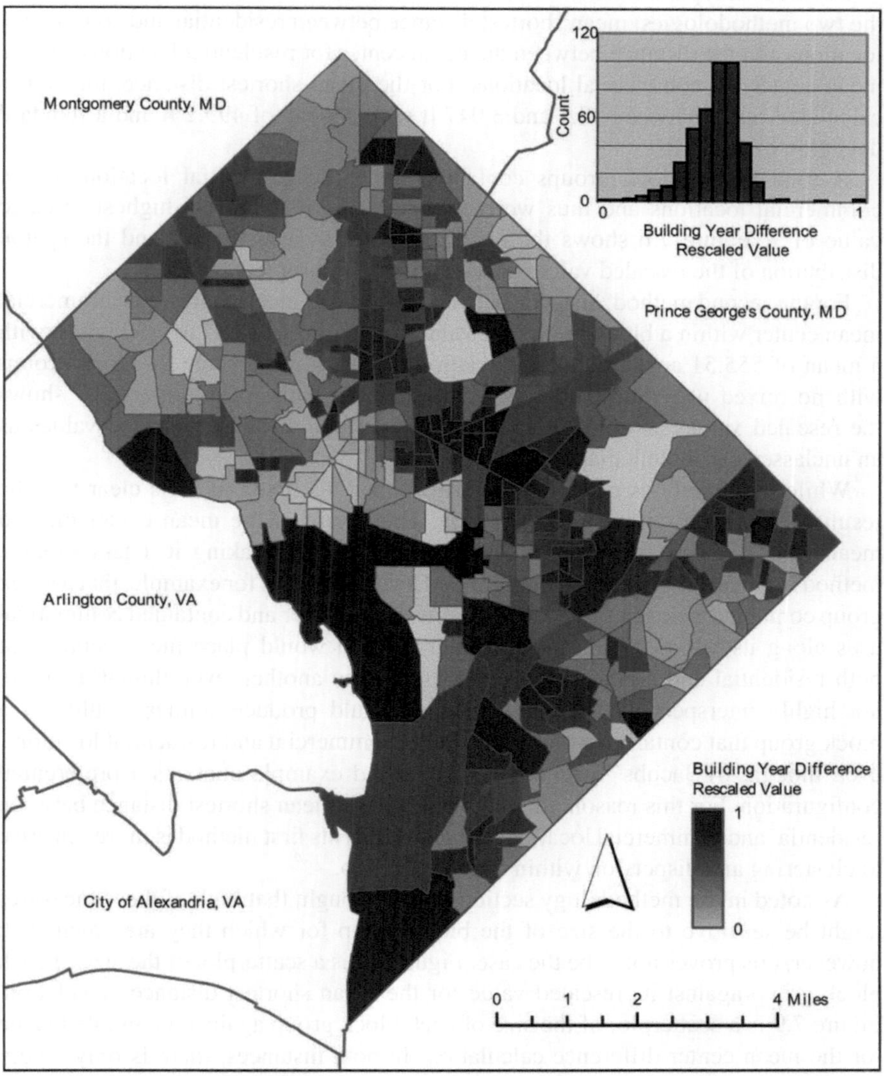

Fig. 7.5 Histogram and map of reclassed values for building year difference

as a histogram and shows the rescaled values as an unclassed choropleth map. Some of the greatest variability is located in the areas around Shaw/Mt. Vernon Square as well as Historic Anacostia and Columbia Heights. These are all areas where are currently undergoing significant new development and seeing noteworthy demographic changes.

The area of the National Mall again stands out as something of an anomaly. This is largely explained by the small number of address points within this block group, thus having even one very new structure can easily alter the standard deviation for this area.

For the fourth parameter, mixed primary uses, the results have been calculated for the two methodologies: mean shortest distance between residential and commercial locations and the distance between the mean center for residential locations and the mean center for commercial locations. For the mean shortest distance, the values calculated range between 18.1 and 3,047 ft with a mean of 499.2 ft and a standard deviation of 482.1 ft.

A total of 94 block groups contained either no residential locations or no commercial locations and thus were automatically assigned the highest rescaled value of 1. Figure 7.6 shows the rescaled values as a histogram and the spatial distribution of the rescaled values as an unclassed choropleth map.

For the second method, distance between residential mean center and commercial mean center within a block group, the values ranged from 14.37 to 3,397.49 ft with a mean of 555.51 and a standard deviation of 506.44. The same 94 block groups with no mixed use were automatically assigned a value of 1. Figure 7.7 shows the rescaled values as a histogram and the distribution of the rescaled values as an unclassed choropleth map.

While both methods produce similarly shaped distributions, it is clear that the results are spatially dispersed differently. The nature of the mean center method means that it is less sensitive to localized conditions, making it a less suitable method for measuring uses at a fine-grained level. Suppose, for example, that a block group contained a cluster of residential units in the center and contained commercial uses along its boundary; the mean center method would place mean centers for both residential and commercial very close to one another, even though they are not highly interspersed. The same method would produce similar results for a block group that contained evenly distributed commercial and residential locations, even though, by Jacobs' estimation, this second example contains a preferential configuration. For this reason, the first method, the mean shortest distance between residential and commercial locations is preferred. This first method is more sensitive to clustering and dispersion within the block group.

As noted in the methodology section, it was thought that both of these measures might be sensitive to the size of the block group for which they are calculated; however, this proves not to be the case. Figure 7.8 is a scatterplot of the size of each block group against its rescaled value for the mean shortest distance calculation. Figure 7.9 is a scatterplot of the size of each block group against its rescaled value for the mean center difference calculation. In both instances, there is only a very

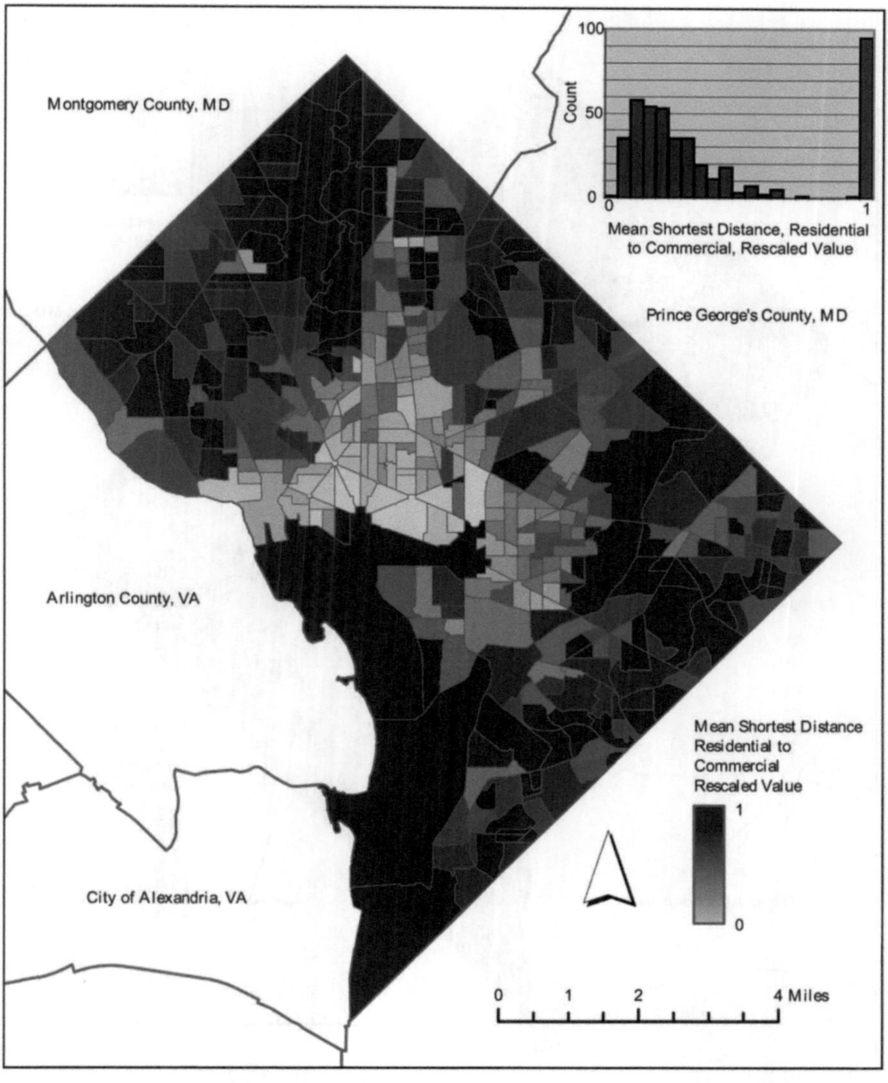

Fig. 7.6 Histogram and map for mixed-use using the mean shortest distance rescaled values

weak correlation between the two parameters, primarily that extremely large block groups have a higher likelihood of not having a sufficient mix of uses – the opposite of what was expected.

In order to examine possible spatial relationships between the two methods for calculating mixed use, a map was created that depicts the difference between the two calculations for each block group. Figure 7.10 shows the difference when the rescaled values for the mean center method are subtracted from the rescaled values

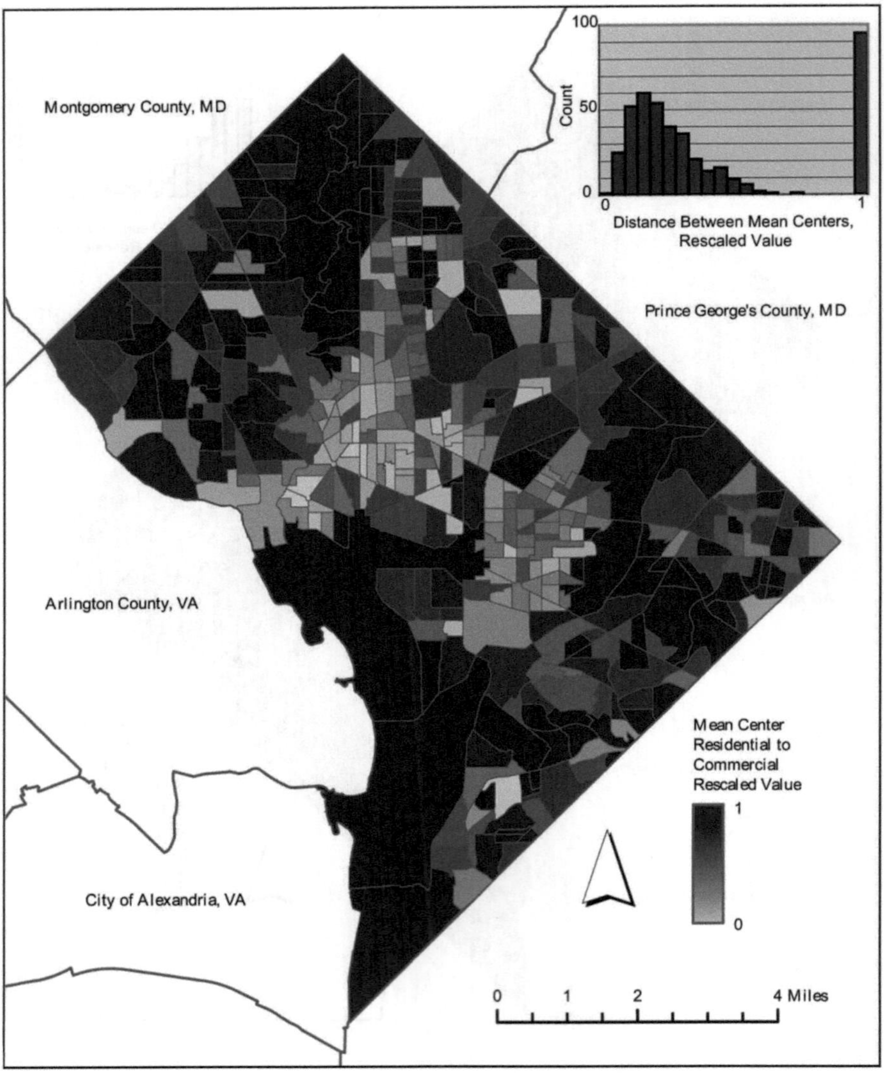

Fig. 7.7 Histogram and map of mixed-use using the mean center difference rescaled values

for the mean shortest distance calculation. Higher numbers indicate that the mean shortest distance method provided a higher value than the mean center method. A visual inspection of this map does not reveal any clear correlations relating to the location of block groups and under which method they perform better. A detailed examination of possible correlations may be warranted as an area of future research, as the possibilities are too numerous to explore within the context of this particular study.

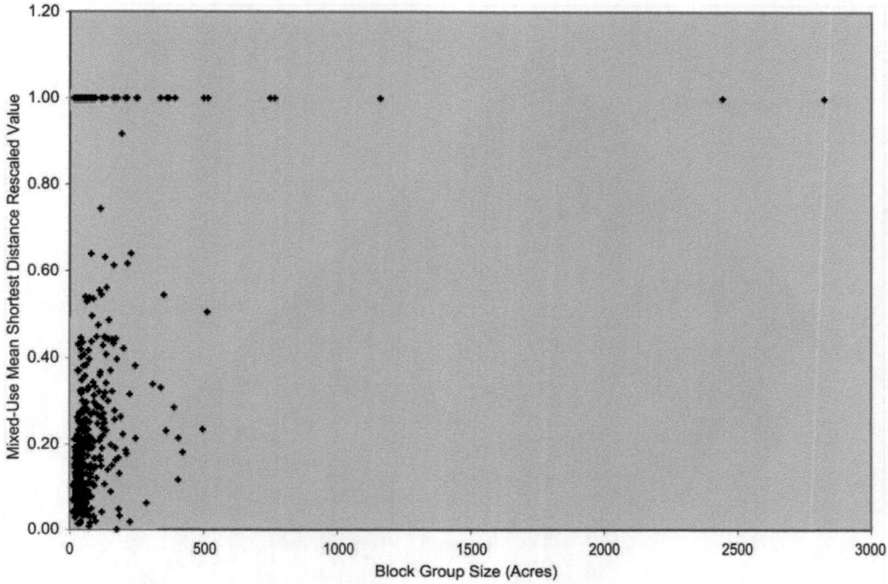

Fig. 7.8 Mixed-use mean shortest distance vs. block group size

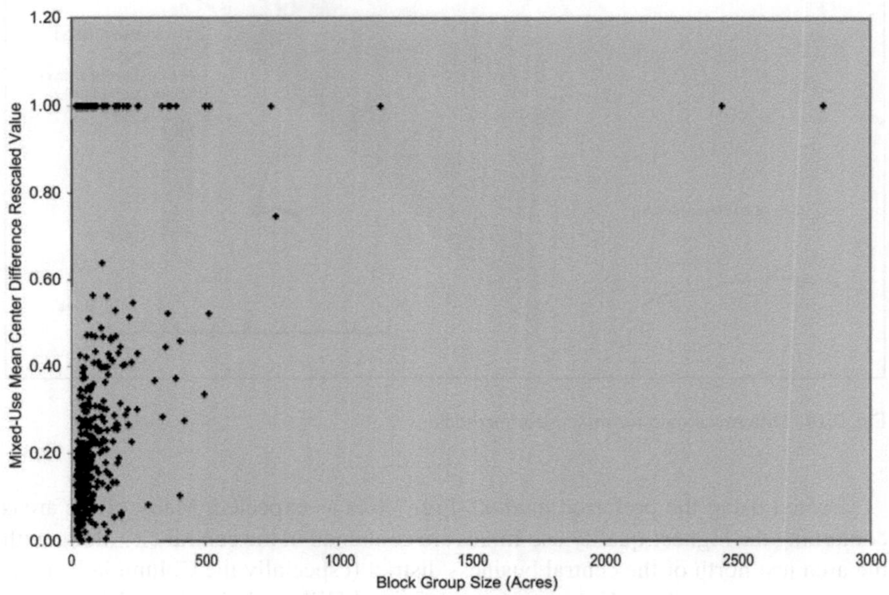

Fig. 7.9 Mixed-use mean center difference vs. block group size

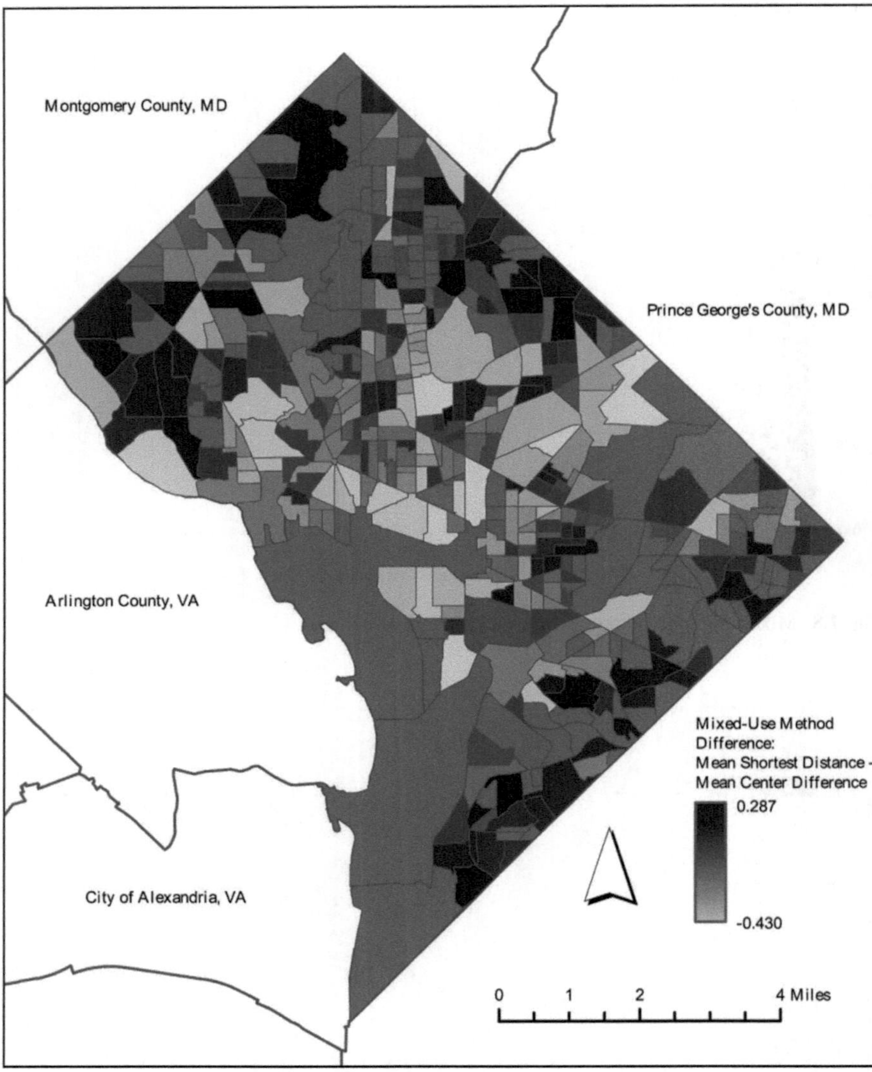

Fig. 7.10 Difference map for mixed-use methods

Created using the preferred method, Fig. 7.6 is as expected. Many of the areas containing the highest-quality use-mixes are contained in the central city along with the area just north of the central business district (especially the Columbia Heights area) as well as parts of Georgetown and Capitol Hill, including the H Street NE Corridor. There are other high-quality pockets, such as the area of Historic Anacostia and the northern section of Connecticut Ave., NW, near Tenleytown/American

University. The poorest performing areas include the National Mall as well as Upper Northwest, which consist primarily of single-family homes.

For the final, composite (*UL*) index, the more sensitive mean shortest distance mixed-use parameter was used. The composite scores range from −1.75 to 1.16, with a mean of −0.005 and a standard deviation of 0.51. Figure 7.11 shows the distribution of the rescaled composite scores as a histogram and as an unclassed choropleth map.

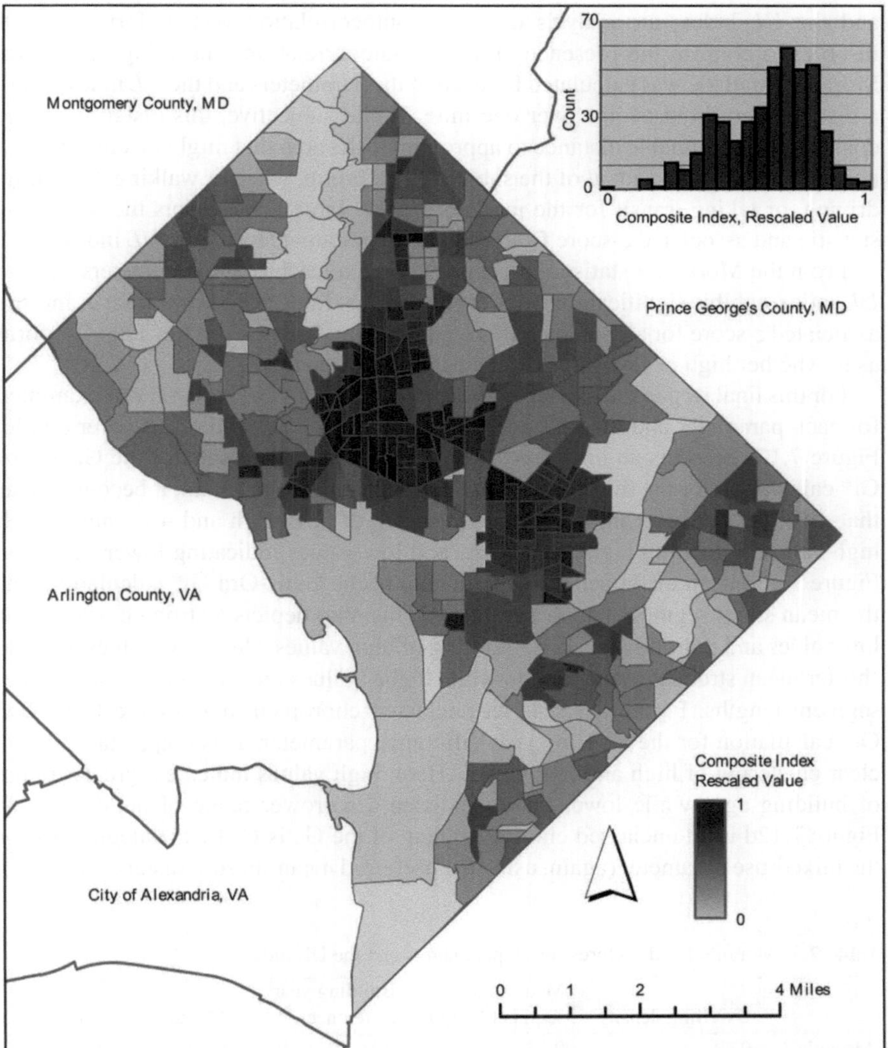

Fig. 7.11 Composite index frequency distribution and map

An examination of Fig. 7.11 shows that there are some well-defined areas of maximum livability within Washington, D.C. The area between Dupont Circle and Shaw/Mt. Vernon, extending north to Columbia Heights, as well as the general area of Capitol Hill clearly appear to be the most livable areas within the city across all four parameters. The *UL* index indicates where all four parameters are performing well. The two high-performing areas in the composite index consistently perform well across parameters. In other areas of the city, such as Georgetown and Historic Anacostia, the *UL* index shows them performing at a mixed level.

In order to better understand the locational relationships of each of the parameters and the *UL* index, an analysis of spatial autocorrelation was performed. First, in order to confirm the presence of spatial autocorrelation (clustering), a general Moran's *I* statistic was calculated for each of the parameters and the *UL* index, using a distance threshold of just over one mile. While subjective, this distance can be considered a reasonable distance to approximate the area that might be considered a neighborhood for a resident of the subject block group, whether walking, bicycling, driving, or taking transit for the mode of travel. Table 7.3 presents the Moran's *I* statistic and associated *z*-score from each of the parameters and the *UL* index.

From the Moran's *I* statistic, it is confirmed that each of the parameters and the *UL* index exhibit significant spatial clustering, as Table 7.3 shows both *I* and the associated *z*-score for all parameters are greater than 0. However, this fails to inform as to whether high or low values are clustering together.

For this final step in the cluster analysis, the Getis-Ord Gi* statistic was examined for each parameter and the *UL* index, again, using a threshold of just over a mile. Figure 7.12a presents an unclassed choropleth map of the results of the Getis-Ord Gi* calculation for the dwelling density parameter. From this map, it becomes clear that dwelling density exhibits strong clustering of both high and low values, with high values indicating higher densities, and low values indicating lower densities. Figure 7.12b is an unclassed choropleth map of the Getis-Ord Gi* calculations for the mean street segment length parameter. This map depicts a strong clustering of low values and a slightly weaker clustering of high values. Here, low values indicate shorter mean street segment length, while higher values indicate longer mean street segment lengths. Figure 7.12c is an unclassed choropleth map of the Getis-Ord Gi* calculation for the building year difference parameter. This map, again, shows clear clustering of high and low values. Here, high values indicate a greater range of building ages while lower values indicate a narrower range of building ages. Figure 7.12d is an unclassed choropleth map of the Getis-Ord Gi* calculations for the mixed-use parameter (again, using the preferred mean shortest distance method).

Table 7.3 Moran's I and z-scores for all parameters and the UL index

	Dwelling density	Mean street segment length	Building year difference	Mixed use	*UL* index
Moran's *I*	0.44	0.25	0.18	0.28	0.4
Z-score	36.55	20.45	15.38	23.2	33.18

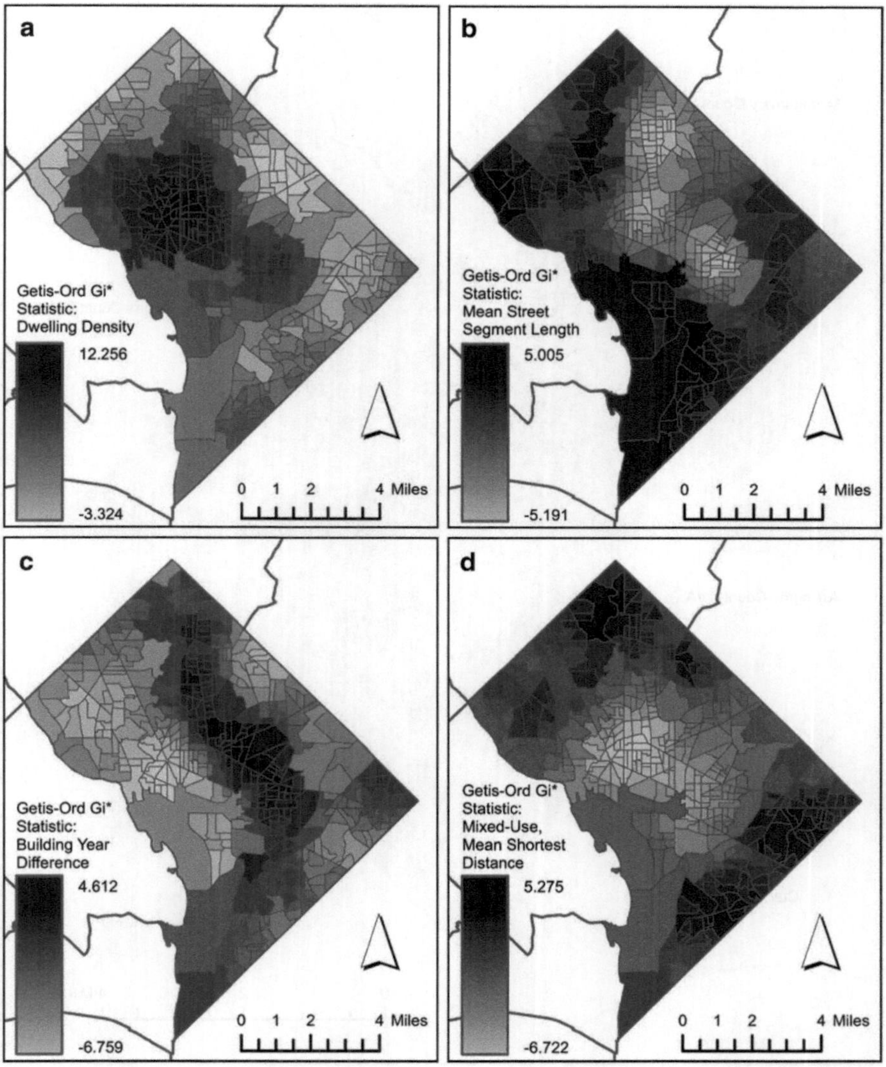

Fig. 7.12 Results of Getis-Ord Gi* statistic for local clustering of each livability measure. (**a**) Getis-Ord Gi* results for dwelling density. (**b**) Getis-Ord Gi* results for mean street segment length. (**c**) Getis-Ord Gi* results for building year difference. (**d**) Getis-Ord Gi* results for mixed-use, mean shortest distance

This map shows a very strong clustering of low values and a weak clustering of high values. Here, low values indicate a more mixed-use environment, while high values indicate a less mixed-use environment.

Finally, Fig. 7.13 shows an unclassed choropleth map of the Getis-Ord Gi* calculation for the *UL* index. This map shows a very strong clustering of high values

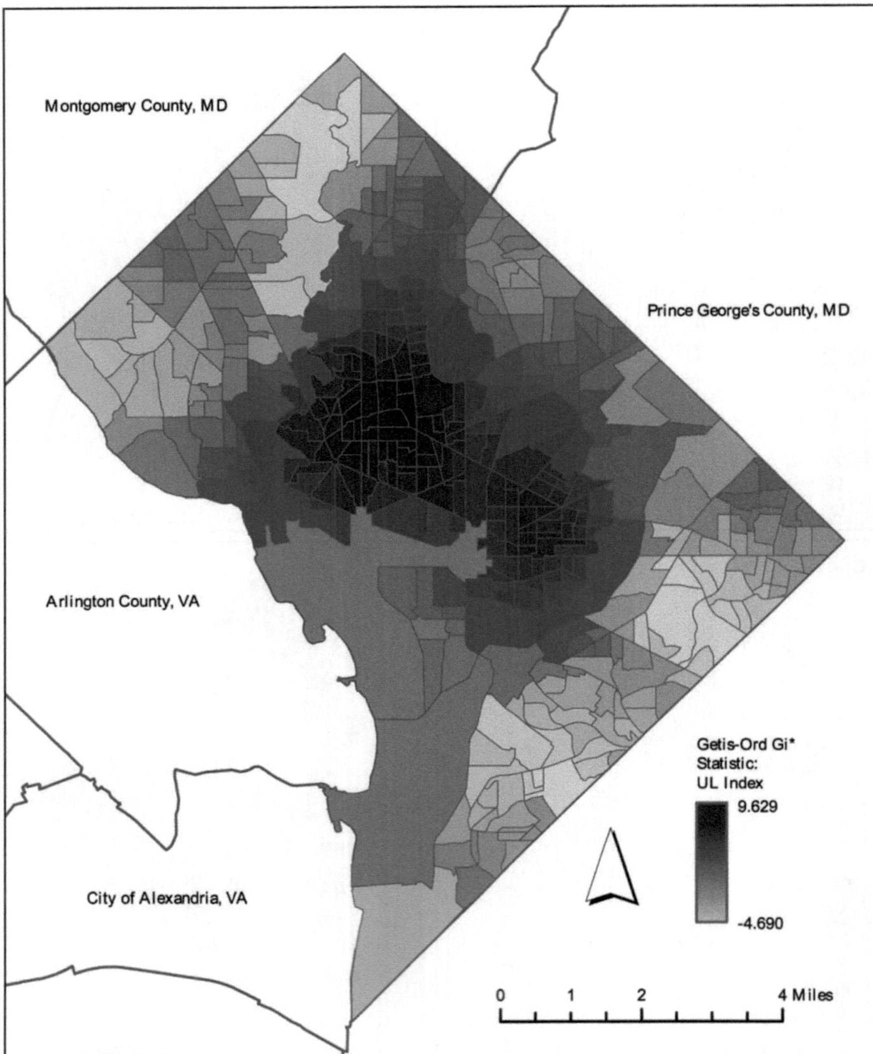

Fig. 7.13 Map of the Getis-Ord Gi* statistic for the composite UL index

as well as a strong clustering of low values. Here, high values indicate a higher composite "livability," while low values indicate a lower "livability" index.

Each of the maps presented in Figs. 7.12 and 7.13 confirm the analysis of high and low-value areas noted earlier in the results, but they also serve to highlight the degree to which the parameters do tend to cluster both at the high and low ends of the spectrum.

Table 7.4 Correlation coefficients between individual measures and the composite UL index

	Dwelling density	Mean street segment length	Building year difference	Mixed use
Correlation coefficient	0.568	−0.627	0.223	−0.863

To further examine the relationships between the *UL* index and the component parameters, correlation coefficients were calculated between the rescaled values for each of the parameters and the *UL* index, as presented in Table 7.4.

Clearly the mixed-use parameter shows the strongest correlation, followed by the mean street segment length parameter and the dwelling density parameter. The building year difference parameter shows the weakest correlation of the four. For both the mean street segment length parameter and the mixed-use parameter, lower values are preferable, so a negative correlation coefficient is to be expected. Based on these coefficients, it is expected that the mixed-use parameter, mean street segment length parameter, and the dwelling density parameter have the greatest influence over the final outcome of the *UL* index. While the building year difference parameter does display a correlation, it is weak in comparison to the other parameters. This is a parameter that has generally been overlooked in the much of the previous research, but is also a parameter that may be important from the perspective of preserving affordable housing and creating a wider socioeconomic demographic within a neighborhood, so this may warrant more detailed analysis in the future.

7.6 Conclusions, Discussion, and Future Research

The research presented here set out to place Jane Jacobs' four generators of diversity into a quantitative and repeatable methodology for use in public discourse of city planning. The methodology employed provides a fine-grained look at each of Jacobs' generators as well as constructing a new Urban Livability (*UL*) Index comprised of all four parameters in concert. Overall, the *UL* index provides a snapshot view of high and low performing areas within the city, while each of the individual parameters can be used to further investigate the level of performance for the four component metrics. These five indices, when taken together, can provide useful information to public agencies and policy-makers as they make planning decisions both at the citywide and neighborhood levels.

Looking first at the results of the dwelling density parameter, the methodology presented here does an acceptable job of presenting the current status in D.C. While there are some shortcomings, such as the possibility that a portion of a block group contains land area that cannot reasonably be expected to contain dwellings (e.g., parkland or large water features), this is more a limitation of the data than the methodology. This type of over-calculation of area can be adjusted for by an astute investigator, provided the appropriate data is available. Additionally, the dwelling location data used here may be among the most difficult to obtain for a researcher in another city. Although such cadastral data are becoming more widely available, that availability is not universal to be sure. In that case, the number of households per census block could be used as a substitute. It is clear that Jacobs was concerned with dwellings themselves, rather than households, and we know that the census can – and not infrequently does – associate more than one household with a single dwelling. The measures are therefore not identical; however, the difference this substitution would create in the results is unknown, it may be small, and it may be place dependent. A test of the effect of this substitution is an area for future research.

The next parameter, block length, is also well represented using the method applied here. It is noteworthy that this method is sensitive to the inclusion of limited-access roadways, such as interstates. This is, from Jacobs' perspective, a benefit of the methodology, as expressways and the like are viewed by Jacobs as detrimental to the city, particularly when they form a barrier that negatively influences neighboring communities (Jacobs 1992, pp. 258–259). Thus, having a methodology that is sensitive to these types of intrusions can be powerful. This is illustrated in the case study presented here, by the lower ranking that is given to the Georgetown area, due to the presence of the Whitehurst Freeway. The presence of the Southeast-Southwest Freeway in the Southern quadrants of the city also plays a significant role in the outcomes in those locations with regard to the block length parameter. It is noted that the block length processing was quite computationally intensive. In the future it may be appropriate to investigate other surrogates for block length, such as number of intersections per unit area.

The third parameter, mix of building age, is also well quantified using the methodology presented here. Perhaps the most challenging aspect of this parameter is that, given few enough structures, it can become excessively sensitive to the range of building ages. Thus, in the case study, the area of the National Mall performed very well, due to the low number of structures within this area. However, the casual observer may disagree with this assessment, particularly given the large geographic area that this particular block group includes. Thus, in terms of the spatial mixing of building ages, the National Mall performs poorly, since buildings of differing ages are not necessarily in close proximity to one another. This is perhaps the parameter of Jacobs that is most open to interpretation, since Jacobs provides little guidance for what range of ages (20 years? 50? 100?) is most beneficial to the district in question. In the future, the exploration of an entropy or diversity index to measure lack of diversity among the building ages may prove fruitful, particularly if a spatial parameter can be included.

The fourth parameter, mixed-primary uses, is the most complex of the four parameters to calculate. However, the method presented here may represent an improvement over other methods that have been utilized in the past. This method provides the level of "mixedness," rather than simply presenting a binary variable, or counting the number of different uses within an area, ignoring their actual proximity to one another within that area. The results of the calculation appear to provide

Table 7.5 Correlation coefficients between all measures

	Dwelling density	Mean street seg length	Building year difference	Mixed use	*UL* index
Dwelling density	–	−0.320	−0.114	−0.388	0.568
Mean street seg length	−0.320	–	−0.139	0.285	−0.627
Building year diff	−0.114	−0.139	–	−0.009	0.223
Mixed use	−0.388	0.285	−0.009	–	−0.863
UL index	0.568	−0.627	0.223	−0.863	–

a result that is close to what the casual observer who is familiar with the city might expect. Future research should explore ways to build on this methodology by effectively incorporating a richer mix of uses.

In addition to the correlations presented in Table 7.4, it is important to examine the correlations between each of individual parameters. Table 7.5 shows the correlation coefficients between each of the individual parameters, as well as the *UL* index.

Looking at each of the coefficients, it is clear that some parameters exhibit stronger correlations than others. The mixed-use parameter and dwelling density parameters appear to exhibit the strongest correlations between each of the other parameters, while the building year difference parameter displays the weakest correlations. Although this analysis is not sufficient to draw strong conclusions, when paired with the strong correlations to the *UL* index for these two parameters, it suggests that mixed-use and dwelling density are perhaps the most important of the four individual parameters in terms of their contribution to the overall livability of the built environment. Jacobs, as noted earlier, would likely dispute the finding that any of the four parameters is more influential than the others.

In order to explore this further, some initial within-city scenario testing was conducted. From the perspective of a city official, it is important to know that if a targeted investment can be made in only one parameter, which parameter would provide the greatest improvement for that investment. The scenario tested is one where the 43 (10 % of the total) bottom performing block groups in a particular parameter are targeted such that their raw parameter input is increased (or decreased) to match the raw value of the best performing block group for that parameter. The subindex is then scaled with these new values and the change in the raw *UL* is observed. This was repeated independently for each parameter. For all block groups outside of the bottom 43 in each parameter, a reduction in the *UL* value is observed. This is due to the fact that this test is in effect shifting the distribution such that the bottom 10 % becomes top performers, making all other values (aside from the one previous top performing block group) lower, because the values are scaled relative to one another. Thus, only the effect of this change on the 43 block groups tested is noted below. This testing does provide some challenges for the mixed-use parameter. This is due to the 94 block groups that contained either no residential or no commercial. Originally, these block groups were automatically

Table 7.6 Results of scenario test for bottom 10 % of block groups

	Dwelling density	Street length	Building year	Mixed use (exclusive)
Mean change in raw *UL*	+0.983	+0.580	+0.795	+0.507
Median change in raw *UL*	+0.980	+0.539	+0.777	+0.447
Minimum change in raw *UL*	+0.969	+0.408	+0.714	+0.382

given the worst value; thus, the bottom 10 % fall into this category. When these block groups are given the best value, they all show a change of plus 1 in the *UL* index. Given the uniform nature of this response for the mixed-use parameter, these 94 block groups were not examined, and the bottom 10 % was chosen selected from the block groups for which a valid distance had been calculated (i.e., they contained both residential and commercial points). Table 7.6 presents the results of this scenario testing.

For all parameters, when the bottom 10 % of block groups are targeted in such a manner that they become equal to the highest performing block group, an investment in dwelling density produces the greatest effect on the *UL* index. This is followed by the building year parameter, the street length parameter, and, finally, the mixed-use parameter. However, as noted earlier, given the high number of block groups that contained no mix of uses, this parameter would hold the greatest influence if those block groups were targeted. This suggests that while adding a mix of uses to areas that currently have none can have a dramatic effect (an increase of 1 in the *UL* index), adding a greater mix to areas that already contain some mixing does not result in a similar improvement. Given the results of this testing, it suggests that while all of the parameters have a significant influence on the *UL*, it may make sense for city planners to target density and (only for areas where no mixing is present) the mixed-use parameters. To a lesser extent, a focus on maintaining a diverse mix of building ages may also be beneficial. This parameter can be difficult to influence since buildings cannot be artificially aged, leaving the options of new construction and preservation of older structures. It is important to note that these conclusions may only be applied to the case study city, Washington, D.C., and not necessarily to other cities. A wider study that includes the same methodology applied to other cities would be needed to draw wider conclusions. This testing highlights the importance of utilizing the *UL* index in conjunction with the four subindex calculations. A skilled planner may locate areas in need of improvement using the *UL* index and then consult the sub-indices to evaluate which parameter will have the greatest impact on that location.

Ideally, city planners (perhaps to Jacobs' chagrin) and policy-makers can utilize the individual parameters, along with the *UL* index to focus revitalization efforts on specific locations within a city. For example, if a particular section of a city is expected to undergo massive redevelopment, and it has also been identified as having excessively long street lengths, the city may choose to make the inclusion of shorter street lengths a prerequisite for any redevelopment or rezoning efforts.

An effort along these exact lines is currently being undertaken in the Crystal City section of Arlington County, just outside of Washington, D.C. This area consists of numerous superblocks and the county has included the addition of new cross-streets to break up these blocks as a part of its transportation plans for this highly urbanized area as it undergoes redevelopment. In a similar vein, using a combination of the mixed-use parameter and *UL* index, city planners may identify large areas with a poor mix of uses. This may lead to a reevaluation of zoning policies within the city, in order to encourage a healthier mix of uses at a fine-grained level. However, as noted by Hirt (2007), even the current efforts to modify US zoning laws fall short of what European communities have achieved with their zoning laws. The use of the methods presented here may open up new ideas regarding zoning laws that encourage a more effective mix of uses under all conditions.

Based on the cluster analysis presented in the results section, it is clear that all of the parameters as well as the *UL* index exhibit clustering at both ends of the spectrum. While cause and effect is difficult to determine, it is possible that this clustering is an effect of spillover from successful neighborhoods. As a particular location becomes more desirable, those individuals and families who wish to live there, but may not have the resources, may be attracted to the edges of that neighborhood and over time begin to emulate the successes of the desired neighborhood within their own. If this is true, it may be possible for city planners to "seed" a neighborhood by focusing on creating a highly livable location that may then influence the surrounding areas over time. Obviously, this is a process that can take decades, so it is difficult to gage the effectiveness of this strategy in the short term. It is also important to note, however, that Jacobs' ideas about livability do not necessarily mirror that of the entire population. Thus, it becomes important to retain areas that are well suited to all segments of a city's population. There are some individuals that are willing to sacrifice the convenience of nearby commerce and entertainment for the urban retreat offered by a single-family home.

While the methods presented here are structured in a manner that allows them to be applied across a number of cities, it is important to remember the unique nature of individual cities. As these methods are applied to other cities, researchers may choose to strengthen the methodology by taking into consideration the unique nature of the subject city and the data that may be available. For example, the Washington, D.C., case study could be strengthened by dealing more effectively with the unique presence of the Federal government within the city. The city contains an inordinately high percentage of federally owned land that is not subject to the ordinances of the local government; indeed Congress may overrule decisions made by the city council. In order to account for this, it may be more useful to exclude all federal properties and land area from the analysis or place a negative value on these particular locations. Similarly, in a city with a significant number of waterways or parkland, these areas may be excluded or treated in a different manner in order to strengthen the results of the analysis. However, it was not the purpose of this research to provide a study strictly of one city, but to provide a framework that can be extended to other cities.

Ideally, this research also lays the groundwork for future research into livability at a detailed intracity level. This includes the examination of possible correlation between the indices presented in this research and the outcomes that Jacobs' sought, such as low crime rates, socioeconomic diversity, and "24-hour" neighborhoods. It may be of particular interest to examine the correlation between these instantiations of Jacob's urban diversity and diversity along racial or ethnic lines. Also, there is more research to be conducted regarding the precise locational nature of the parameters; do some parameters, such as dwelling density and street segment length have a tendency to co-locate (as the correlation coefficients presented here suggest)? Furthermore, it would be useful to have this methodology applied to other major world cities and have corresponding subjective methods (such as collecting resident's perceptions of their neighborhoods) applied to these cities in order to begin building a broad-based assessment of Jacobs' theories. We encourage researchers to incorporate the characteristics of their own city – where the city lies in the urban life cycle, what are its economic advantages or physical characteristics, what are its citizens' beliefs regarding livability – in order to more holistically determine how these measures could contribute to policy and practice.

References

Apparicio P, Seguin A, Naud D (2008) The quality of the urban environment around public housing buildings in Montreal: An objective approach based on GIS and multivariate statistical analysis. Soc Indic Res 86(3):355–380. doi:10.1007/s11205-007-9185-4

Bramley G, Power S (2009) Urban form and social sustainability: the role of density and housing type. Environ Plan B: Plan Des 36(1):30–48. doi: http://dx.doi.org/10.1068/b33129

Cervero R (2002) Built environments and mode choice: toward a normative framework. Transp Res D 7(4):265–284. doi: http://dx.doi.org/10.1016/S1361-9209(01)00024-4

Cozens P, Hillier D (2008) The shape of things to come: new urbanism, the grid and the Cul-De-Sac. Int Plan Stud 13(1):51–73. doi: http://dx.doi.org/10.1080/13563470801969962

Day K (2003) New urbanism and the challenges of designing for diversity. J Plan Educ Res 23(1):83–95. doi: http://dx.doi.org/10.1177/0739456X03255424

Doi K, Kii M, Nakanishi H (2008) An integrated evaluation method of accessibility, quality of life, and social interaction. Environ Plann B 35(6):1098–1116. doi:10.1068/b3315t

Filion P, Hammond K (2003) Neighbourhood land use and performance: the evolution of neighbourhood morphology over the 20th century. Environ Plan B: Plan Des 30(2):271–296. doi: http://dx.doi.org/10.1068/b12844

Grant J (2002) Mixed use in theory and practice: Canadian experience with implementing a planning principle. J Am Plan Assoc 68(1):71–84

Hirt S (2007) The devil is in the definitions: contrasting American and German approaches to Zoning. J Am Plan Assoc 73(4):436–450. doi: http://dx.doi.org/10.1080/01944360708978524

Jacobs J (1992) The death and life of great American cities. Vintage Books, New York

Krizek K (2003) Operationalizing neighborhood accessibility for land use – travel behavior research and regional modeling. J Plan Educ Res 22(3):270–287. doi: http://dx.doi.org/10.1177/0739456X02250315

Laurence PL (2006) The death and life of urban design: Jane Jacobs, the Rockefeller foundation and the new research in urbanism, 1955–1965. J Urban Des 11(2):145–172. doi: http://dx.doi.org/10.1080/13574800600644001

Martinez J (2009) The use of GIS and indicators to monitor intra-urban inequalities. A case study in Rosario, Argentina. Habitat Int 33(4):387–396. doi:10.1016/j.habitatint.2008.12.003
Mellon JG (2009) Visions of the livable city: reflections on the Jacobs-Mumford debate. Ethics Place Environ 12(1):35–48. doi: http://dx.doi.org/10.1080/13668790902753047
Miles R, Song Y (2009) "Good" neighborhoods in Portland, Oregon: Focus on both social and physical environments. J Urban Aff 31(4):491–509. doi: http://dx.doi.org/10.1111/j.1467-9906.2009.00457.x
Nasar JL (2003) Does neotraditional development build community? J Plan Educ Res 23(1):58–68. doi: http://dx.doi.org/10.1177/0739456X03256224
Openshaw S (1983) The modifiable areal unit problem. Geo Books, Norwick
Pacione M (2003) Quality-of-life research in urban geography. Urban Geogr 24(4):314–339
Rodrigue J-P, Comtois C, Slack B (2009) The geography of transport systems, 2nd edn. Routledge, New York
Song Y (2005) Smart growth and urban development pattern: a comparative study. Int Reg Sci Rev 28(2):239–265. doi:http://dx.doi.org/10.1177/0160017604273854
Talen E (1999) Charter of the new urbanism, 2nd edn. McGraw Hill Education, New York
Taylor R, Koons B, Kurtz E, Greene J, Perkins D (1995) Street blocks with more non-residential land-use have more physical deterioration – evidence from Baltimore and Philadelphia. Urban Aff Rev 31(1):120–136
Tesfazghi E, Martinez J, Verplanke J (2010) Variability of Quality of Life at Small Scales: Addis Ababa, Kirkos Sub-City. Soc Indic Res 98(1):73–88. doi:10.1007/s11205-009-9518-6
Van Kamp I, Leidelmeijer K, Marsman G, de Hollander A (2003) Urban environmental quality and human well-being – Towards a conceptual framework and demarcation of concepts; a literature study. Landscape Urban Plan 65(1–2):7–20
Wansborough M, Mageean A (2000) The role of urban design in cultural regeneration. J Urban Des 5(2):181–197
Wood L, Frank L, Giles-Corti B (2010) Sense of community and its relationship with walking and neighborhood design. Soc Sci Med 70(9):1381–1390. doi:http://dx.doi.org/10.1016/j.socscimed.2010.01.021

Part III
Urban Transportation and Mobility

Chapter 8
Everyday Cycling in Urban Environments: Understanding Behaviors and Constraints in Space-Time

Godwin Yeboah, Seraphim Alvanides, and Emine Mine Thompson

Abstract Cycling in British cities is increasing but at a slow rate nationally. The ultimate realizations of cycling benefits in urban areas, such as cities in North East England, are hampered by lack of appropriate data to aid in our understanding of cycling behaviors to inform policy strategies and improve cycling uptake as well as data processing methodologies. Several efforts are being made to enhance data availability to understand cycling behaviors to inform policy strategies for which this research aims to contribute by providing evidence on the use of the area's cycling infrastructure by utility cyclists. A proposed corridor space analytical approach was used to analyze the newly collected 7-day GPS data from 79 utility cyclists to estimate the extent to which respondents used the area's cycling infrastructure. The data was used together with the area cycling infrastructure data from Newcastle City Council. Findings from the corridor space analysis suggest that 57.4 % of cyclists from sample prefer cycling on the cycle network, while 33.8 % cycle outside the cycle network with 8.8 % near the cycle network. Also, for all cycle trips, men tend to dominate in cycling *on* and *near* the cycle network. Both the males and females tend to use the cycle network more than *off* the network for utility trips. With 42.6 % of cyclists still cycling outside the designated cycle network, it is imperative that policy initiatives are aimed towards investing in cycling research and infrastructure to further deepen our understanding to encourage cycling around the study area. It was also suggested that the captured detailed actual route choice preferences could serve as input to the development of agent-based models towards understanding cycling behaviors around the study area.

Keywords Cycling behaviors • Corridor space analysis • Built environment • GPS tracking • Time geography • Spatial analysis • Sustainable transport policy

---

G. Yeboah (✉)
The Centre for Transport Research, University of Aberdeen, Aberdeen, UK
e-mail: godwin.yeboah@abdn.ac.uk

S. Alvanides • E.M. Thompson
Faculty of Engineering and Environment, Northumbria University, Newcastle upon Tyne, UK

M. Helbich et al. (eds.), *Computational Approaches for Urban Environments*, Geotechnologies and the Environment 13, DOI 10.1007/978-3-319-11469-9_8

8.1 Introduction

There is demand for sustainable ways of living due to problems such as traffic congestion, population growth, climate change, low physical activity, sedentary lifestyles, and health-related issues (e.g., obesity & noncommunicable diseases) to name a few (GAPA 2010a, b). Motorized transport contributes to greenhouse gas emissions which also impacts on climate change. Cycling as a means of transport has the potential to contribute to sustainable way of living, thereby ameliorating these problems; hence, understanding cycling as means of transport is paramount. This research is partly a response to calls from UK National Institute for Health and Clinical Excellence 2012 recommendations as well as urban transport literature for further research to incorporate the investigation and discovery of cyclists' perception and experiences (Forsyth and Krizek 2011; Skinner and Rose 2007; NICE 2012), to support urban designers as well as cycling policy interventions and transportation engineers, and thereby to increase cycling uptake to ensure sustainable means of transport with low impact on environment. The ultimate realizations of cycling benefits by cities – such as cities in North East England – are hampered by lack of appropriate data to inform policy strategies to improve cycling uptake as well as data processing methodologies. Moreover, several efforts are being made to enhance data availability to understand cycling behaviors to inform policy strategies for which this research aims to contribute by providing evidence on the use of the area's cycling infrastructure by utility cyclists. *Utility cycling* is defined as any cycling not done primarily for fitness, recreation (such as cycle touring), or sport (such as cycle racing) but as a means of transport and covers activities such as traveling to work and to shops, running errands, seeing friends and family, and going to locations of other social activities. This definition extends Skinner and Rose's (2007, p. 84) definition which suggests that utility cycling can be defined as "day-to-day cycling for mundane trips to local shops, to work or to school." In this chapter, the term *utility* is viewed as "practical, day to day" purposeful trips (LTSA 2004, p. 10) and encompasses cycling for commuting purposes.

The purpose of this research is to provide evidence on the use of the area's cycling infrastructure by experienced utility cyclists. This research has for the first time facilitated the collection and analysis of detailed adult utility bicyclists' route choice preferences in the UK, bringing substantive empirical evidence for understanding daily cycling behaviors. Sener et al. (2009, p. 513) argue that bicyclists' demographics such as gender, among other bicyclist characteristics, influence bicyclists' route choice. Dill and Voros (2007) in their literature review also point out gender as one of the main demographic factors to consider in understanding cycling behavior. Since the main intent is to demonstrate how corridor space analytical approach could be used to understand such variables, gender is conveniently selected as an explanatory variable for the practical demonstration of implementing the proposed technique.

The chapter is divided into six parts including this introductory part. The next two parts give contextualized background to the research and brief description of the study area. The third part discusses the approaches used in data collection and processing and presents descriptive statistics about the primary sample. The concept of corridor space analysis is introduced in the fourth part followed by application of the concept using real data in Part V. The last part discusses the results and concludes the chapter.

8.2 Contextual Background

Cycling in British cities is on the increase but at a slow rate (CTC 2014), partly due to the increased costs of private and public transport as a result of the recession (Allen 2012; Brignall and King 2012) and partly due to the 2012 London Olympics as the UK government – taking advantage of the success stories in the 2012 Olympics such as the *Paralympics* and the *Tour de France* – has unveiled the largest investment of about £77 million to make cycling more visible and useful to encourage everyday cycling (DfT 2013b) and political backing (Briggs 2012; Charlesworth 2012; Shankleman 2012; Hill 2012) and to a lesser extent as a result of active transport interventions (Cope et al. 2011). The vast majority of these initiatives and interventions tend to be less top-down with an urgent call for those at the top to take bold initiatives as pointed out in the recent Get Britain Cycling report (APPCG 2013; Goodwin 2013), driven by environmental issues (DfT 2007; UNECE 2009, 2011), health (NIHR 2012), and political agendas, usually followed by improvements in the current national cycling facilities for recreation (Sustrans 2012) and commuting, for example, the London Barclays Cycle Superhighways (TfL 2012), but hardly ever matched by significant investments in the infrastructure which will go a long way to densify existing cycle networks to enable "everyday practical journeys" (Jones 2012, p. 148).

The lack of significant investments in cycling infrastructure and limited consultation with commuter cyclists, employers, and bicycle user groups (BUGs) are evidenced by the mushrooming of cycling campaigners and activist groups in many British cities. Examples of these groups are Newcastle Cycling Campaign, Gateshead Cycling Forum, and London Cycling Campaign, to name a few (NCC 2012; GCF 2013; LCC 2013). These groups question whether this wave of perceived cycling uptake presents a genuine mode shift in British commuting patterns or whether it is a temporary phenomenon hyped by the media and politicians with vested interests. There remains the danger that cycling will fail to reach *critical mass*, as a result of societal attitudes towards cyclists, coupled with lack of real investment in cycling infrastructure and gaps in the current cycling facilities. This means that following the Olympic summer and current recession, cycling uptake is likely to implode once the Olympic hype is over and we see past recession returning to the old habits and reliance of private and public transport even for the most cycleable of short trips (i.e., less than 5 miles). As cycling in British cities increases, so do conflicts between cyclists and other road users, as well as debates with city planners who are trying to balance cities' transport infrastructures in the face of public spending cuts and limited investment. In Tyne and Wear

(North East England), incremental steps are taken by the local authorities to provide cycling infrastructure, albeit at a slow rate compared to the uptake of cycling in the commuting area. As a result, there appears to be increasing dissatisfaction from local campaign groups about the lack of a coherent cycling strategy, negative attitudes towards cyclists, and even the refusal of the local public transport providers to allow bicycles on the metro system (MacMichael 2012).

Although cycling is increasing in Britain, it still remains relatively low, about 2 % of trips shorter than 2.5 km, compared to other European countries like 37 % for the Netherlands and 27 % for Denmark (Pucher and Buehler 2008, p. 498). Even in cities like Copenhagen with high cycling uptake, further research is being undertaken to deepen understanding of what constrains or enables cycling (Meyer 2011; Snizek et al. 2013). The last decade has seen annual increase of pedal cycle traffic in Britain from 4.4 billion kilometers in 2002 to 5 billion kilometers in 2012 with some fluctuations in between (DfT 2013a). The consensus on the benefits of cycling is growing among researchers, and the means – *the how* – by which cycling uptake could be realized is the confronting issue (Pucher et al. 2010, p. S107). Given an emphasis on *how*, the importance of infrastructure (i.e., the built environment) and planning for cycling together with cyclists' movement behavior is inescapable. Docherty and Shaw (2008, p. 125) argue that the major factors constraining everyday cycling have a primary relationship with the environment in which it takes place. Cycling is a safe, convenient, practical, and healthy means of transport (Pucher and Buehler 2008) and relatively cheaper than other modes. British cities such as Newcastle upon Tyne lack appropriate spatial data along with associated processing methods to support local transport planning. An example is the decision by UK Department for Transport (DfT) not to use GPS to collect travel data in the national travel survey 2013 onwards partly as a result of lack of GPS data processing methodologies (Guell et al. 2012, p. 4). This work is, in part, a response to existing calls, for further research to incorporate the investigation and discovery of cyclists' perceptions and actual experiences (NICE 2012; Forsyth and Krizek 2011; Skinner and Rose 2007), to support urban designers and transportation engineers, and for cycling policy interventions.

One of the major goals for transportation and spatial planning is the provision of accessibility (Dijst and Vidakovic 2000). The freedom of accessibility is rooted in the nature and architecture of the built environment. This freedom, one way or the other, has been compromised and historically been offered to motorized road transportation. The dominance of motorized road transportation policies has rendered and given little or no room for cycling transportation policies in the UK. As Pucher and Buehler (2008) note, "... Government policies are at least as important: transport policies, land-use policies, urban development policies, housing policies, environmental policies, taxation policies and parking policies. In many respects, the UK and the USA have given the green light to the private car, almost regardless of its economic, social and environmental costs. In sharp contrast, cycling has prospered in the Netherlands, Germany and Denmark over the past three decades precisely because these countries have given the red light, or at least the yellow warning light, to private cars."

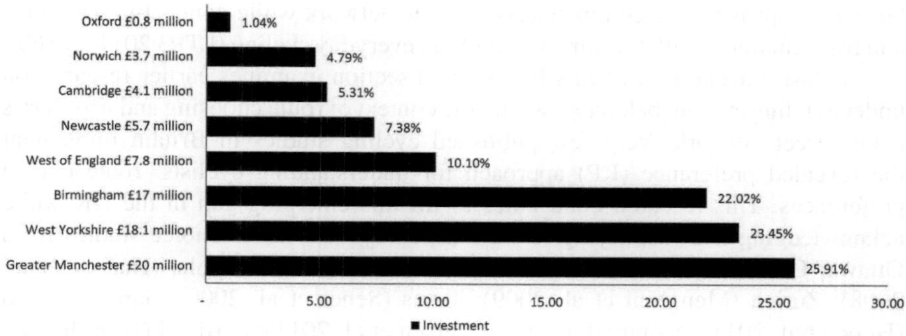

Fig. 8.1 Share of investment in cycle transport infrastructure upgrade for 8 cities in England (*Original data source: Press release – Government shifts cycling up a gear* (DfT 2013b))

It is only recently that major investment in cycling infrastructure is supported by politicians through an investment of about £94 million of public money into cycling in England with the aim of undertaken transport network upgrades to help cyclists at 14 locations on trunk road networks along with identified national parks (Walker 2013a; Siddique 2013). Although the funds have been announced, knowing where to invest in a practical sense is key to getting better value for the money. This became necessary because it was found out that major roads pose as obstacles to journeys by bike (Siddique 2013). The eight cities mentioned to benefit from the huge investment were Newcastle, Leeds, Manchester, Oxford, Norwich, Birmingham, Bristol, and Cambridge. The shares of investment of the public funds for the cities are shown in Fig. 8.1. Additionally, £17 million investment funds are to cover national parks: South Downs (£3.8), New Forest (£3.6 m), Peak District (£5 m), and Dartmoor (£4.4) (Westcott 2013). Although policy shift to investment is positive, more is needed as the declared investment still does not match other EU countries such as the Netherlands where they fund about £24 per person per year (Peck 2013). With the entire funding for investment in cycling combined in the UK, the budget per person per year is estimated to be about £18 (DfT 2013c), a shortfall of £6 compared to Netherlands. Despite the effort and commitment by the UK government in improving cycling, not everyone agrees that the government is doing enough. Walker (2013b) argues that the government is not doing enough for the realization of UK becoming one of the *cycling nations* in Europe. The basis of argument is that the long awaited DfT response to the *Get Britain Cycling inquiry* recommendations was far below expectation and seems not to offer any hope by linking responses to already existing nationwide projects (Walker 2013b; DfT 2013c; Goodwin 2013). Moreover, the MP for Newcastle Central, Chi Onwurah, has also argued for better national leadership and backing for cyclists around Newcastle upon Tyne (the study area) during a recently held (Get Britain Cycling) debate in parliament (Pearson 2013). This political campaign suggests that there is more to be done than just a well-formulated local plan as in the Tyne and Wear region, where strategic policies are in place to address the improvement of on- and off-road cycle lanes by improving the completeness of cycle network while connecting key hubs and trip generators with the aim of improving everyday cycling (LTP3 2011, p. 160).

The last but one part of this background section examines earlier research on understanding cycling behaviors within the context of route choosing and trip shares using street network. Very few published cycling studies in Britain implement the revealed preference (RP) approach for understanding cyclists' route choice preferences. This research contributes to fill this emerging gap in the UK while acknowledging and identifying few published cyclists' route choice studies from Ottawa, Guelph, and Toronto (Aultman-Hall 1996); Minneapolis (Harvey et al. 2008); Zurich (Menghini et al. 2009); Texas (Sener et al. 2009); San Francisco (Hood et al. 2011); Montreal, Quebec (Larsen et al. 2011); Portland (Broach et al. 2012); Auckland (Ehrgott et al. 2012); and ongoing work in Denmark (www.bikeability.dk). Almost all of these studies have some form of stated preference (SP) component as part of the research design, with the exception of Zurich where only GPS secondary data without additional stated preferences of the sample was used for the research. Moreover, Duncan and Mummery (2007) in comparing GIS measures with data from GPS conclude that the use of GPS in active transport research is encouraged, enabling further work to be undertaken especially in cycling. These aforementioned studies have used a variety of techniques in the quest of understanding cycling behaviors which some are not suitable for analyzing detailed quantitative cycling data. For example, Larsen et al. (2011) study concludes that the grid-cell method is not appropriate for detailed analysis of cyclists' actual route choice preferences. Additionally, they emphasized the importance of cycling infrastructure and the fact that methods assisting objective revelation of priority areas are essential to provide the evidence needed as input to effective use of finite resources allocated to the building and improvement of cycling infrastructure. The proposed technique in this study is therefore to be considered as an addition to existing basket of techniques for understanding route choice in cycling research. The reader may refer to study by Prato (2009) reviewing alternative solutions in determining preferences of various travelers with the aim of increasing route heterogeneity but in the context of general route choice modeling.

This study also has potential to provide some empirical bases for modelers in cycling research to reasonably validate models based on understanding cycling behaviors. The ability of ABM to capture emergent phenomena providing a natural description of a system and its flexibility are some of its strengths (Bonabeau 2002), which could be of benefit in using it to understand cycling behavior either at the city or route levels. The uses of such simulations are numerous: (1) to inquire better understanding of some aspect of the real social world (Axelrod 1997a, b), (2) to enable prediction or forecasting (Gilbert and Troitzsch 1999, p. 4–5), (3) to fabricate new tools that substitute weakness in human capabilities (Gilbert and Troitzsch 1999, p. 5), (4) for training, (5) for entertainment, and (6) for potential in facilitating discovery and formalization. Due to some of its numerous uses, suggestive call has been placed for building of community of social scientists who promote simulation as a field (Axelrod 2003). Examples of ABM applicable areas have been categorized but in a business context, namely, (1) *flows* which comprise crowd evacuation, traffic flow management; (2) *markets*; (3) *organization*; and (4) *diffusion* (Bonabeau 2002, p. 7281). Despite the great potential for the use of ABM in understanding various societal problems, very little is known about how ABM can be used for understanding cycling behavior. The only unpublished or ongoing works identified which include the use of ABM in cycling-related research are the Danish Bikeability project (Bikeability.dk); a personal short-term scientific mission to explore how an ABM toolkit called NetLogo can be used to model cycle tracks (Yeboah 2012); and, to some extent, the recent ESRC seminar series *Modelling on the Move* (modellingonthemove.org) focusing on transport modeling. Although this is the case, more empirical evidence on cyclists' route choice behavior as well as step-by-step approach together with concepts is needed to enable full implementation of agent-based models, especially in relation to using actual route choice parameters and cycle networks and facilities as inputs. The output of this study will provide some gender-based evidence on cyclists' route choosing around the study area for further analysis in subsequent studies.

8.3 Study Area

The selection of the study area was based on assumptions focusing on practicality, convenience, and the fact that the central part of Tyneside conurbation had the potential of registering traces of cyclists for the study. The latter will be shown in the next section. The Tyneside conurbation comprises the four main local authorities that make up the conurbation: Newcastle upon Tyne, Gateshead, North Tyneside, and North Tyneside. This central part of Tyneside conurbation which is around the southeastern path of Newcastle upon Tyne also has a high potential of traffic congestion (LTP3 2011). Traffic congestion tends to increase the operation costs of employers in situations where travel demand increases (Aditjandra et al. 2013, p. 55). According to the next decade and third strategy for the local transport plan (LTP3) for the area, the promotion of sustainable and safe communities has been considered as one of the challenges to address. Given that the area is reported to be the most deprived in England, there are motivations to improve active travel modes such as cycling while promoting healthy living lifestyles (LTP3 2011, p. ii; Rogers 2011). The deprivation ranking is based on the 2010 Index of Multiple Deprivation (IMD) data and summary measures constructed by Social Disadvantage Research Centre at Oxford University, which ranks districts in Tyne and Wear as Newcastle upon Tyne (40), Gateshead (43), Sunderland (44), South Tyneside (52), and North Tyneside (113) in increasing order, respectively. The IMD consists of indicators such as income, employment, health deprivation and disability, education skills and training, barriers to housing and services, crime, and living environment.

Moreover, understanding cycling behaviors can help transportation engineers to devise strategies to control traffic flow in and around the area. Given the design of the research, the criteria for the sample for data collection were around the central part of the Tyneside conurbation, whereas the observed route choices of cyclists

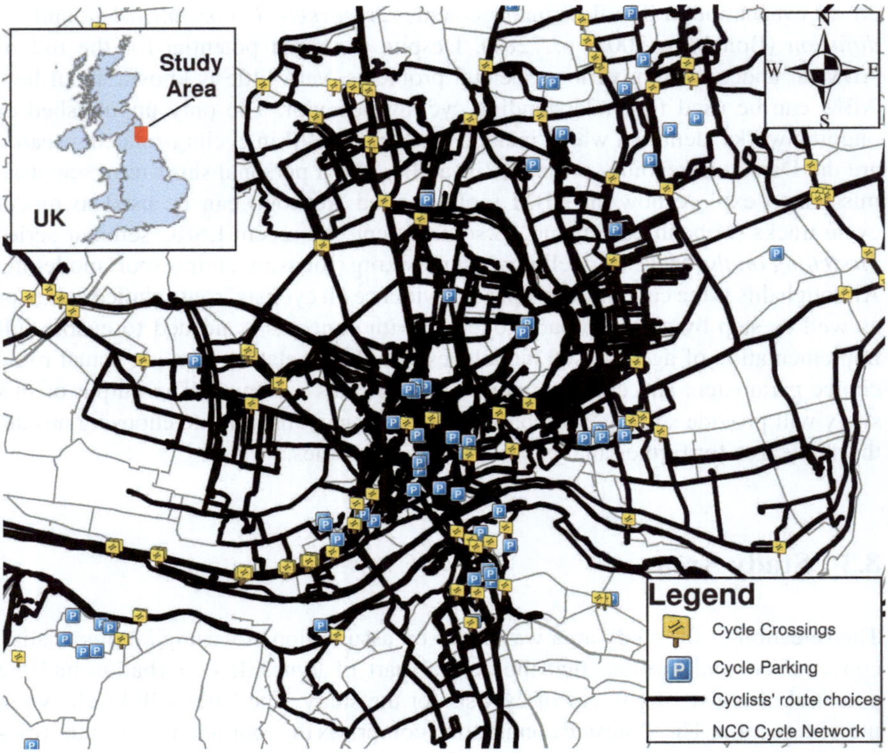

Fig. 8.2 Map of study area with cyclists' route choices and cycle infrastructure data

were more flexible but constrained to the northeast region of England. Taking this approach allowed the spatial characteristics of the primary sample to reflect the spatial dynamics identified prior to the primary data collection. Figure 8.2 shows the study area with cyclists' route choices and cycle infrastructure data. The left side shows the major transport networks which appear to have its major hub around Newcastle Central Rail Station.

8.4 Methods Used in This Study

8.4.1 Data Collection, Processing, and Sample Characteristics

8.4.1.1 Survey Instruments

Three survey instruments were developed, in addition to the use of GPS devices and materials for the field campaign. Four GPS devices were evaluated and QStarz

BT-Q1000XT selected for the data collection (Yeboah et al. 2012). First, a self-reported travel diary was designed which is adopted from questionnaire forms from UK Department for Transport (DfT 2011) and used to collect detailed information on the mode and duration of each daily trip by participants. Second is a self-reported form, named Form A, for the collection of further information of participants such as demographic and socioeconomic characteristics, experience with use of GPS device for the data collection, and confirmation of collected data. Third is a self-reported travel behavior form, named Form B, for the collection of cognitive and attitudinal data on participants' travel environment, attitude, behavior, norm, intention, and habit. Form B is adopted from Lemieux and Godin's (2009) work, but results are not presented here. The detailed information from travel diaries was used in order to clean up the collected GPS tracks and identify the cycling tracks for further analysis using GIS methods. Additional materials, mainly for the field campaign, were prepared: flyers, brochures, and posters; web pages using SurveyMonkey (http://en.wikipedia.org/wiki/Survey_monkey); and a leaflet containing frequently asked questions as well as important issues on the use of the GPS device. The leaflet was added to the travel diaries and given to participants during the data collection phase. A consent form and research statement were also prepared and added to the instruments in accordance with Northumbria University policy on ethics (Northumbria 2010).

8.4.1.2 GPS Tracking

An online and offline campaign strategy was used to approach potential participants. The campaign period for the field data collection was in September 2011. For the one online, an email was sent to 350 email contacts among which included email lists of bicycle user groups of both Northumbria and Newcastle Universities after securing approval from the moderators. An email address was created and introduced to the potential participants in the first call message and used thereafter. The idea of using a separate email address was to allow ease of management of email responses from participants. Another innovation introduced in the campaign is the use of twitter service to solicit for participation. Other campaign techniques were the use of flyers, brochures, and A4 size posters on notice boards.

Participants carried the GPS device for 7 days while filling the forms described earlier. The data collection wave is from October to November 2011. Literature suggests some variation of duration for GPS-based data collection, but most of the studies are about 1 week (Anderson et al. 2009; Van der Spek et al. 2009). Reasons for the choice of duration as well as time or distance logging interval depend on several things, for example, memory capacity, battery life, as well as the research design and scope. The log interval used in the design of this research is 5 s.

A total of 118 responses were received from the field campaign with 111 screened as utility cyclists making the response rate about 34 %. The idea behind the screening and sample size is also to ensure that the expected wide range of activity and movement patterns (Van der Spek et al. 2009, p. 3052) are captured that will be relevant to the investigation. Sampling cyclists for a particular cycling research project is a challenging task for researchers (Krizek et al. 2009). The sampling criteria for the study are articulated as follows: any adult *utility cyclist* who is more than 19 years and willing to freely volunteer as a participant; should be a *utility cyclist* and commute by bicycle at least once a day in a week; must have home, work, or school location within Newcastle upon Tyne geographic area; and must be willing to carry a personal GPS device continuously for 1 week. A week is defined as 7 days of the week which is from Monday to Sunday.

8.4.1.3 Space-Time Cube-Based Data Processing

Space-time cube (STC)-based processing is the idea of exploiting and using the STC space construct, as originally proposed by Hägerstrand (1970) and adopted by Kapler and Wright (2004), to edit raw GPS data by mainly visual inspection with additional data (Yeboah 2014; Yeboah and Alvanides 2013; Yeboah et al. 2012). By using the STC in this way, the applicability and usability of the cube achieve a complete cycling of data cleaning, analysis, and visualization (Fig. 8.3). The raw GPS data – after some basic checks on time zone, column headers, and coordinate system using the QTravel Software which comes with the GPS device – is imported into an STC using GeoTime software (www.geotime.com) and its link into the ArcMap environment. The basic check is also to ensure that the raw GPS data is properly downloaded with correct time zone settings. In the process of export/transfer, the attributes of the data are mapped to enable GeoTime to load all attributes of the raw data to enable further selection and visual inspection in STC space in GeoTime. Additional secondary information, such as OpenStreetMap (http://wiki.openstreetmap.org) basemap, is loaded in ArcMap, and this reflects automatically in the GeoTime STC space. It is at this point that travel diaries are consulted to identify cycle trips in a particular date and time. The identified information is visually

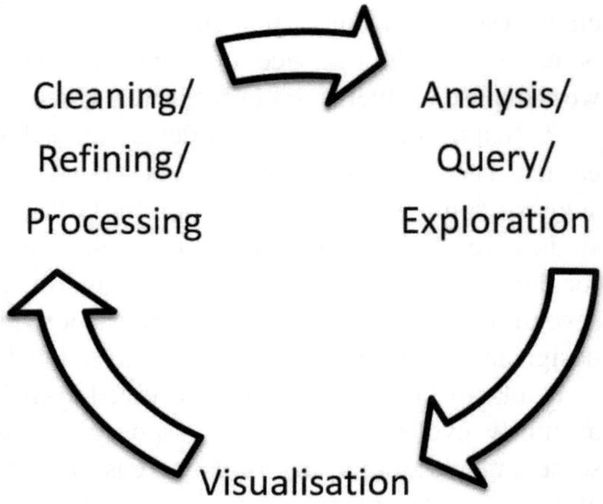

Fig. 8.3 Space-time cube usability cycle (*Source: adapted from* Yeboah (2014))

inspected in the STC by categorizing the loaded week data by date and selecting that date and subsequently navigating to find particular journey in that day as recorded in the diary. Once particular journey is identified, visual inspection continues using the loaded secondary data vis-à-vis the diary information such as start/stop location name to support decision making in visually detecting outliers and redundant points. The outcome is a refined point set (i.e., processed data) describing a particular trip. The average STC-based manual processing time recorded *per cycle trip only* was about 7 min. This duration is expected to vary based on experience of the analyst as well as the file size (e.g., week-/month-long data) per participant.

8.4.1.4 Sample Characteristics

There were a total of 941 cycle trips by the sample of 79 adult cyclists, 319 trips of these made by females and 622 made by males. Weighted average distances per trip for both female and male were 3.5 km and 5.4 km, respectively (Table 8.1). The average body mass index (BMI) for all the 79 cyclists was 23.3 kg/m^2 with a standard deviation of 3.4 kg/m^2. Table 8.1 shows average BMI for only female cyclists as 23.1 kg/m^2 and that of males as 23.5 kg/m^2. The BMI average of 23.3 kg/m^2 with standard deviation (SD) of 3.4 kg/m^2 for the sample (age range: 23–67) in this study is 1.3 points higher than Lemieux and Godin (2009) who reported BMI average of 22.0 kg/m^2 (SD = 3.4 kg/m^2) for age range from 19 to 48 years and mean age of 24.0 ± 4.9 years. The mean ages for the two samples were however different. This study had mean age of 38 years (SD = 10.3) with a minimum of 23 years and a maximum of 67. The older sample of our study explains the overall higher BMI; nevertheless, these findings suggest that cyclists tend to have an ideal weight according to the World Health Organization (WHO) weight groupings according to BMI. According to Yang and French (2013), the WHO weight groupings according to BMI values could be defined as follows: $BMI < 18.5\ kg/m^2$ for underweight, $18.5\ kg/m^2 \leq BMI < 25.0\ kg/m^2$ for ideal weight, $25\ kg/m^2 \leq BMI < 30\ kg/m^2$ for overweight, and $BMI \geq 30\ kg/m^2$ for obese. These values along with the findings clearly position cycling as a very attractive way of tackling obesity and promoting healthy lifestyles.

Table 8.1 Gender versus number of cycle trips and distance traveled

Gender (number)	Over 1-week period per person					
	Number of trips	Total[a] distance (km)	Average[a] distance (km/trip)	Average[a] distance (km/person)	Min/max trip distance (km)	Average BMI (kg/m^2)
Female (27)	319	2,137.4	3.5	41.1	0.25/13	23.1
Male (52)	622	3,373	5.4	64.9	0.12/36	23.5
79	941	5,510.4	4.5	53		

[a]Weighted by gender. Female distance value is weighted to control for gender imbalance using a factor of 52/27

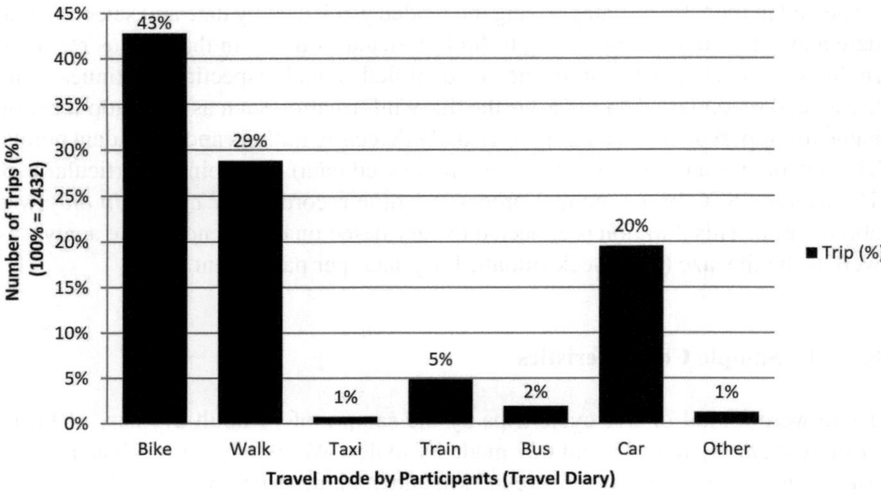

Fig. 8.4 Reported travel mode by participants from travel diary

Table 8.2 Summary of reported trips and logged GPS points

Item	Participants	All diary trips	All GPS points (Raw)	Only cycle trips (OCT)	OCT GPS points
Total	79	2,432	2,415,666	941	205,987

Out of the 2,432 reported trips, 43 % of participants' trips were actually reported to be cycle trips which are slightly less than combined trips for walking and car use (Fig. 8.4). The other trips' mode comprised mainly motor bike, scooter, and running/jogging.

The total GPS logged points for the sample for the 7-day period were almost two and half million points using a log interval of 5 s. Thus, the smaller the logging interval, the higher the likelihood of more point data which, if no careful consideration is given to data handling, can prove difficult in when it comes to data management and editing. For example, given a total 2,415,666 points in the case of this research, all merged data was impossible to be stored in MS Excel workbook given its limitation of 1,048,576 rows suggesting that without awareness, almost half of the point data will be lost in the course of editing or storage (Table 8.2).

Given the fact that 941 trips were identified from the revealed route choice preferences (i.e., the GPS-measured trips), 9.6 % of the cycle trips reported in the travel diary were not recorded by the GPS device. The trend for reasons in data loss for cycle trips follows similar observations made by Dill (2009), suggesting that there are still challenges in the use of GPS technologies in capturing movement behaviors.

8.4.2 *The Concept of Corridor Space Analysis*

Corridor-based analysis has been used in several transport-related studies: the location of high crash concentrations within the context of bridging the gap in highway safety analyses (Smith et al. 2001), prehistoric cultural activity (Hazell and Brodie 2012), as well as modeling and identification of species migration corridors (Hargrove and Westervelt 2012). This section discusses the concept of corridor space analysis (CSA) for exploring movement patterns as well as some merits and potential drawbacks of such an approach for data analysis.

8.4.2.1 Corridor Space Analysis for Exploring Movement Patterns

Data collection, alone, is not enough for analysis; matching GPS-tracked data to other spatial datasets is also necessary albeit a difficult task. It is in this difficulty that the concept of corridor space is introduced to address how spatial analysis could be done when collected data and available datasets do not fit properly due to data inaccuracy issues. The *corridor space* is defined as a buffer zone around cycle lanes/paths used for detecting cycle trips/cycle trip sections/other available cycle infrastructure. The analysis associated with the use of corridor space in determining an area of interest and further inquiry therein is what is termed the corridor space analysis here. A cycle trip is defined here as any journey by an adult cyclist bounded by origin and destination and identifiable in both a travel diary by purpose(s) and GPS data by geometry. The concept is used to distinguish cycle trips *off*, *on*, or *near* the *official* cycle network in the study area.

For a given buffer distance *a*, half of that would equal *b* as shown in Fig. 8.5. Mathematically, let us assume $B = (B_1, B_2, \ldots, B_n)$ where B_1 is a trip or trip segment within the blue region (i.e., the on region) as shown in Fig. 8.5. Similarly, $G = (G_1, G_2, \ldots, G_n)$ where G_1 is a trip or trip segment within the green region (i.e., the near region), whereas $R = (R_1, R_2, \ldots, R_n)$ where R_1 is a trip or trip segment within the off region. The total distances for each region (i.e., BT, RT, and GT) could be represented in Eqs. (8.1), (8.2), and (8.3), such that the total distance for all cycle trips should approximate $\mathrm{BT} + \mathrm{RT} + \mathrm{GT}$ and that $\mathrm{BT} \neq \mathrm{GT}$.

$$\mathrm{BT} = \sum_{k=0}^{n} B_k \tag{8.1}$$

$$\mathrm{RT} = \sum_{k=0}^{n} R_k \tag{8.2}$$

$$\mathrm{GT} = \sum_{k=0}^{n} G_k \tag{8.3}$$

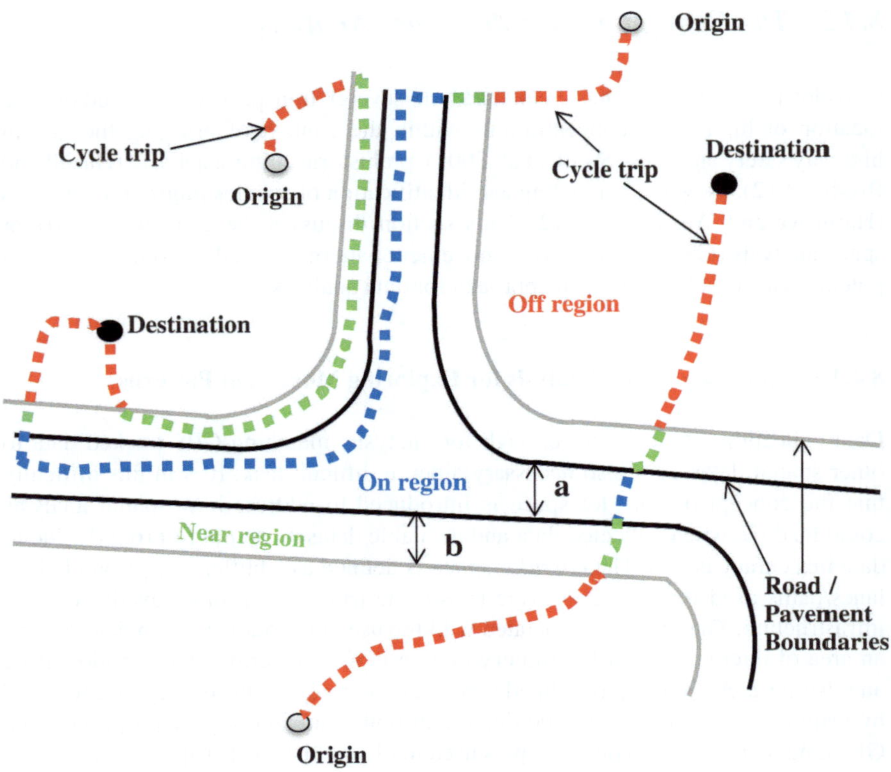

Fig. 8.5 Conceptual mapping of corridor space analytical frame

The *near* fuzzy region, the green region in Fig. 8.5, and its estimate serve as a trade-off and a measure of whether the *on* as well as the *off* estimates make comparative sense. Thus, a *near* estimate equal to or close to the *on* estimate may give clues to the reliability of the latter. The decision on best estimates with associated multiple buffers is left to the analyst but with some reasonable justification. The assumption here is that a *near* estimate equaling or less than one third of an *on* estimate is a meaningful trade-off. The notion of *trade-off* here is similar, for example, to argued case for using a calibration process as a trade-off between bias and standard error in the mechanics of geographically weighted regression (GWR) method (Fotheringham et al. 2002, p. 52). In other words, the trade-off serves as a tolerable measure for the acceptance of *on* and *off* estimates in our case.

8.4.2.2 Corridor Space Analysis: Advantages and Disadvantages

Understanding route choice preferences for people's travel behavior is important in the context of transportation research. Both the built environment in which the travel behavior occurs and the behavior itself need to be measured to allow for comparisons and further analysis. In doing so, the measured data for both realms will need to match; otherwise, comparative analysis becomes difficult. Where captured data for both realms are matched, the difficulty in analysis is reduced. On the contrary, if captured travel data does not match the available transport network data, then the introduction of new analytical approaches becomes imperative and is more useful in applications where uncertainties are difficult to eliminate, for example, estimating travel behavior in terms of trips *on*, *near*, and *off* a given transport network.

A limitation of corridor space analysis is associated with the needed accuracy of outcome and more importantly the context of usage. For example, in highly precise applications where millimeter to centimeter levels of output are needed, the need for such approaches to data analysis may not be as demanding as tolerable meter lever estimates will not be acceptable. Corridor analysis, however, is very useful in furthering understanding in data analysis and exploration.

The next best approach to route choice-related data analysis demands prior alignment (i.e., map matching) to the road network. Such an approach, however, comes with further demands: a high (topologically correct) resolution right-of-way network which has a high probability of containing most, if not all, of the potential route choices. Given the time-consuming nature of data integration, in a case where such a high-resolution network is not available or disparate or somewhat questionable in terms of well-connected end-to-end points in the network, careful consideration of the research scope and feasibility is necessary to avoid excessive delay of progress. This depends largely on time constraints and data availability; the decision is in the hands of the researcher to decide. It is in this spirit that the concept of corridor space was developed and the method used in this research.

8.5 Analysis and Results

The aim of this section is to present a comparative geographical analysis of primary tracks on everyday utility cycling, in comparison to *official* cycling network data of the study area. Rather than using 300 m grid-like corridors for analysis, as was in the case of Larsen et al. (2011) study, *corridor space analysis* already explained is used here. The analysis was intended to provide substantive empirical evidence on the use of the area's cycling infrastructure by utility cyclists, by estimating the cycle kilometers on the cycling network as a percentage of the total, for a given sample.

In adopting the concept of corridor space to compare route choice preferences of participants, two spatial analytical functions were used: buffer and overlay (*identity*) analyses. The following steps were taken for the analysis in ArcGIS. First, two sets of buffers were created around the area's cycle network, at multiples of 10 m on each side of the network, as shown in Fig. 8.5 as widths *a* and *b*. These values were used based on the outcome of the *on/near/off* estimates with the near value estimate used as a trade-off measure in the settlement of both on and off estimates. Second, the ArcGIS *identity* function was applied to overlay the buffers with the primary GPS data. Both the selected point sets forming cycle tips and the subsequently generated cycle trips in the form of polylines were used. The *identity* function computes the spatial intersection of features (points, lines, buffers) and merges their respective attributes. Finally, the ArcGIS spatial statistic functions were used to compute various distances, and database queries were used to extract trips of interest for further space-time cube categorization and visualization. What follows are the actual results from the corridor space analysis covering only home-to-work cycle trips, all cycle trips except home-to-work trips, and all cycle trips. All percentages were derived from weighted distances.

Figure 8.7 shows trip shares for the off/on/near regions for all cycle trips. It also shows the temporal dimension of all the cycle trips per the on/off/near regions. Figure 8.7 further suggests that longer trips occurred in November than in October. Despite the possibility to infer some meanings from such a static 3D visualization, practical experience with Kapler and Wright's (2004) implementation of STC visualization suggests that an interactive form of visualization is more useful. Although such an implementation of STC can be argued to partly satisfy questions raised by Kraak (2003), there is the possibility, like in our case, that such 3D visualization for cycling data can only show a brief overview of activities but may still be possible to sparkle the mind with ideas as suggested by Kraak (2003).

Findings from the corridor space analysis suggest that 57.4 % of cyclists' bike trips were found on the cycle network, while 33.8 % cycles were found outside the cycle network with 8.8 % near the cycle network (Fig. 8.6). Also, for all cycle trips, men tend to dominate in cycling *on* and *near* the cycle network. Both male and female tend to use the cycle network more than *off* the network for utility trips.

Figure 8.6 shows a more summarized form of Table 8.3 using a histogram. Table 8.3 shows the details of the weighted distances for each of corridor regions

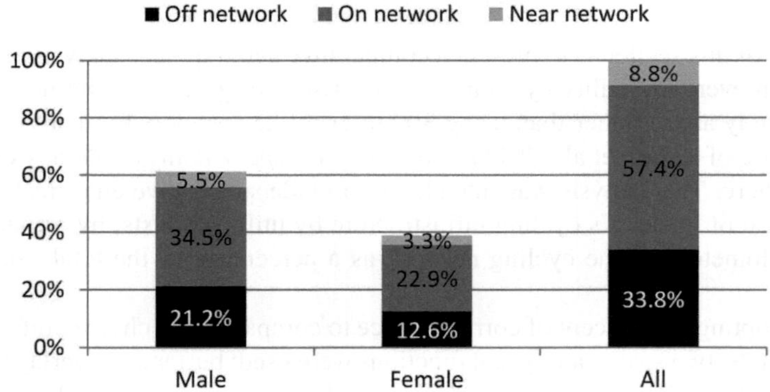

Fig. 8.6 All trips – gender and cycling with off/on/near network – corridor space characteristics

Table 8.3 All trips: gender and cycling with off/on/near network – corridor space characteristics

All trips	Off/on/near network – corridor space characteristics (weighted distance traveled in km/%)			Total (km)
	Outside buffers (off network) km	10 m buffer (on network) km	10–20 m buffer (near network) km	
Female	695 (32.5 %)	1,262 (59.1 %)	179 (3.3 %)	2,136
Male	1,167 (34.6 %)	1,900 (56.3 %)	305 (5.5 %)	3,372
Total	1,862 (33.8 %)	3,162 (57.4 %)	484 (8.8 %)	5,508

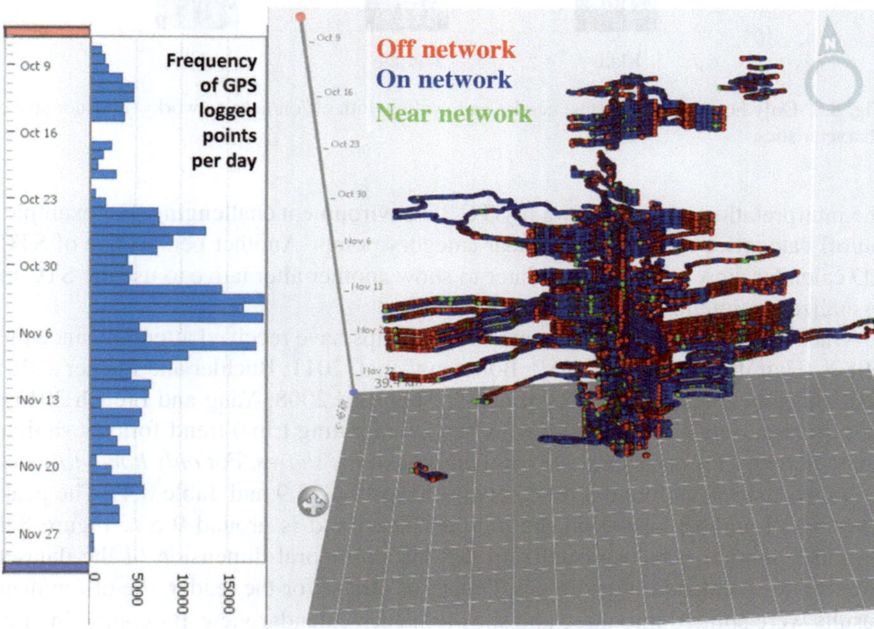

Fig. 8.7 All trips in space and time: cycling with off/on/near network – corridor space characteristics

(i.e., on/off/near regions). The percentage differences between male and female bike trips falling within the *on* regions (34.5−22.9 = 8.6%) are almost the same as the percentage of bike trips falling within the *off* region (21.2−12.6 = 11.6%) suggesting some similarities in cycling behaviors based on gender. The addition of the trade-off region, the *near* region (5.5−3.3 = 2.2)%, estimates to the *on* estimates, making it about 10.8 %, even increases the degree of similarities between *on* and *off* intra-differences. Figure 8.7 shows the temporal dimension of all the cycle trips per the on/off/near regions and suggests that longer trips occurred in November. It also shows the frequency of GPS logged points per day pointing out low cycling uptake on Sundays (i.e., dates shown on scale like October 16, 23, etc.). Depending on how the STC 3D environment is oriented, different conclusions could be drawn making

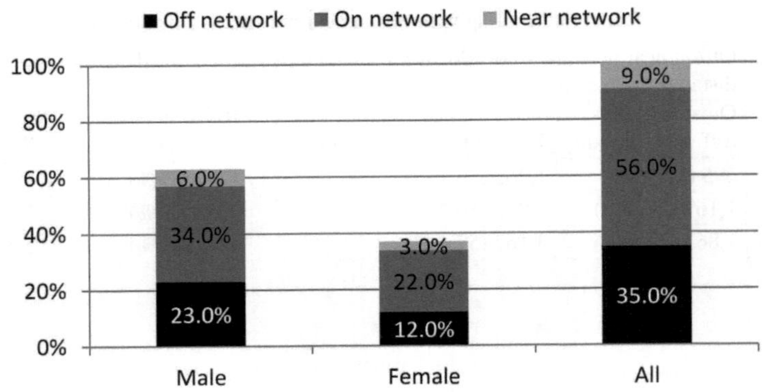

Fig. 8.8 Only home-to-work trips: gender and cycling with off/on/near network – corridor space characteristics

the interpretation of static views in STC 3D environment challenging. For example, on/off categories overlap almost near category points. Another perspective of STC 2D calendar view will be shown later to show another alternative to use the STC to visualize spatiotemporal data.

One-way commute or only home-to-work trips have received attention since the 1980s (Hamilton and Röell 1982; Boussauw et al. 2011; Buehler and Pucher 2012; Dill and Carr 2003; O'Kelly et al. 2012; Shephard 2008; Yang and French 2013). In our case, *only home-to-work trips* (i.e., commuting trips) trend follows similar characteristics for *all trips shares* and *all other trips shares*. For *only home-to-work cycle trips*, 56 % is found *on* the network (Figs. 8.8, 8.9 and Table 8.4). The peak hour was found to be within the *ante meridian* and is around 9 am. Figure 8.9 provides another way of visualizing the spatiotemporal dimension of the dataset referred to as calendar view in GeoTime. For clarity for the reader, the off/on/near results were splitted into three and shown in each calendar view. By comparing the presentation in Figs. 8.9 and 8.7, it seems clearer using 2D calendar view than the 3D view in GeoTime. The 3D view appears to be more useful when visualizing one category, for example, if all trips are considered by imagining that all the colors are one and therefore just represent all trips without any categories.

Figures 8.10 and 8.11 show the corridor space characteristics for *all other trips* except only *home-to-work* cycle trips in the form of histogram and space-time representation, respectively. Figure 8.11 shows the rush-hour time being moved to the post meridiem where peak is around 5.30 pm. The present of all three categories (i.e., off/on/near) exists, but only very general inferences can be made. Figure 8.11 shows that the 3D view can also be oriented in a way that if static view could resemble the calendar view, this will help the interpreter to examine the timings in detail. But still, visualizing more than one category may be difficult to present. Table 8.5 shows the actual computed but weighted distance values for all other trips.

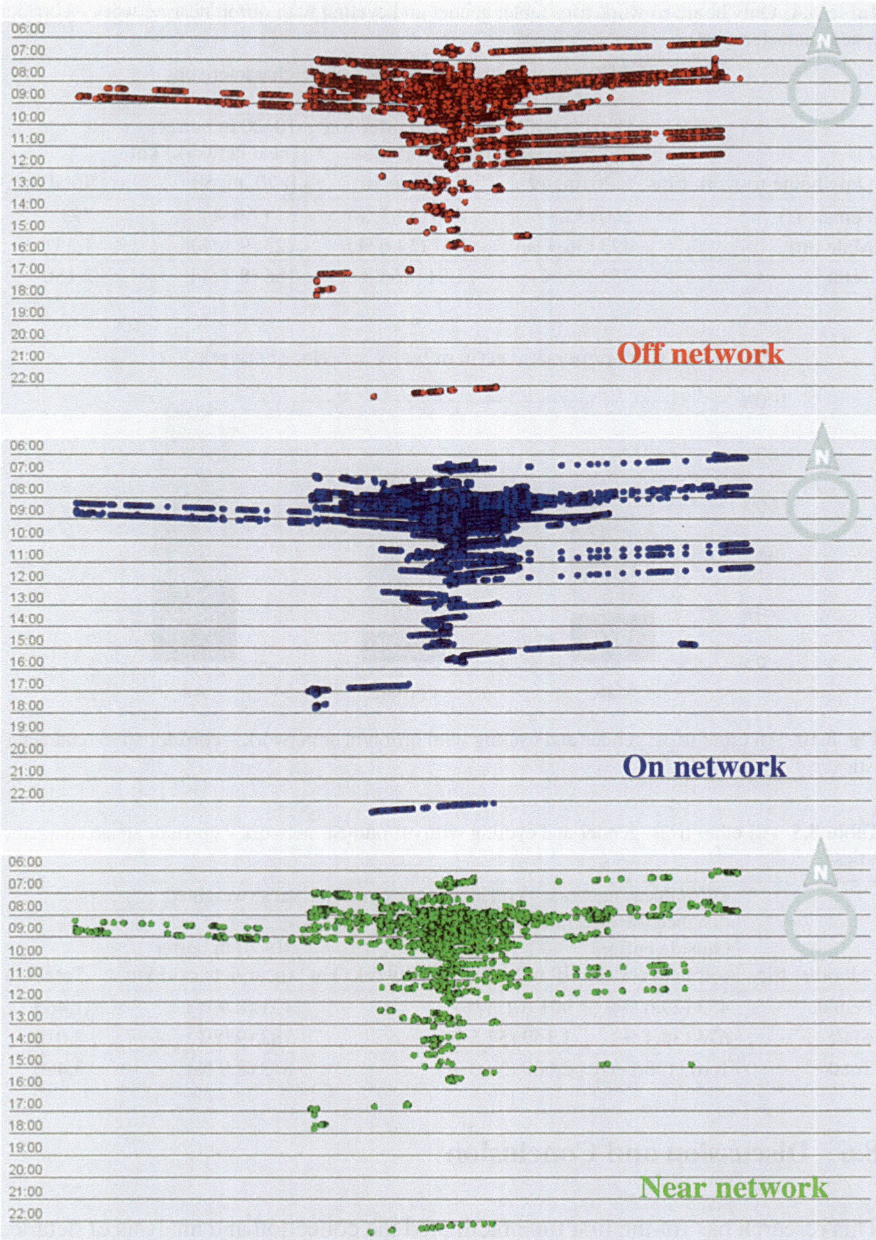

Fig. 8.9 Only home-to-work trips in space and time: cycling with off (*top panel*), on (*middle panel*), near network (*bottom panel*) – corridor space characteristics

Table 8.4 Only home-to-work trips table: gender and cycling with off/on/near network – corridor space characteristics

	Off/on/near network – corridor space characteristics (weighted distance traveled in km/%)			
	Outside buffers (off network) km	10 m buffer (on network) km	10–20 m buffer (near network) km	
Only home-to-work trips	f: 20, m: 49	f: 20, m: 50	f: 20, m: 50	Total (km)
Female (f)	260 (32.8 %)	470 (59.3 %)	63 (7.9 %)	793
Male (m)	493 (36.3 %)	741 (54.6 %)	123 (9.1 %)	1,357
Total	753 (35 %)	1,211 (56.3 %)	186 (8.7 %)	2,150

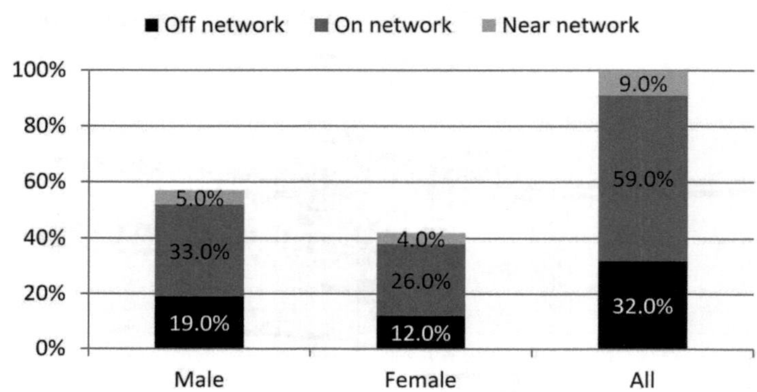

Fig. 8.10 All other trips: gender and cycling with off/on/near network – corridor space characteristics

Table 8.5 All other trips: gender and cycling with off/on/near network – corridor space characteristics

	Off/on/near network – corridor space characteristics (weighted distance traveled in km/%)			
All other trips	Outside buffers (off network)	10 m buffer (on network) km	10–20 m buffer (near network) km	Total km
Female	433 (29.6 %)	901 (61.5 %)	131 (8.9 %)	1,465
Male	674 (34.4 %)	1,159 (57.5 %)	182 (9.0 %)	2,015
Total	1,107 (33.1 %)	2,060 (58.0 %)	313 (8.9 %)	3,480

8.6 Discussion and Conclusion

This research has for the first time facilitated the collection and analysis of detailed bicyclists' route choices in the UK, bringing substantive empirical evidence for understanding daily cycling behaviors. Few published studies related to cycling have the uniqueness of the research design developed for this research study about utility cyclists which show actual route choice preferences within the study area (Menghini et al. 2010; Broach et al. 2012), but not in the UK. The triangulation of

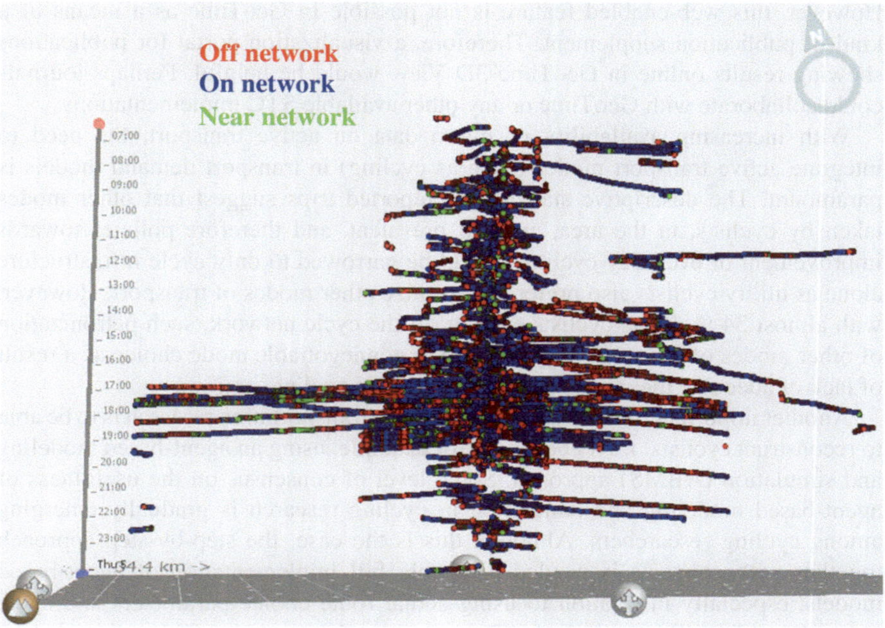

Fig. 8.11 All other trips in space and time: gender and cycling with off (*red*), on (*blue*), near network (*green*) – corridor space characteristics

evidence using available secondary data further deepens the existing understanding of cycling patterns and infrastructure within the study area. Additionally, the STC-based data processing visual technique has the potential of allowing anyone who is familiar with the neighborhood to easily clean a messy GPS dataset without any algorithmic knowledge of complex toolkits. Another novel concept introduced is the corridor space spatial analysis approach which offers unique ways and means for understanding cyclists' interactions with the built environment, which partly constitutes the cycling infrastructure.

Findings from the corridor space analysis suggest that 57 % of cyclists from sample prefer cycling on the cycle network, while 34.1 % cycles prefer outside the cycle network with 8.9 % near the cycle network. Also, for all cycle trips, it was found out that males cycle more on and off the network than females with the gender difference in *off network* trips 3 % lower than the gender difference in *on network* trips (11.6 %). For only home-to-work cycle trips, gender difference in *on network* 12 % and that of *off network* is 11 %. With 43 % of cyclists still cycling outside the designated cycle network, it is imperative that policy initiatives are aimed towards investing in cycling research and infrastructure (i.e., lanes, parking, crossings, etc.). Visualization experience using the STC suggests that for 3D STC visualization, it is more appropriate to present the result in an interactive manner in GeoTime. A general overview of the spatiotemporal dimension of cycling data can be presented using 3D STC View, but even better symbolization for better presentation is still needed.

However, this web-enabled feature is not possible in GeoTime as a means of a kind of publication supplement. Therefore, a visualization portal for publications showing results online in GeoTime 3D View would be helpful. Perhaps journals could collaborate with GeoTime or any other available STC implementations.

With increasing availability of micro data on active transport, the need to integrate active transport modes (such as cycling) in transport demand models is paramount. The descriptive statistics of reported trips suggest that other modes taken by cyclists, in the area, are still prevalent, and therefore policies towards improvement of everyday cycling should be narrowed to only cycle infrastructure alone as utility cyclists also prefer to patronize other modes of transport. However, with almost 34 % of the cyclists cycling off the cycle network, such patronization of other modes of transport might be due to nonnegotiable mode choice as a result of lack of adequate means to get to destination by cycling.

Another important future research prospect in transportation research is to be able to reconstruct cyclists' travel behavior, for example, using an agent-based modeling and simulation (ABMS) approach. Some level of consensus on the usefulness of agent-based modeling and simulation in cycling research is gradually emerging among cycling researchers. Although this is the case, the step-by-step approach together with concepts is needed to enable full implementation of agent-based models, especially in relation to using actual route choice parameters and cycle networks and facilities as inputs. The emerging importance of bicycling and its associated nonlinear travel behavior has now prompted researchers to turn to the application of agent-based modeling and simulation techniques to create platforms to further understand behavior of cyclists. The ongoing development of Copenhagen agent-based model of bicyclists' experiences (in short CopenhagenABM) as part of the Bikeability.dk project in Denmark, at the time of writing, is a typical example. Future research should use datasets in different geographical contexts and adopt some or all of the methodological approaches used in the CopenhagenABM to enable easy comparison of models, thereby allowing testing of variants of hypotheses and existing theories. It is planned that the detailed dataset collected in this research will be used as input to the development of NewcastleABM together with comparative analysis of common variables alongside the CopenhagenABM.

Acknowledgments The authors would like to thank Northumbria University at Newcastle for sponsoring this research. Also, the authors thank all anonymous participants who helped in collecting the primary data for this research. We also acknowledge the constructive comments made by two anonymous reviewers and the book editors to improve the content.

References

Aditjandra PT, Mulley C, Nelson JD (2013) The influence of neighbourhood design on travel behaviour: empirical evidence from North East England. Transp Policy 26(0):54–65. doi:http://dx.doi.org/10.1016/j.tranpol.2012.05.011

Allen E (2012) Petrol prices will rise again as motorists must accept higher taxes to help clear Britain's debt, insists Transport Minister. DailyMail

Anderson T, Abeywardana V, Wolf J, Lee M (2009) National travel survey GPS feasibility study – Final report final edn. NatCen & GeoStats for DfT, London. Accessed 2 Mar 2011

APPCG (2013) Get Britain cycling – summary and recommendations. Cyclists' Touring Club (CTC). http://www.ctc.org.uk/sites/default/files/get_britain_cycling.pdf. Accessed 24 Apr 2013

Aultman-Hall LM (1996) Commuter bicycle route choice: analysis of major determinants and safety implications. Doctor of Philosophy (PhD), McMaster University, Hamilton, Canada

Axelrod R (1997a) Advancing the art of simulation in the social sciences. In: Conte R, Hegselmann R, Terna P (eds) Simulating social phenomena. Springer, Berlin, pp 21–40

Axelrod R (1997b) Advancing the art of simulation in the social sciences. Complexity 3(2):16–22. doi:10.1002/(sici)1099-0526(199711/12)3:2{\mathsurround=\opskip$<$}16::aid-cplx4{\mathsurround=\opskip$>$}3.0.co;2-k

Axelrod R (2003) Advancing the art of simulation in the social sciences (2003 updated version). Special Issue on Agent-Based Modeling 12(3)

Bonabeau E (2002) Agent-based modeling: methods and techniques for simulating human systems. Proc Natl Acad Sci U S A 99(Suppl 3):7280–7287. doi:10.1073/pnas.082080899

Boussauw K, Neutens T, Witlox F (2011) Minimum commuting distance as a spatial characteristic in a non-monocentric urban system: the case of Flanders. Pap Reg Sci 90(1):47–65. doi:10.1111/j.1435-5957.2010.00295.x

Briggs M (2012) London mayor election: who is the greenest choice? Ecologist. http://www.theecologist.org/News/news_analysis/1337944/london_mayor_election_who_is_the_greenest_choice.html. Accessed 25 June 2012

Brignall M, King M (2012) Budget 2012: Osborne disappoints motorists over fuel price increase. Guardian. http://www.guardian.co.uk/money/2012/mar/21/budget-2012-osborne-fuel-price-increase. Accessed 25 June 2012

Broach J, Dill J, Gliebe J (2012) Where do cyclists ride? A route choice model developed with revealed preference GPS data. Transp Res A Policy Pract 46(10):1730–1740. doi:10.1016/j.tra.2012.07.005

Buehler R, Pucher J (2012) Cycling to work in 90 large American cities: new evidence on the role of bike paths and lanes. Transportation 39(2):409–432. doi:10.1007/s11116-011-9355-8

Charlesworth A (2012) Boris' bike scheme goes live tomorrow. BusinessGreen. http://www.businessgreen.com/bg/news/1805995/boris-bike-scheme-goes-live-tomorrow. Accessed 25 June 2012

Cope A, Nair S, Torrance J, Allan W, Wilson A (2011) Cycling in the city regions: delivering a step change (Final report). Sustrans. http://www.sustrans.org.uk/assets/files/rmu/110411_Cycling_in_the_city_regions_Sustrans_PTEG_report_final.pdf. Accessed 29 June 2011

CTC (2014) CTC cycling statistics. http://www.ctc.org.uk/resources/ctc-cycling-statistics. Accessed 18 Feb 2014

DfT (2007) Towards a sustainable transport system: supporting economic growth in a low carbon world. In: Presented to Parliament by the Secretary of State for Transport, by Command of Her Majesty. The Stationery Office (TSO), London. Accessed 12 Sept 2012

DfT (2011) In confidence – national travel survey (Questionnaire form). Dft with National Centre for Social Research (NatCen), London. Accessed 29 June 2011

DfT (2013a) Pedal cycle traffic (TRA04). UK Department for Transport (DfT). https://www.gov.uk/government/statistical-data-sets/tra04-pedal-cycle-traffic. Accessed 18 Feb 2014

DfT (2013b) Press release – Government shifts cycling up a gear. https://www.gov.uk/government/news/government-shifts-cycling-up-a-gear?dm_i=6EB,1RI71,3DY6U8,6AE0L,1. Accessed 12 Aug 2013

DfT (2013c) Her Majesty's Government response to the 'Get Britain cycling' report published by the All Party Parliamentary Cycling. https://www.gov.uk/government/uploads/system/uploads/attachment_data/file/232611/appcg-response.pdf. Accessed 24 Apr 2013

Dijst M, Vidakovic V (2000) Travel time ratio: the key factor of spatial reach. Transportation 27(2):179–199. doi:10.1023/a:1005293330869

Dill J (2009) Bicycling for transportation and health: the role of infrastructure. J Public Health Policy 30(S1):S95–S110

Dill J, Carr T (2003) Bicycle commuting and facilities in major U.S. cities: if you build them, commuters will use them – another look (Revised version). TRB 2003 annual meeting CD-ROM, TRB

Dill J, Voros K (2007) Factors affecting bicycling demand: initial survey findings from the Portland, Oregon, region. Transp Res Rec: J Transp Res Board 2031(−1):9–17

Docherty I, Shaw J (2008) Traffic jam: ten years of 'sustainable' transport in the UK. Policy Press, Bristol

Duncan MJ, Mummery WK (2007) GIS or GPS? A comparison of two methods for assessing route taken during active transport. Am J Prev Med 33(1):51–53

Ehrgott M, Wang JYT, Raith A, van Houtte C (2012) A bi-objective cyclist route choice model. Transp Res A Policy Pract 46(4):652–663. doi:10.1016/j.tra.2011.11.015

Forsyth A, Krizek K (2011) Urban design: is there a distinctive view from the bicycle? J Urban Des 16(4):531–549. doi:10.1080/13574809.2011.586239

Fotheringham AS, Brunsdon C, Charlton M (2002) Geographically weighted regression: the analysis of spatially varying relationships. Wiley, Hoboken

GAPA (2010a) NCD prevention: investments that work for physical activity. Global Advocacy Council for Physical Activity (GAPA), Toronto

GAPA (2010b) The Toronto charter for physical activity: a global call to action. Global Advocacy Council for Physical Activity, Toronto

GCF (2013) Gateshead cycling forum. http://www.gatesheadcycling.org.uk/.

Gilbert N, Troitzsch KG (1999) Simulation for the social scientist. Open University Press, Philadelphia

Goodwin P (2013) Get Britain cycling – full report from the inquiry. All Party Parliamentary Cycling Group (APPCG). http://allpartycycling.files.wordpress.com/2013/04/get-britain-cycling_goodwin-report.pdf. Accessed 25 April 2013

Guell C, Panter J, Jones NR, Ogilvie D (2012) Towards a differentiated understanding of active travel behaviour: using social theory to explore everyday commuting. Soc Sci Med 75(1):233–239. doi:10.1016/j.socscimed.2012.01.038

Hägerstrand T (1970) What about people in regional science? Pap Reg Sci 24(1):6–21. doi:10.1007/bf01936872

Hamilton BW, Röell A (1982) Wasteful commuting. J Polit Econ 90(5):1035–1053. doi:10.2307/1837132

Hargrove W, Westervelt J (2012) An implementation of the Pathway Analysis Through Habitat (PATH) algorithm using NetLogo. In: Westervelt JD, Cohen GL (eds) Ecologist-developed spatially-explicit dynamic landscape models, Modeling Dynamic Systems. Springer, New York, pp 211–222. doi:10.1007/978-1-4614-1257-1_12

Harvey F, Krizek KJ, Collins R (2008) Using GPS data to assess bicycle commuter route choice. Transportation Research Board, 20 p

Hazell LC, Brodie G (2012) Applying GIS tools to define prehistoric megalith transport route corridors: Olmec megalith transport routes: a case study. J Archaeol Sci 39(11):3475–3479. doi:10.1016/j.jas.2012.05.015

Hill D (2012) Ken Livingstone is streets ahead of Boris Johnson in mayoral cycling policy race. The Guardian. http://www.guardian.co.uk/politics/davehillblog/2012/apr/06/ken-livingstone-jenny-jones-boris-johnson-london-cycling-policies. Accessed 25 June 2012

Hood J, Sall E, Charlton B (2011) A GPS-based bicycle route choice model for San Francisco, California. Transport Lett: Int J Transp Res 3(1):63–75. doi:10.3328/tl.2011.03.01.63-75

Jones T (2012) Getting the British back on bicycles—the effects of urban traffic-free paths on everyday cycling. Transp Policy 20:138–149. doi:10.1016/j.tranpol.2012.01.014

Kapler T, Wright W (2004) GeoTime information visualization. In: Information visualization, 2004. INFOVIS 2004. IEEE Symposium on 10–12 Oct 2004, pp 25–32. doi:10.1109/INFVIS.2004.27

Kraak M-J (2003) The space-time cube revisited from a geovisualization perspective. In: Paper presented at the 21st International Cartographic Conference (ICC), Durban, South Africa, 10–16 August 2003

Krizek KJ, Handy SL, Forsyth A (2009) Explaining changes in walking and bicycling behavior: challenges for transportation research. Environ Plan B Plan Des:725–740. doi:citeulike-article-id:8424062

Larsen J, Patterson Z, El-Geneidy A (2011) Build it. But where? The use of geographic information systems in identifying locations for new cycling infrastructure. Int J Sustain Transp 7(4): 299–317

LCC (2013) London cycling campaign. http://lcc.org.uk. Accessed 17 July 2013

Lemieux M, Godin G (2009) How well do cognitive and environmental variables predict active commuting? Int J Behav Nutr Phys Act 6(1):12

LTP3 (2011) LTP3: the third local transport plan for Tyne and Wear trategy 2011–2021. Tyne and Wear Integrated Transport Authority, Tyne and Wear. Accessed 1 Aug 2011

LTSA (2004) Cycle network and route planning guide. Land Transport Safety Authority, New Zealand. Accessed 24 Aug 2011

MacMichael S (2012) Tyne & Wear Metro operator rejects calls to allow trial of letting bikes on trains. http://road.cc/content/news/52483-tyne-wear-metro-operator-rejects-calls-allow-trial-letting-bikes-trains. Accessed 1 Oct 2012

Menghini G, Carrasco N, Schüssler N, Axhausen K (2009) Route choice of cyclists in Zurich: GPS-based discrete choice models. Arbeitsberichte Verkehrs- und Raumplanung 544, IVT, ETH Zurich, Zurich

Menghini G, Carrasco N, Schüssler N, Axhausen KW (2010) Route choice of cyclists in Zurich. Transp Res Part A Policy Pract 44(9):754–765. doi:http://dx.doi.org/10.1016/j.tra.2010.07.008

Meyer MS (2011) Presentations from the Bikeability Seminar: December 2 2010. bikeability.dk. http://www.bikeability.dk/. Accessed 4 July 2011

NCC (2012) Newcastle cycling campaign. http://www.newcycling.org/. Accessed 17 July 2013

NICE (2012) Walking and cycling: local measures to promote walking and cycling as forms of travel or recreation. NICE public health guidance 41. Accessed 28 Nov 2012

NIHR (2012) Improving the health and wealth of the nation through research. www.nihr.ac.uk. Accessed 12 Sept 2012

Northumbria (2010) Northumbria research ethics and governance handbook – third edition 2009–2010, 3rd edn. Northumbria University at Newcastle, Newcastle upon Tyne. Accessed 7 Feb 2014

O'Kelly M, Niedzielski M, Gleeson J (2012) Spatial interaction models from Irish commuting data: variations in trip length by occupation and gender. J Geogr Syst 14(4):357–387. doi:10.1007/s10109-011-0159-3

Pearson A (2013) Newcastle MP calls for national backing for city's cyclists. Chronicle Live, 4 Sep 2013, 10:26 am. Accessed 5 Sept 2013

Peck C (2013) Get Britain cycling report recommends £10 per head, per year funding for cycling. http://www.ctc.org.uk/news/get-britain-cycling-report-recommends-%C2%A310-head-year-funding-for-cycling. Accessed 23 Apr 2013

Prato CG (2009) Route choice modeling: past, present and future research directions. J Choice Model 2(1)

Pucher J, Buehler R (2008) Making cycling Irresistible: lessons from the Netherlands, Denmark and Germany. Transp Rev 28(4):495–528

Pucher J, Dill J, Handy S (2010) Infrastructure, programs, and policies to increase bicycling: an international review. Prev Med 50(Supplement 1):S106–S125

Rogers S (2011) Indices of multiple deprivation: find the poorest places in England. The Guardian, Tuesday 29 March 2011, 10.00 BST. Accessed 1 Apr 2011

Sener IN, Eluru N, Bhat CR (2009) An analysis of bicycle route choice preferences in Texas, US. Transportation 36(5):511–539. doi:10.1007/s11116-009-9201-4

Shankleman J (2012) Boris Johnson: 'I delivered every single one of my green pledges'. BusinessGreen – Incisive Media. http://www.businessgreen.com/bg/interview/2171102/boris-johnson-delivered-single-green-pledges. Accessed 25 June 2012

Shephard R (2008) Is active commuting the answer to population health? Sports Med 38(9):751–758

Siddique H (2013) David Cameron to announce largest ever investment in cycling. The Guardian. Accessed 12 Aug 2013

Skinner D, Rose P (2007) Hell is other cyclist: rethinking transport and identity. In: Horton D, Rosen P, Cox P (eds) Cycling and society. Ashgate Publishing Ltd, Aldershot, UK, pp 83–96

Smith RC, Harkey DL, Harris B (2001) Implementation of GIS-based highway safety analyses: bridging the gap. http://www.fhwa.dot.gov/publications/research/safety/1039.pdf. Accessed 11 Aug 2012

Snizek B, Sick Nielsen TA, Skov-Petersen H (2013) Mapping bicyclists' experiences in Copenhagen. J Transp Geogr (0). doi: http://dx.doi.org/10.1016/j.jtrangeo.2013.02.001

Sustrans (2012) Connect2: enabling millions of people to make everyday, local journeys by foot or by bike. http://www.sustransconnect2.org.uk. Accessed 12 Sept 2012

TfL (2012) Barclays cycle superhighways. TfL. http://www.tfl.gov.uk/roadusers/cycling/11901.aspx. Accessed 12 Sept 2012

UNECE (2009) Amsterdam declaration – making THE link, transport choices for our health, environment and prosperity. United Nations Economic Commission for Europe (UNECE), Denmark. Accessed 12 Sept 2012

UNECE (2011) Green and healthy jobs in transport: launching a new partnership under THE PEP. United Nations Economic Commission for Europe (UNECE), Denmark. Accessed 12 Sept 2012

Van der Spek S, Van Schaick J, De Bois P, De Haan R (2009) Sensing human activity: GPS tracking. Sensors 9(4):3033–3055

Walker P (2013a) David Cameron's 'cycling revolution' is barely off the starting blocks. TheGuardian, Monday 12 August 2013 11.39 BST

Walker P (2013b) Despair, cyclists: Britain will not be a 'cycling nation' in your lifetime. The Guardian, Friday 30 August 2013, 10.43 BST

Westcott R (2013) Cycling gets £94m push in England. BBC. http://www.bbc.co.uk/news/uk-23657010. Accessed 12 Aug 2013

Yang J, French S (2013) The travel – obesity connection: discerning the impacts of commuting trips with the perspective of individual energy expenditure and time use. Environ Plan B Plan Des 40(4):617–629

Yeboah G (2012) Exploring agent-based modelling technique for cycle track modelling. European Union MOVE COST Grant ID: ECOST-STSM-IC0903-051112-023655-23655. University of Copenhagen, Denmark

Yeboah G (2014) Understanding urban cycling behaviours in space and time (Unpublished Doctoral Thesis). Unpublished, Northumbria University at Newcastle, Newcastle upon Tyne

Yeboah G, Alvanides S (2013) Everyday cycling in urban environments: understanding behaviours and constraints in space-time. In: Poster presented at the GIS Research UK 2013 conference, University of Liverpool, UK

Yeboah G, Alvanides S, Thompson EM (2012) Methodological perspective on understanding cycling behaviours of commuters. In: Poster presented at the cycling and society symposium 2012, University of East London Stratford Campus, UK, 3–4 Sept 2012

Chapter 9
Performance Improvements for Large-Scale Traffic Simulation in MATSim

Rashid A. Waraich, David Charypar, Michael Balmer, and Kay W. Axhausen

Abstract In contrast to aggregated macroscopic models of traffic simulation, multi-agent microscopic models, such as MATSim, enable modeling of individual behavior and facilitate more detailed traffic analysis. However, such detailed modeling also leads to an increased computational burden, such that simulation performance becomes critical.

This paper looks specifically at the MATSim simulation framework and proposes several ways to improve its performance. This is achieved through a combination of several approaches, including reducing disk access, decoupling computational tasks, and making use of parallel computing. Additionally, for the traffic simulation, an event-based model is adopted instead of a fixed-increment time advance approach.

Experiments show that by applying these methods, a simulation speedup of four times and more is achieved (depending on the scenario) when compared to the current Java micro-simulation in MATSim.

Initial simulation experiments on a high-resolution navigation network of Switzerland – containing around one million roads and 7.3 million agents – demonstrate that real-world scenarios can now be executed in around one-and-a-half weeks using the improved model. Ways to further shorten the computational time of MATSim are also described.

Keywords Large-scale traffic simulation • Agent-based modeling • Parallelization

---

R.A. Waraich (✉) • D. Charypar • K.W. Axhausen
Institute for Transport Planning and Systems (IVT), ETH Zurich, Zurich, Switzerland
e-mail: waraich@ivt.baug.ethz.ch; charypar@ivt.baug.ethz.ch; axhausen@ivt.baug.ethz.ch

M. Balmer
Senozon AG, Schaffhauserstrasse 331, 8050 Zürich, Switzerland
e-mail: balmer@senozon.com; http://www.senozon.com

M. Helbich et al. (eds.), *Computational Approaches for Urban Environments*, Geotechnologies and the Environment 13, DOI 10.1007/978-3-319-11469-9_9

9.1 Introduction

Traffic simulations can be performed at different levels of detail. One common technique is to model traffic as flows consisting of aggregated number of cars between different areas (de Dios Ortúzar and Willumsen 2011). While this technique is quite fast, it does not allow modeling of individual preferences or temporally and spatially detailed analysis.

In contrast, *agent-based travel demand* models like MATSim represent each person in the simulation as an individual agent (MATSim 2009). The travel demand by each agent in this case is based on an *activity-based* model, where activity times and durations are time dependent (Axhausen and Gärling 1992). Furthermore, a dynamic micro-simulation model is used to model detailed traffic interactions, which are again time dependent.

This enables various kinds of new applications, which are not possible with the first approach – e.g., detailed modeling of car sharing (Ciari et al. 2008) or modeling the charging behavior of electric vehicles (Waraich et al. 2009). Also, commercial applications of such detailed micro-simulations can be envisioned. For example, companies owning advertising space could offer a more sophisticated service to customers, where not only the traffic volume along a road determines the price but also the target audience of an advertisement is considered.

While more powerful, such detailed micro-simulation models are more expensive, in terms of computation time, than aggregated models. This chapter describes efforts to improve the performance of an agent-based micro-simulation model called Multi-Agent Transport Simulation Toolkit (MATSim 2009). This model is aimed at the simulation of large travel demand scenarios. But in order to perform a simulation of a full population run of Switzerland, with 7.3 million agents on a high-resolution navigation network, it is estimated that the existing Java-based micro-simulation in MATSim would require around 3–4 weeks. In the direction of reducing the computational time of MATSim, two of its central components are redesigned. The first component is the mobility simulation where the traffic dynamics are modeled. In order to make the mobility simulation faster, a new micro-simulation is implemented based on the ideas of an existing event-based micro-simulation (Charypar et al. 2007a). While multiple distributed and parallel traffic simulations in the C++ programming language have been implemented in the past (Barceló et al. 1998; Nökel and Schmidt 2002; Nagel and Rickert 2001), to the best of the authors' knowledge, this chapter presents the first large-scale implementation of such a simulation in the Java programming language (implemented mid-2009). Therefore, this chapter also discusses specific challenges for large-scale traffic simulation in Java, which has not been discussed in the related C++ literature.

A second major performance improvement achieved in this work is related to another core component of the MATSim framework called event handling. This component is needed to process the results of the mobility system and is therefore essential for integration with other MATSim internal components and also for extension of the MATSim model. Parallel computing is used to make event handling faster.

This chapter is structured as follows: in the next section, MATSim is described together with several mobility simulation implementations available for MATSim and open issues in this regard. Thereafter, the implementation of the various performance improvements with regards to the micro-simulation and the event-handling model are described. This is followed by experiments which assess the performance gains due to the newly implemented models. Before concluding, open issues are discussed together with possible future work.

9.2 Related Work

In the following section, a description of MATSim is given followed by a presentation of the different micro-simulation models and the event-handling module in MATSim.

9.2.1 MATSim

In MATSim, individuals are modeled as agents who want to perform activities throughout the day, such as being at home or work and going shopping. But due to the spatial separation of the corresponding activity locations, agents need to travel. This leads to many additional choices for the agent, such as the mode of travel, the activity duration, the location, and the route choice. The goal of MATSim is to find a plan for each agent, which maximizes the overall utility of the agent, including items as travel time, ticket fares, or street toll prices. This optimization needs to be performed while keeping constraints of the agent's environment in mind, such as street network capacities or opening times at shops and working hours. This optimization of agents' plans in MATSim is achieved by applying an iterative process, which is depicted in Fig. 9.1.

In the beginning, the simulation starts with an initial plan for each agent depicted as *initial demand*. A simple plan for an agent, who wants to leave home at 7:25 a.m. in the morning, work for 8 h and 20 min, and then drive back home, might look as follows:

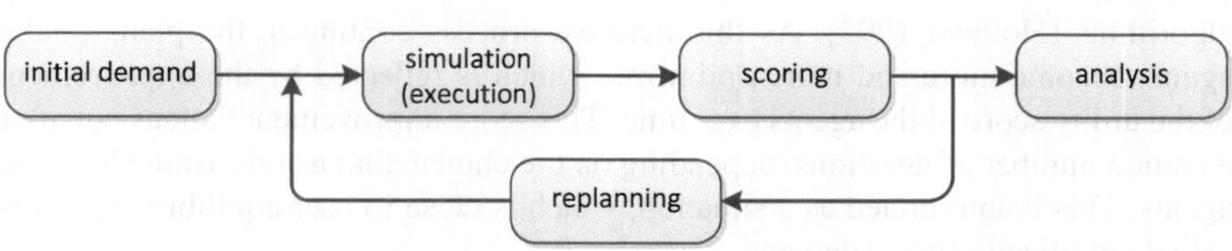

Fig. 9.1 Coevolutionary simulation process of MATSim

```
<plan>
        <act type="home" x="125" y="256" link="122" end_time="07:25:00" />
        <leg mode="car">
                <route> 122 19 59 </route>
        </leg>
        <act type="work" x="-500" y="100" link="59" dur="08:20:00" />
        <leg mode="car">
                <route> 59 10 7 122 </route>
        </leg>
        <act type="home" x="" y="0" link="122" />
</plan>
```

The plan also contains information about the road (link), where the home and work activities are located, and the route (link ids) the agent wants to drive. The micro-simulation in MATSim follows the instructions in the plan and executes it step by step. This means that the agent leaves home at the time specified in the plan and, thereafter, is routed through a virtual road network throughout the day. As an agent's vehicle is typically not traveling alone on the road network, interactions with other vehicles and road capacity constraints come into play. While the simulation is running, information about the performance of the plan is also collected. For example, did the agent need to pay a toll, how long was the travel time, and how long did the agent work? This information is used to calculate different *utility* components for the various aspects of the plan. These are added up during the *scoring* step, such that each executed plan has a score assigned to it. The next step in the iteration is called *replanning*. In this step, either an old plan, which was generated in a previous iteration, is reselected for execution in the next iteration or a new plan is generated, possibly by adapting a previous plan. This allows alternative choices for the agent such as the mode of travel or the travel route. Often, the probability of reselecting a plan for execution in the next iteration is based on its score, meaning that a plan with a higher score has a higher chance of reselection. Due to memory limitations, only a small number of plans are kept, and the plan with the lowest score is deleted whenever a new plan is generated. This corresponds to mutation, selection, and survival of the fittest within the context of evolutionary algorithms (Holland 1992). As this iterative process continues, the plans of the agents become more and more optimized, which is reflected by the improvement of the utility score of the agents over time. This score improvement flattens out after a certain number of iterations, depending on the choice dimensions available to the agents. This is interpreted as a situation, which is close to user equilibrium, and is called optimized/relaxed demand.

After this brief description of the MATSim model, in the following section the different traffic micro-simulation models available for MATSim are presented in more detail.

9.2.2 *Traffic Simulation Model*

Traffic flow simulations range from detailed physical simulations to macroscopic models. Detailed physical simulation attempt to capture as many traffic flow phenomena as possible, e.g., car following and lane changing, by representing space continuously and simulating very small time steps, e.g., Fellendorf and Vortisch (2010). A second less detailed approach is represented by cellular automata, where cars move between fixed-sized cells, e.g., Nagel and Rickert (2001). Interactions between cars in neighboring cells are present, such that travel speeds and densities can be modeled as disaggregated. Less details are present in mesoscopic models, where detailed traffic demand is present however only aggregated supply, e.g., Ben-Akiva et al. (1998). Macroscopic models represent the highest abstraction level, where supply and demand is modeled on an aggregated level, e.g., Cayford et al. (1997). For a more detailed literature review related to traffic simulations, see Charypar et al. (2007a).

There is a trade-off between computation time and model detail. As MATSim aims for large-scale simulation, it uses a queue-based approach, which in terms of detail is located somewhere between the cellular automata and mesoscopic approach. While the details in terms of implementation differ slightly, in general all of MATSim's traffic simulators consider roads as active elements, which move around cars. Each road link contains a queue which stores the entry time of each car. Adjacent links collaborate with each other to assure that link capacity, free speed travel time, intersection precedence, and space availability are taken into account during the simulation. There are several implementations of the traffic simulation available, which are presented in the following section.

9.2.3 *QueueSim and JQueueSim*

The first micro-simulation developed for MATSim is called QueueSim and is based on a fixed-increment time advance model (Raney et al. 2003). In this model, vehicles are moved along links in fixed time steps of 1 s. Although the model is quite flexible, for larger simulations it is too slow because of the fixed simulation time step. A parallel version of QueueSim was implemented leading to a significant speedup (Cetin 2005). Both of these simulations are implemented in the C++ programming language. In order to improve the maintainability of the code, MATSim was later re-implemented in Java, including the nonparallel version of QueueSim, which is called JQueueSim here. In recent years, the performance of JQueueSim has improved, but the underlying simulation method remains the same.

9.2.4 DEQSim

A major performance breakthrough, within the MATSim context, is achieved with a more recent micro-simulation called Deterministic Event-Driven Queue-Based Traffic Flow Micro-Simulation (DEQSim, Charypar et al. 2007a), which is implemented in C++. Instead of performing the simulation along fixed time steps, an event-based model is used performing only discrete actions which are relevant to the model, i.e., entering and leaving roads. Furthermore, DEQSim has been parallelized making it one of the fastest large-scale transport micro-simulations currently available (Charypar et al. 2007b).

A major drawback of DEQSim within the MATSim context is that it is implemented in C++, while the other modules of MATSim are implemented in Java. This means that the communication of DEQSim with the other MATSim modules is bridged by a slow file input/output (I/O) interface.

9.2.5 Graphical Processing Units

Yet a different approach to accelerate the micro-simulation was tried using graphical processing units (GPUs) on computer graphic cards. These GPUs perform many more operations in the same amount of time than central processing units (CPUs) on computer boards. A first successful implementation of QueueSim on GPUs rendered a speedup of 67 times (Strippgen and Nagel 2009) compared to JQueueSim. The main drawback of GPUs is similar to that of DEQSim, as the interface between the graphic card and the rest of MATSim modules poses a bottleneck. Furthermore, current GPUs have a limited amount of memory. For example, the traffic simulation of Switzerland, with 7.3 million agents, requires around 60 GB of memory. Graphic cards today have often less than 4 GB of memory. Also, maintainability of the implementation is an issue, as the program code is not written in Java.

In order to provide a faster micro-simulation in Java than QueueSim, here a redesign and re-implementation of DEQSim in Java is proposed, called *JDEQSim*. A second feature, for which an improvement is also proposed, is the event-handling module in MATSim, which is described in the next section.

9.2.6 Event Handling

The output of the traffic simulation contains detailed information about the course of the simulation. It describes, e.g., when an agent's vehicle enters a road or arrives at an activity location. This information is embedded within a data structure called "events," which contains information such as the identification of the agent and the link where the event occurred and its time. These events are used to communicate the output of the micro-simulation to other modules in MATSim, such as scoring which can use it to calculate the performance score of an agent's plan. These events can also be utilized to extend the simulation, e.g., modeling the state of charge of the battery in an electric car or updating the capacity of a parking lot.

After their creation by the micro-simulation, events are handed over to the *event-handler* module. Modules interested in certain types of events can register with the event handler. The event handler then processes each event according to the instructions of the registered modules.

It is clear that the performance of event handling is critical to the performance of MATSim and its extensions. In the next section, the implementation of the JDEQSim and the performance improvements for the event handler are described.

9.3 Implementation

9.3.1 JDEQSim

The re-implementation of DEQSim in Java provided the opportunity to rethink and redesign its code structure. The C++ DEQSim code is used to understand the internal structure of the DEQSim traffic model but is not used for the implementation of JDEQSim itself. The design of JDEQSim is influenced by OMNeT++ (OMNeT++ 2009), which is a modular and open-architecture discrete-event communication network simulator. To a certain extent, many elements used in JDEQSim are similar to concepts presented by Axhausen (1988).

The JDEQSim implementation consists of three parts: simulation units, messages, and a scheduler. Vehicles and links are the basic building blocks of the traffic simulation and are called *simulation units*. These simulation units communicate with each other by exchanging different kinds of *messages,* which can be thought of as internal events, which stay inside the traffic simulation. These are different from the external events, which are forwarded to the event handler by the micro-simulation. Each message contains a time stamp, e.g., when a vehicle is allowed to enter the next link or when a car should start a trip. Each newly created message is sent to the *scheduler.* The scheduler contains a message priority queue, which is ordered by message time and message type. At the beginning of the simulation, the end time of the first activity of each agent is scheduled in the queue. When the micro-simulation is started, the scheduler fetches the first message and delivers each message to its intended target simulation unit, where it gets executed, e.g., a car leaves one road and enters a new one. Often, execution of the instructions written in a message leads to the creation of new messages, which are then added to the queue of the scheduler. The processing of the messages also leads to the generation of external events, which includes the passing of these events to the event handler, where they are processed further. The scheduler will always only process the first message in the queue until all of the messages have been processed and the micro-simulation ends.

As mentioned above, the messages are not only ordered according to time but also according to the message type, which is not the case in most event-based simulations. This is required to solve the situation, in which two events happen at the same time and a deterministic order needs to be maintained. For example, it is logical that an agent first has to arrive home before any activity can be performed there. Therefore, if the arrival at home and the start of an activity there happen at the same time due to a missing delay in-between, priority needs to be given to the arrival event for it to be processed first. This is important both for the internal and external consistency of the traffic simulation, including event handling.

9.3.1.1 Simulation of Transportation Modes Besides Cars

While DEQSim only supports the simulation of cars, a simple and general model for other transport modes is present in JQueueSim. The model allows to define a constant travel speed for a new transportation mode. This model has also been implemented in JDEQSim.

Two other features, which distinguish the DEQSim and QueueSim model, are described in the following section, as they have been implemented in JDEQSim.

9.3.1.2 Traveling Gaps in a Queue

When the front car in a traffic queue moves, it leaves behind a gap which travels backwards. Cars behind in the queue have to wait until such a gap reaches them before they can start driving. Such gaps, traveling backwards as a traffic queue is dissolving, are implemented in DEQSim (Charypar et al. 2007b). They have also been implemented in JDEQSim as this makes the model more realistic.

9.3.1.3 Prevention of Gridlock

Gridlock, which can occur in all of the micro-simulation models, is a problem. For example, if links are full and there is a circular flow relationship between vehicle movements on two or more links, it can lead to vehicles waiting for each other forever. In order to address this issue in DEQSim, if a vehicle waits for a long time at the front of a link, it is moved to the next link. This introduces a minimum flow at network links. Furthermore, the space available on a link is also temporarily modified.

An alternative to this mechanism could be to remove the agent and its vehicles from the simulation if it does not move after a certain maximum duration at the front of a road. This mechanism is implemented in JQueueSim but has the disadvantage that, as the agent is suddenly removed during the simulation, the scoring of the agent and further processing is stopped. This measure is rather abrupt and can lead to temporary wrong results at modules processing events, e.g., wrong traffic counts. For this reason, the JDEQSim approach to avoid gridlock is implemented.

Besides the re-implementation and extension of DEQSim in Java, an attempt is also made to parallelize JDEQSim, which is described in the next section.

9.3.2 *Parallelization of JDEQSim*

In Charypar et al. (2007b), the parallelization of DEQSim is described. This is achieved by partitioning the traffic network into several pieces, which are assigned to separate CPUs of a shared memory machine. The message passing interface (MPI, Snir et al. 1995) is used for communication between CPUs. When an agent travels from the network area assigned to one CPU to a different one, MPI is used for passing the agent's data between CPUs. This operation includes periodically synchronizing the state of links at the border of each network partition.

In Java, *threads* are used as a basis for parallel programming (Lea 1999), where such instructions which need to be executed in parallel are distributed to different threads. In order to pass data between two threads, the *synchronized* keyword is used. The advantage of the Java synchronized keyword compared to MPI in C++ is that no explicit data structures have to be built for transferring data between threads. This means that all data within the Java virtual machine (JVM, Lindholm and Yellin 1999) are accessible to all threads. But in the context of parallelizing JDEQSim, this turns out to be a major disadvantage; whereas MPI allows to explicitly specify which data to transfer between CPUs, it is not always obvious what data will be exchanged due to a synchronized statement in Java. While many elements of data transfer between threads are hidden and handled by the JVM, which simplifies programming, this also means that the programmer has little control over them for performance optimization.

Before describing a successful implementation of the parallelization of JDE-QSim, two straightforward, but failed, attempts in this regard are described. This helps to better understand the path taken and the various issues involved.

9.3.2.1 Failed Attempt 1: Single Scheduler Queue

As mentioned earlier, in Java each thread can access all data within the JVM. A simple parallelization solution is therefore to maintain a single scheduler queue within the JVM. The network links are assigned to different threads, e.g., using a network partitioning algorithm as used in DEQSim, such that most collocated links are assigned to the same thread.

In order to keep the state of the scheduler queue synchronized between threads, each access of the queue needs to be synchronized. Unfortunately, this leads to too many synchronizations between the threads, thus deteriorating performance. An attempt was also made to improve the performance by introducing message buffers, which are attached to the scheduler, but this did not help to resolve the problem.

9.3.2.2 Failed Attempt 2: Multiple Scheduler Queues

In order to try to solve the problem with the single scheduler queue, separate scheduler queues per thread are defined. Different threads still need to synchronize to ensure data consistency, e.g., when vehicles move between network partitions. Furthermore, periodical synchronization between border links at predefined intervals is needed to ensure that one thread does not advance the simulation too much. This synchronization interval is determined by the travel time needed to travel between two partitions. This method is almost identical to how the parallelization of DEQSim is performed.

Unfortunately, this approach does not perform well because of the periodical synchronization between threads and the waiting time involved; whenever one thread is too far ahead, it has to wait on the other thread. This issue could be improved by applying a different method, which is described in the next section.

9.3.2.3 Successful Parallelization: Decoupling of Executor Threads

The main problem detected with the multiple scheduler queues approach is that the synchronization between threads happens too often, thus hindering parallelization. A new approach in this regard is implemented which successfully decouples the threads, as shown in Fig. 9.2. Instead of periodically synchronizing between network

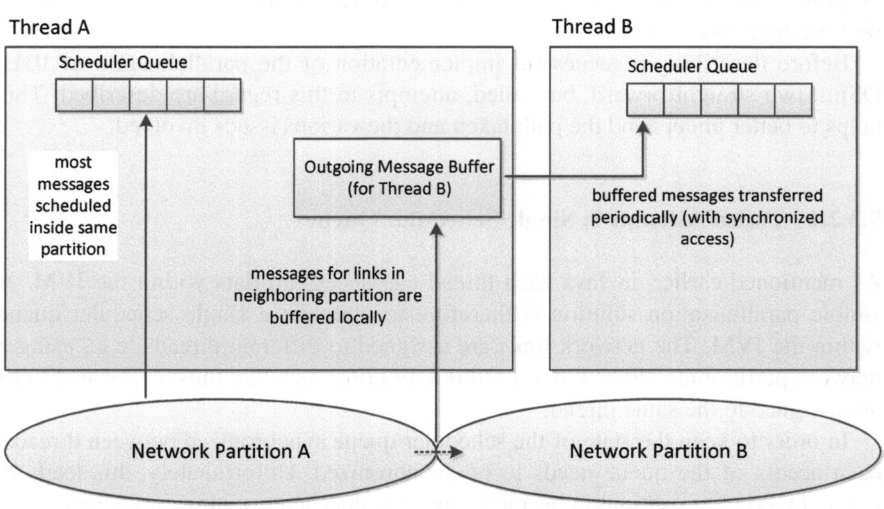

Fig. 9.2 Parallelization of JDEQSim

partitions on a link level, as in DEQSim, a periodic synchronization for the whole partition is performed periodically after a fixed interval δ. This means that all messages generated during this interval, which need to be added to the scheduler queue of an adjacent network partition, are buffered and then added at once using synchronized access. The maximum allowed δ between two threads is determined by the link with the shortest travel time, which resides at the border of the two network partitions. As a larger δ is better for simulation performance, possible adaptations to the network portioning algorithm are discussed further in the future work section.

9.3.3 Parallel Event Handling

During the development of JDEQSim and its parallelization, it was observed that event handling is executed on the same thread as the micro-simulation itself. But as the event-handling process is independent of the micro-simulation and can be further split into multiple event-handling tasks, it is ideal for parallelization. Such a parallelization allows us not only to run the micro-simulation faster but it also improves the performance of extensions of MATSim, which use the event-handling interface.

Currently, five default event handlers are present in MATSim. Figure 9.3 shows the relative time proportions of these five handlers to each other. Two of these event handlers are needed for gathering and communicating information between the micro-simulation and other MATSim modules: *EventsToScore* for accumulation of the utility score components and *TravelTimeCalculator* for estimating the travel time. Two additional event handlers generate statistics and graphs of the simulation (*LegHistogram* and *CalcLegTimes*). The most time-consuming event handler is *EventWriterTXT* which writes all events produced by the micro-simulation to a file, allowing later post processing and analysis of simulation results.

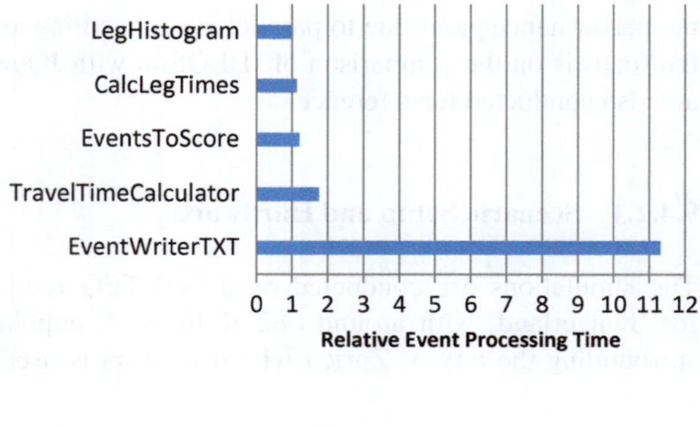

Fig. 9.3 The execution time of default event handlers

9.3.3.1 Implementation

The parallelized version of event handling is called parallel event handling and allows the user to specify how many threads should be dedicated to this module during the simulation. The current implementation applies a round-robin approach (Hahne 1991) to assign event handlers to threads. This means it tries to assign the same number of event handlers to each thread. It is obvious from Fig. 9.3 that this approach is suboptimal because it would be best to put the event handler writing out events to a separate thread and the other handlers on a second one.

Fortunately, writing out events to a file is not part of the communication interface between JDEQSim and MATSim, which is different in DEQSim. This means the event file is needed only in the last iteration as a backup for further analysis. Additionally, events can be written out at predefined intervals, e.g., each 10th iteration, in order to analyze intermediate results of a bigger simulation while it is still running. This means that during most iterations, a quite balanced parallelization of event handling is possible using up to four cores if considering the default simulation without any other scenario-specific event handlers.

While the implementation of parallel event handling could be improved in several ways, as discussed in the future work section, the reason to start with a simple implementation is that the existing interfaces did not need to be changed.

9.4 Experiments and Results

9.4.1 Single Thread JDEQSim with Parallel Event Handling

While the parallel version of JDEQSim is still experimental work, the nonparallel version of JDEQSim is planned to be run with parallel event handling in the near future for large-scale scenarios.[1] Therefore, experiments presented here are conducted in the latter configuration to demonstrate the performance gains when comparing it to the current state-of-the-art micro-simulation in Java (JQueueSim). Furthermore, the performance gains due to parallel event handling are also demonstrated. While the focus is on the comparison of JDEQSim with JQueueSim, runs with DEQSim are also conducted for reference.

9.4.1.1 Scenario Setup and Hardware

The simulations are conducted on a NAVTEQ road network (NAVTEQ 2009) for Switzerland, with around 882 K links. A population sample of the people surrounding the city of Zurich who drive cars is used, containing around 614 K

[1]Have been conducted in 2010 (see, Meister et al. 2010)

agents. The hardware used for this experiment is a Sun Fire X4600 M2, with 16 cores (8 dual core CPUs) and 128 GB of memory. Due to the large number of simulations, not all runs could be repeated multiple times. To give a sense of the variability of the JDEQSim results for a similar JDEQSim run over 50 iterations, the standard deviation for the computation time amounted to 13 % of its mean value (for micro-simulation and event handling).

9.4.1.2 Runtime: Micro-simulation and Event Handling

In the first experiment, the runtime of the three micro-simulations is compared using nine configurations, where also the number of threads used is varied (see Fig. 9.4). Configurations include runs with and without parallel event handling (abbreviated as PEH in Fig. 9.4). Furthermore, runs with and without the EventWriterTXT event handler are conducted (abbreviated as EWT in Fig. 9.4).

Both JDEQSim and JQueueSim runs with a single thread use the default event handling, while runs with more than one thread use parallel event handling. For example, the JDEQSim run with two threads uses one thread for the micro-simulation and one for parallel event handling. In the DEQSim runs, first the micro-simulation is run using the indicated number of threads, thereafter the events are written to a file and read in by the nonparallel version of the event handler for further processing.

The number of threads usable in the different configurations cannot be chosen arbitrarily: The number of threads "n" which can be used by DEQSim is constrained to $n = 2^i$, where $i \geq 0$. This is due to the network partitioning algorithm used. For JDEQSim, the maximum number of usable threads in the presented scenario is five without EWT and six with EWT. In these cases one thread is used for the

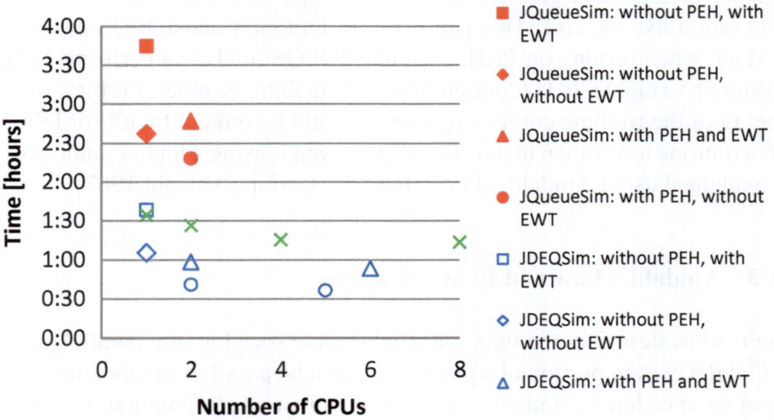

Fig. 9.4 The computation time for the three micro-simulations as a function of number of threads

micro-simulation and one thread for each of the default event handlers. However, only a selected number of configurations and thread combinations are simulated and discussed here.

The first two configurations look at JQueueSim without PEH and with and without EWT. The comparison of these two configurations highlights the overhead due EWT (43 %). Configuration three and four look at JQueueSim for the case where PEH is turned on, again with and without EWT. The experiments show that the newly implemented parallel event handling reduces the time of an iteration by around 26 % for the case where writing out events is turned on (configuration one vs. three) and by around 13 % for the case where writing out events is turned off (configuration two vs. four). The higher gain for the case where events are written out to the hard drive is expected; as in this case parallel event handling successfully decouples the micro-simulation from I/O operations.

Configurations one to four and five to eight correspond to each other, and only the latter uses JDEQSim instead of JQueueSim. In order to measure possible performance gains due to the implementations made in this chapter, JQueueSim and JDEQSim need to be compared to the case where writing out events is turned off and the latter uses PEH (configuration four vs. eight). The reason for comparing the case where event writing is turned off is important because in most iterations, this configuration is run. In this case, the runtime is reduced by around four times for the given scenario. The major part of this speedup (ca. 76 %) is due to the differences in models of JDEQSim and JQueueSim (event-based vs. fixed time steps). This can be seen when comparing configuration two and six where only a single thread is used both for JDEQSim and JQueueSim. The remaining performance gain is due to the parallelization of event handling (configuration six vs. eight).

Configuration nine shows DEQSim runs for various numbers of CPUs used. While the runtimes of JDEQSim and DEQSim are similar, when PEH is not used and EWT is turned on (configuration five vs. nine), already turning off EWT leads to a major performance gain for JDEQSim compared to DEQSim (50 %, compare configuration five vs. six). This gap between DEQSim and JDEQSim even builds up further, when turning on PEH, such that JDEQSim always performs better than DEQSim up to eight CPUs (configuration seven/eight vs. nine). Furthermore, as the flattening of the runtime curves suggests, it might be quite difficult for DEQSim to reach a runtime lower than that of JDEQSim, even if using a higher number of CPUs as is explained using Amdahl's law in the next section (Amdahl 1967).

9.4.1.3 Amdahl's Law and Its Implications

Amdahl's law describes the maximum achievable speedup of a parallel program. It says that if a certain portion of a program cannot be parallelized, then the maximum achievable speedup is limited – even with unbounded computation power. The maximum achievable speedup with n threads for a program where fraction b of the program cannot be parallelized can be calculated using Eq. (9.1).

$$S(n) = \frac{1}{b + \frac{1-b}{n}} \tag{9.1}$$

To give an example of Amdahl's law, if 5 % of a program cannot be parallelized, then it is not possible to achieve a speedup of more than 20. The implication of Amdahl's law is present both for DEQSim and JDEQSim runs but at different points. For DEQSim, the interface between the micro-simulation and MATSim is the bottleneck. Because of the I/O overhead of the communication between the micro-simulation and MATSim, a speedup of even two seems impossible. This means that more than 50 % of the micro-simulation consists of parts which have to be executed sequentially. A second and smaller part of nonparallelizable code present in the DEQSim runs is the event handling, which cannot run in parallel mode for DEQSim at the moment. In case of parallel event handling, the maximum achievable speedup is limited by the slowest event handler.

This first experiment suggests that to make most efficient use of CPUs, JDEQSim should be run with one parallel event-handling thread. As the machine used in this experiment has around 128 GB RAM and 16 cores and the scenario uses less than 15 GB of RAM, several JDEQSim runs could run in parallel, which is useful especially during the calibration phase.

9.4.2 Influence of Network Size

In the previous experiment, JDEQSim performed around four times faster than JQueueSim. But this cannot be generalized, because if the network is congested, then JDEQSim can be much faster than JQueueSim. Such congestion can happen especially during the initial iterations, in which the routes are far from optimal and can lead to a simulation period stretching far beyond 24 h. This can lead to long run times for JQueueSim as its runtime is directly correlated to the simulation period.

Furthermore, different ratios of network size and population can widen the gap between the speedup of JQueueSim and JDEQSim, which is highlighted here. In this experiment, all micro-simulations are run using two threads. Both JDEQSim and JQueueSim runs are performed with parallel event handling, using a single thread and no events are written to the hard drive. The network capacity is chosen in such a way that no congestion should happen in order to remove possible adverse influence of this on JQueueSim. The three scenarios which are considered are:

- Scenario A: Network with 882 K links and 61 K agents (36 M events)
- Scenario B: Network with 61 K links and 616 K agents (58 M events)
- Scenario C: Network with 882 K links and 614 K agents (363 M events)

The results of the experiments in Fig. 9.5 show that DEQSim and JDEQSim scale linearly with the number of events. Only in Scenario A, in the case of DEQSim, the I/O overhead of loading the network is immense compared to the actual simulation time.

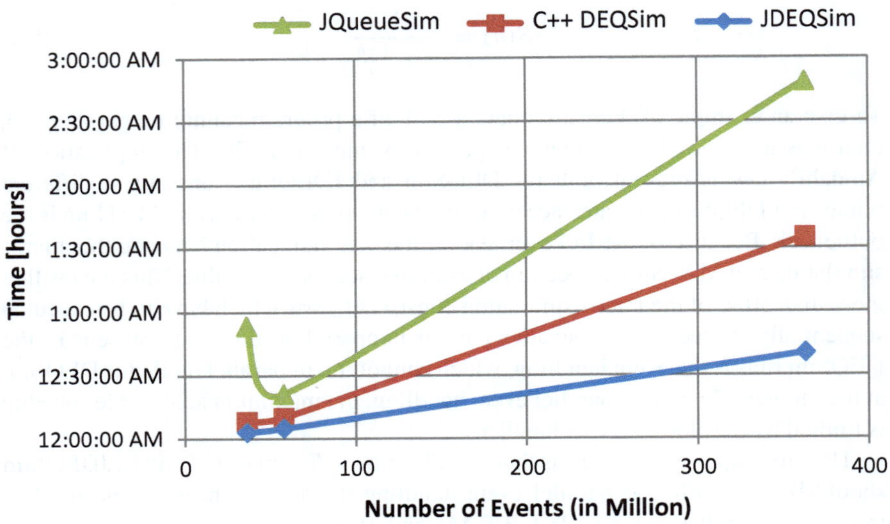

Fig. 9.5 The influence of network size on the three micro-simulations

For JQueueSim the situation is different. While Scenarios A and B generate the same magnitude of events, the run times are substantially different. This has to do with the substantially different ratio of network-to-population size. Therefore, in Scenario A, JQueueSim performs extremely badly compared to JDEQSim. In fact, JDEQSim is more than 17 times faster than JQueueSim in Scenario A, while for Scenarios B and C, it is around four times faster.

9.4.3 Scalability of Event Handlers

While the first two experiments look at the overall performance increase, due to both the micro-simulation and parallel event handling, in this section we only look at the latter. As mentioned earlier, the event handler with the longest computation time defines the runtime of parallel event handling. Therefore, in order to test how parallel event handling would scale with multiple event handlers and threads, identical test handlers are added to the simulation. The test handler performs computationally intensive tasks and does not involve any disk I/O. This is important because event handlers requiring I/O are inherently difficult to parallelize due to the speed limit of the hard drive and are therefore not suitable for this experiment.

Figure 9.6 shows the speedup for the different runs, where different numbers of handlers are involved. This experiment shows that parallel event-handling scales linearly up to 4 threads, but the speedup with 8 threads already drops to around 6. This drop is severe if we consider that only little use of Java's synchronized keyword

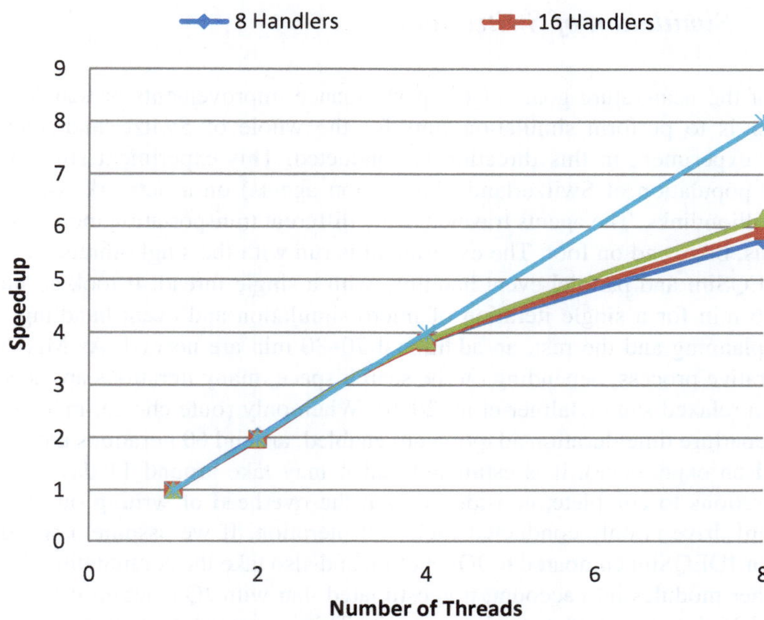

Fig. 9.6 Performance of parallel event handling

is made. Even more complex parallel programs written in C++ with MPI achieve speedups of around 8 in this case (Charypar et al. 2007a).

While only a small number of event handlers are present by default in MATSim, many applications are under development and planned in the MATSim community (MATSim 2009) which require additional event handlers. The good news is that if more handlers are added to parallel event handling, the speedup gets slightly better, as can be seen in the 8-thread scenario. This is expected because adding more work to the handler reduces the relative penalty of synchronization between threads.

9.4.4 Speedup for Parallel JDEQSim

As described earlier, a first prototype of the parallelization of JDEQSim for two threads has been implemented. As event handling has not been adapted yet to properly function with parallel JDEQSim, only measurements of the micro-simulation are reported here where event handling is turned off. The experiment consists of 1.62 million agents residing in the surroundings of Zurich city. The network contains 163 K links. This experiment required 29 min and 40 s when running with JDEQSim, while on the parallel JDEQSim (2 threads), the experiment only took 18 min and 37 s. This is a speedup of 1.6, which is encouraging, but many problems remain unresolved, as is discussed in the future work section.

9.4.5 *Simulation of Switzerland*

One of the near-future goals of the performance improvements presented in this chapter is to perform simulation runs for the whole of Switzerland; therefore, a first experiment in this direction is conducted. This experiment simulates the whole population of Switzerland (7.3 million agents) on a network with around one million links. The agents traveled using different transportation modes, such as car, bus, bike, and on foot. The experiment is run with the single-threaded version of JDEQSim and parallel event handling with a single thread. It took around 3 h and 16 min for a single iteration of micro-simulation and event handling, while for replanning and the rest, an additional 70–80 min are needed. As MATSim is an iterative process, depending on the search space, many iterations are needed to reach a relaxed state (Balmer et al. 2009). When only route choice, mode choice, and departure time/duration adaption are enabled, around 60 iterations are required (based on experience). It is estimated that it may take around 11 days for these 60 iterations to complete, considering that the overhead of writing out events to the hard drive is only conducted each 10th iteration. If we assume a speedup of four for JDEQSim compared to JQueueSim and also take the computational time of the other modules into account, it is estimated that with JQueueSim it would take around 36 days to calculate such a scenario. This is a speedup of around 3.2 for the overall simulation. While the performance gains achieved by the work presented in this chapter are important, this means that more progress related to performance is still needed, *which is discussed in the next section.*

9.5 Discussion and Future Work

While two ways to significantly shorten the runtime of the MATSim simulation are presented in this chapter, it seems like the "low-hanging fruits have been picked" and additional improvements will not be as straightforward and may possibly lead to not as much performance improvement, as well as requiring major changes, to the existing models and interfaces. In the following section, issues involved are discussed together with possible solutions.

9.5.1 *Parallel Micro-simulation*

In order to achieve a major breakthrough with regards to the micro-simulation performance, making use of multiple threads seems to be crucial. Although a first success in the direction of a parallelization of DEQSim has been made, a successful integration of this into MATSim needs more work and would also require fundamental changes to the existing models.

Two points seem central to a successful integration of parallel JDEQSim in MATSim: decoupling of threads and integration of parallel JDEQSim with parallel event handling. In both cases also synchronization overheads between threads in Java contribute to the increased computation time which is further discussed in the following sections.

9.5.1.1 Load Balancing

For decoupling of different micro-simulation threads, the workload needs to be distributed evenly among them, and the time period δ for two periodical synchronizations between threads should be as long as possible.

For assigning the same amount of work to all threads, currently the network is partitioned at the beginning of the iteration. However, as traffic load changes over the day, this may lead to a major imbalance of workload among threads during the course of the iteration. This causes faster threads to wait for slower ones due to the periodic synchronization. This situation could be improved by changing the network partition assignment to threads during the iteration to correct for the imbalance. This operation could be performed at the time when synchronization between threads happens.

Besides the workload, the coupling of two threads is affected by the time period δ after which two threads need to synchronize. This time is determined by the link with the shortest travel time at the border of two network partitions. Therefore, a way forward might be to partition the network in a way which maximizes the duration between two consecutive thread synchronizations. But it is unclear how much coupling between threads is caused due to an imbalance of workload between threads and how much improvement could be achieved by partitioning the network in a way that δ is maximized. Therefore, extensive experiments would be required in this regard to be able to make recommendations on how to proceed in this regards.

9.5.1.2 Trade-Off: Preciseness Versus Performance

If the synchronization time period between two threads is above its maximum value (defined by the link with the shortest travel time at the partition border), race conditions will occur, meaning that cars at the border links could enter the neighboring partition too early or too late, leading to a distortion of traffic patterns. It would be interesting to investigate up to what value of δ a significant performance increase can be achieved and how much the traffic patterns are affected due to this. According to initial experiments in this regard, a possible trade-off between preciseness and performance could be a viable way to increase performance.

9.5.1.3 Adaptation of Event Handling

The events processed by event handling must have an ascending time stamp. However, this is not naturally the case when events are generated by different threads of the parallel micro-simulation. Therefore, neither the current single-threaded event handler nor the new parallel event handler is suitable for use with the parallel version of JDEQSim. Buffering and sorting of events need to be implemented between the micro-simulation and the event handler in order to make this possible.

In the following section, additional issues and possible improvements related to event handling are discussed.

9.5.2 Event Handling

9.5.2.1 Performance

According to the experiments conducted, parallel event-handing scales well if the load is well balanced. But this is often not the case when one or several event handlers involved contain many disk I/O operations. In this case, these event handlers pose a bottleneck to the current parallelization approach in several ways: firstly, as I/O operations are slow, event handlers involving many I/O operations are the slowest event handlers; secondly, as writing to a hard disk is limited by the speed of the hard disk, this becomes even more of an issue because such handlers slow each other down even further or might even be influenced by other I/O operations on the same computer.

In order to solve this problem, one might distinguish event handlers based on the criterion if other modules in the simulation depend on their output. Only such modules which fulfill this criterion require that event handling is completed before the next step in the MATSim loop is executed. Of the five default event handlers, this is only the case for two of the event handlers (EventsToScore and TravelTimeCalculator). The other three event handlers only produce output for later analysis. This means that these three event handlers could continue their processing while the rest of the MATSim iterations continue. This approach could certainly be used to reduce the overall computational time of MATSim.

A second performance improvement could be achieved by replacing the current round-robin algorithm, which assigns an even number of event handlers to all threads with one which performs better load balancing. This could be achieved in the following way: the average runtime of the different event handlers could be tracked and a load balancing could be performed, every couple of iterations, while taking this information into account.

9.5.2.2 Usability

While event handling is a simple way for users to access the output of the MATSim simulation and also to extend the simulation itself, with parallel event handling and other performance improvements, the possibility of making errors especially for novice users of MATSim increases.

A possible pitfall of using parallel event handling could arise in the following way: while with nonparallel event handling it is not a problem if different event handlers access each other's data, this could cause race conditions in the case of parallel event handling. This means that if no synchronization is used, data written by one thread could not be visible to another thread and, as such, data inconsistencies could occur. While the default event handlers are all implemented in a way that this problem cannot occur, users not familiar with the intricacies involved could make such a mistake.

Also, the proposed performance improvement – where a distinction between event handlers is made, whether they produce output for other MATSim modules or not – adds to the complexity of how event handling is used till now. This means that there is a trade-off between usability and performance with regards to parallel event handling which users need to be informed of.

9.5.3 *Replanning Modules*

Besides the micro-simulation and event handling, replanning is the third module in MATSim which requires major computational time. Apart from reducing the computational time of the various modules involved, work that helps to reduce the number of iterations required to reach a relaxed state is also important. Ongoing work related to Meister et al. (2006) where new optimization methods and heuristics are applied in order to reduce the number of iterations is interesting in this regard.

Another research strand to investigate is by how much the runtime can be reduced by making more dynamic use of the replanning module than at the moment. Currently, the probability for applying a replanning strategy in MATSim is fixed at the beginning of the simulation. For example, in each iteration 10 % of the agents try to reroute. While systematic research in this regard is limited, experience suggests that some search dimensions do not need a constant replanning share throughout the simulation. For example, the share of rerouting could be reduced over time, as optimal routes are often found within the first 10–20 iterations with a 10 % reroute share. This could accelerate the overall simulation as the freed up processing power could be used by other replanning modules.

9.6 Conclusions

This chapter presents to the best of the authors' knowledge the first implementation of a large-scale traffic simulation in Java while making use of parallel computing. Several issues discussed in this chapter have been raised for the first time in traffic simulation literature and might be useful for the implementation of other traffic simulation models in Java as well.

The main contribution of this chapter is that two methods are proposed and implemented to enhance the performance of the agent-based travel demand simulation MATSim. This is achieved by refining the micro-simulation and by parallelizing the processing of its output. As a result of these performance improvements, larger runs can be simulated in less time and using fewer CPUs/cores than possible before.

Experiments show that, through the proposed changes, the runtime of the current Java-based micro-simulation is improved by a factor of four and more, depending on the scenario. As MATSim is aimed at the simulation of large-scale scenarios and simulation runs of the whole of Switzerland are planned in the near future on high-resolution networks, it is shown that the computational time for the whole MATSim run is reduced by a factor of around 3 to about 4.5 h per iteration.

While this is a significant performance enhancement, further improvements of various modules of the MATSim simulation are also proposed, especially with regards to the parallelization of the micro-simulation.

References

Amdahl G (1967) Validity of the single processor approach to achieving large scale computing capabilities. In: Spring joint computer conference. ACM, New York, pp 483–485

Axhausen KW (1988) Eine ereignisorientierte simulation von Aktivitätenketten zur Parkstandswahl. Ph.D. thesis, University of Karlsruhe, Karlsruhe (in German)

Axhausen KW, Gärling T (1992) Activity based approaches to travel analysis: conceptual frameworks, models and research problems. Transp Rev 12(4):323–341

Balmer M, Rieser M, Meister K, Charypar D, Lefebvre N, Nagel K (2009) MATSim-T: architecture and simulation times. In: Bazzan ALC, Klügl F (eds) Multi-agent systems for traffic and transportation engineering. Information Science Reference, Hershey, pp 57–78

Barceló J, Ferrer JL, García D, Florian M, Le Saux E (1998) Parallelization of microscopic traffic simulation for ATT systems analysis. In: Equilibrium and advanced transportation modelling. Springer, New York, pp 1–26

Ben-Akiva M, Bierlaire M, Koutsopoulos H, Mishalani R (1998) DynaMIT: a simulation-based system for traffic prediction. In: DACCORS short term forecasting workshop, The Netherlands, February 1998.

Cayford R, Wie-Hua L, Daganzo CF (1997) The NETCELL simulation package: technical description. California PATH research report UCB–ITS–PRR–97–23, University of California, Berkeley, CA

Cetin N (2005) Large-scale parallel graph-based simulations. Ph.D. thesis, ETH Zurich, Zurich.

Charypar D, Axhausen KW, Nagel K (2007a) An event-driven queue-based traffic flow microsimulation. Transp Res Rec 2003:35–40

Charypar D, Axhausen KW, Nagel K (2007b) An event-driven parallel queue-based microsimulation for large scale traffic scenarios. In: The 11th world conference on transportation research, Berkeley, June 2007

Ciari F, Balmer M, Axhausen KW (2008) Concepts for a large scale car-sharing system: modelling and evaluation with an agent-based approach, Arbeitsberichte Verkehrs und Raumplanung, 517. IVT, ETH Zürich, Zürich

de Dios Ortúzar J, Willumsen LG (2011) Modelling transport, 4th edn. Wiley, Chichester

Fellendorf M, Vortisch P (2010) Microscopic traffic flow simulator VISSIM. In: Barceló J (ed) Fundamentals of traffic simulation. Springer, New York, pp 63–93

Hahne E (1991) Round-Robin scheduling for max-fin fairness in data networks. IEEE J Select Areas Commun 9(7):1024–1039

Holland JH (1992) Adaptation in natural and artificial systems: an introductory analysis with applications to biology, control, and artificial intelligence. MIT Press, Cambridge

Lea D (1999) Concurrent programming in Java: design principles and patterns. Addison-Wesley Longman Publishing Co., Inc., Boston

Lindholm T, Yellin F (1999) Java virtual machine specification. Addison-Wesley Longman Publishing Co., Inc., Boston

MATSim (2009) Multi agent transportation simulation toolkit. Webpage: http://www.matsim.org, August 2009

Meister K, Balmer M, Axhausen KW, Nagel K (2006) planomat: a comprehensive scheduler for a large-scale multi-agent transportation simulation. Paper presented at the 11th international conference on travel behaviour research, Kyoto, August 2006

Meister K, Balmer M, Ciari F, Horni A, Rieser M, Waraich RA, Axhausen KW (2010) Large-scale agent-based travel demand optimization applied to Switzerland, including mode choice. In: The 12th world conference on transportation research, Lisbon, July 2010

Nagel K, Rickert M (2001) Parallel implementation of the TRANSIMS micro-simulation. Parallel Comput 27(12):1611–1639

NAVTEQ (2009) NAVTEQ. Webpage: http://www.navteq.com. June 2009

Nökel K, Schmidt M (2002) Parallel DYNEMO: meso-scopic traffic flow simulation on large networks. Netw Spatial Econ 2(4):387–403

OMNeT++ (2009) OMNeT++. Webpage: http://www.omnetpp.org

Raney B, Cetin N, Völlmy A, Vrtic M, Axhausen K, Nagel K (2003) An agent-based microsimulation model of Swiss travel: first results. Netw Spatial Econ 3(1):23–41

Snir M, Otto S, Walker D, Dongarra J, Huss-Lederman S (1995) MPI: the complete reference. MIT Press, Cambridge, MA

Strippgen D, Nagel K (2009) Using common graphics hardware formulti-agent traffic simulation with cuda. In: Simutools'09: proceedings of the 2nd international conference on simulation tools and techniques. ICST, Brussels, pp 1–8

Waraich RA, Galus MD, Dobler C, Balmer M, Andersson G, Axhausen KW (2009) Plug-in hybrid electric vehicles and smart grid: investigations based on a micro-simulation. In: Proceedings of the 12th international conference on travel behaviour research (IATBR), Jaipur, December 2009

Part IV
Remote Sensing

Chapter 10
Recent Advances on 2D and 3D Change Detection in Urban Environments from Remote Sensing Data

Konstantinos Karantzalos

Abstract Urban environments are dynamic and complex by nature, evolve over time, and constitute the key elements for currently emerging environmental and engineering applications in global, regional, and local spatial scales. Their modeling and monitoring is a mature research field that has been extensively studied from the remote sensing, computer vision, and geography scientific communities. In this chapter, a comprehensive survey of the recent advances in 2D and 3D change detection and modeling is presented. The analysis is structured around the main change detection components including the properties of the change detection targets and end products; the characteristics of the remote sensing data; the initial radiometric, atmospheric, and geometric corrections; the core unsupervised and supervised methodologies and the urban object extraction and reconstruction algorithms. Experimental results from the application of unsupervised and supervised methods for change detection and building detection are given along with their qualitative and quantitative evaluation. Based on the current status and state of the art, the validation reports of relevant studies, and the special challenges of each detection component separately, the present study highlights certain issues and insights that may be applicable for future research and development, including *(i)* the need for novel multimodal computational frameworks and *(ii)* for efficient unsupervised techniques able to identify "from-to" change trajectories, along with the importance *(iii)* of automation, *(iv)* of open data policies, and *(v)* of innovative basic research in the core of the change detection mechanisms.

Keywords Urban growth • Building detection • (Un)supervised classification • Monitoring • Modeling

K. Karantzalos (✉)
Remote Sensing Laboratory, National Technical University of Athens, Heroon Polytechniou 9, Zographos 15780, Greece
e-mail: karank@central.ntua.gr; http://users.ntua.gr/karank/

M. Helbich et al. (eds.), *Computational Approaches for Urban Environments*, Geotechnologies and the Environment 13, DOI 10.1007/978-3-319-11469-9_10

10.1 Introduction

Understanding and modeling in detail the dynamic 3D urban scenes can enable effectively urban environment sustainability. In particular, the efficient spatiotemporal urban monitoring in large scale is critical in various engineering, civilian, and military applications such as urban and rural planning, mapping, and updating geographic information systems, housing value, population estimation, surveillance, transportation, archeology, architecture, augmented reality, 3D visualization, virtual tourism, location-based services, navigation, wireless telecommunications, disaster management, and noise, heat, and exhaust spreading simulations. All these subjects are actively discussed in the geography, geoscience, and computer vision scientific communities both in academia and industry. Organizations like Google and Microsoft are trying and seeking to include extensively up-to-date 2D and 3D urban models in their products (Microsoft Virtual Earth and Google Earth).

The prohibitively high cost of generating manually such 2D and 3D dynamic models/maps explains the urgent need towards automatic approaches, especially when one considers modeling and monitoring time-varying events within the complex urban areas. In addition, there is an emergence for algorithms that provide generic solutions through the automated and concurrent processing of all available data like panchromatic, multispectral, hyperspectral, radar, and digital elevation data. However, processing multimodal data is not straightforward (He et al. 2011b; Longbotham et al. 2012; Berger et al. 2013) and requires novel, sophisticated algorithms that on the one hand can accept as an input multiple data from different sensors, data with different dimensions, and data with different geometric, spatial, and spectral properties and on the other hand can automatically register and process them.

Furthermore, despite the important research activity during the last decades, there are, still, important challenges towards the development of automated and accurate change detection algorithms (Lu et al. 2011c; Longbotham et al. 2012; Hussain et al. 2013). It has been generally agreed and is verified by the quantitative evaluation of recent research efforts that there isn't, still, any specific single, generic, automated methodology that is appropriate for all applications and/or all the case studies. The maximum accuracy of the 2010 multimodal change detection contest was just over 70 % (Longbotham et al. 2012). This is in accordance and closely related with Wilkinson's earlier report on the minor improvement during the last decade on the performance of classification algorithms (Wilkinson 2005). Even the latest machine learning techniques haven't contributed much on the remote sensing data classification problem. Standard approaches usually result in similar levels of accuracy with the newer more advanced ones. Therefore, several aspects of the change detection process, towards the efficient 2D and 3D updating of geospatial databases, possess emerging challenges.

The aforementioned need for more intensive research and development is, furthermore, boosted by the available and increasing petabyte archives of geospatial (big) data. Along with the increasing volume and reliability of real-time sensor observations, the need for high performance, big geospatial data processing, and analysis systems, which are able to model and simulate a geospatially enabled content, is greater than ever. Both in global and local scales, the vision towards a global human settlement layer (Craglia et al. 2012) with multiscale volumetric information describing in detail our planet in 4D (spatial dimensions plus time) requires generic, automated, efficient, and accurate new technologies.

Towards this end, a significant amount of research is still, nowadays, focusing on the design, development, and validation of novel computational change detection procedures. Among them, those concentrating on forest change detection are holding the biggest share (Hansen and Loveland 2012) due to the importance on climate change, biodiversity and the suitability of past and current satellite remote sensing sensors, their spatial and spectral properties, and operational monitoring algorithms (Phelps et al. 2013). Cropland, vegetation, and urban environments are the other change detection and monitoring targets that benefit more from the current and upcoming very high-spatial-resolution, very high-spectral-resolution, and very high-temporal-resolution remote sensing data.

This chapter is focusing on the recent advances on change detection computational methods for monitoring urban environments from satellite remote sensing data, with emphasis on the most recent advances in the domain. In order to study change detection methodologies, their main key components are identified and studied independently. The most recent techniques are presented in a systematic fashion. In particular, publications during the last 6 years are reviewed and recent research efforts are classified in certain categories regarding the type of the algorithm employed, the type of geospatial data used, and the type of the detection target. Earlier reviews (Lu et al. 2004, 2011c; Radke et al. 2005) give a detailed summary of the efforts during the last decades (Singh 1989). Moreover, the focus here is on change detection methods applied to medium-, high-, and very high-resolution data, since for urban environments smaller scales do not provide spatial products with suitable accuracies for local geospatial database update. In the following sections several aspects of the change detection targets, end products, the relevant remote sensing data, preprocessing, and core change detection algorithms are detailed and discussed.

10.2 Change Detection Targets and End Products

The main detection targets in urban environments are land cover, land use, urban growth, impervious surfaces, man-made objects, buildings, and roads. With the same order one can indicate a suitable spatial accuracy from regional to more local scales. Therefore, each query for monitoring specific phenomenon, terrain classes, or terrain object poses specific constrains that describe the end product of the procedure. Which is the detection target and the desired location and size, which is the desired time period, and which is the required spatial accuracy?

Table 10.1 Change detection and monitoring targets

Land cover/ land use	Urban growth	Impervious surfaces and man-made objects	Buildings	Slum/ damaged buildings
Boulila et al. (2011)	Bagan and Yamagata (2012)	Chini et al. (2008)	Benedek et al. (2012)	Brunner et al. (2010)
Chen et al. (2012)	Michishita et al. (2012)	Leinenkugel et al. (2011)	Bouziani et al. (2010)	Dong and Shan (2013)
Chen et al. (2013)	Shafizadeh-Moghadam and Helbich (2013)	Lu et al. (2011a)	Champion et al. (2010)	Kit and Lüdeke (2013)
Del Frate et al. (2008)	Taubenböck et al. (2012)	Weng (2012)	Crispell et al. (2012)	Klonus et al. (2012)
Deng et al. (2009a)	Villa (2012)	Xian and Homer (2010)	Doxani et al. (2012)	Wang and Jin (2012)
Dos Santos Silva et al. (2008)	Zhang and Seto (2011)		Hebel et al. (2013)	
Hansen et al. (2014)			Du et al. (2012)	
He et al. (2011a)			Poulain et al. (2011)	
Hu and Zhang (2013)			Taneja et al. (2013)	
Lu et al. (2011c)			Tang et al. (2013)	
Schneider (2012)			Tian et al. (2013)	
Sexton et al. (2013)				
Sjahputera et al. (2011)				
Xian et al. (2009)				
Zanotta and Haertel (2012)				
Zhang et al. (2013)				

The answer to the aforementioned questions indicates various parameters and sorts significantly the required approaches and algorithms that should be employed. Table 10.1 summarizes the recent research activity on change detection and monitoring of urban environments according to the desired product and target that each recent study has been focusing on. Land cover/land use, urbanization, impervious surfaces and man-made objects, building, and slum or damaged buildings compromise the five dominant categories.

These categories are not referring to different terrain objects but rather on a hierarchical terrain object relation like in most model-based descriptions (ontologies, grammars, *etc.*). This categorization depicts both the different end-product requirements like their spatial scale and the type of urban objects/terrain classes are required for detection and monitoring. Along with the different specifications of the currently available remote sensing data, this is, actually, the main reason why these categories seem to form different groups in the literature including data, methods, and validation practices. In particular, the biggest share are holding the efforts which focus either on land cover/land use or either on building change detection.

On the one hand, the opening of the United States Geological Survey's Landsat data archive (Woodcock et al. 2008; Wulder et al. 2012) and the newly launched Landsat Data Continuity Mission (LDCM) enabled the easy access to a record of historical data and related studies on monitoring mainly land-cover/land-use changes, updating land national cover maps, and detecting the spatiotemporal dynamics, the evolution of land-use change, and landscape patterns. With this increased data availability and the increasing open data policies both in the USA and EU, similar studies can correspond to the current demand for improving the capacity to mass process big data and enable the efficient spatiotemporal modeling and monitoring.

On the other hand, a significant amount of research was focused on local scales and building change detection. Novel promising automated algorithms were developed which allow one to automatically detect, capture, analyze, and model efficiently single buildings in dynamic urban scenes. Mainly model-based approaches, like parametric, structural, statistical, procedural, and grammar-based ones, have been design to detect, both in 2D and in 3D, buildings and spatiotemporal changes. Google Earth, Virtual Earth, and other government applications and databases must be/remain updated, and therefore, the motivation on automated algorithms instead of costly manual digitization procedures is, still, high.

Apart from the requirements regarding the multiple properties of the desired product and detection target, the change detection procedure is affected by a number of parameters including spatial, spectral, thematic, and temporal constraints; radiometric, atmospheric, and geometric properties; and soil moisture conditions. Therefore, a sophisticated methodology should be able to address in a preprocessing step all the various constrains and conditions that will enable an effective and accurate core spatiotemporal analysis. In the following two subsections, certain important aspects regarding the multiple properties of the remote sensing data are detailed along with a brief description on the required preprocessing steps.

10.3 Remote Sensing Data

During the last decades important technological advances in optics, photonics, electronics, and nanotechnology allowed the development of frame and push-broom sensor with high spatial and spectral resolution. New satellite mission have been scheduled continuously and gradually remote sensing data of higher quality from either passive or active sensors will be available. However, today data with high spatial and spectral resolution is either for military or commercial use. In Table 10.2, a summary of the currently available satellite remote sensing sensors,

Table 10.2 Summary of the currently available satellite remote sensing data, which have been employed in recent studies, and their major specifications and cost

	Resolution			Cost (*archive data*)		
Satellite	Spectral	Spatial	Temporal	Per scene (*300 km²*)	Per km²	Comments
LANDSAT-8	VNIR, SWIR and LWIR	Pan: 15 m; MS: 30 m	16 days	–	–	*Free, open data*
		Thermal: 100 m				
SENTINEL 1-6	Radar, Optical, etc	>10 m	2/5 days	–	–	*Free, open data*
SPOT-5	Pan, R,G,NIR	Pan: 5 m, MS: 10 m	3 days	225€ (Pan and MS)	0,75€	
ALOS	Pan, R,G,B,NIR	Pan: 2,5 m, MS: 10 m	46 days	291€ (Pan and 4MS)	0,97€	*Low temporal resolution*
CARTOSAT-1	1 Pan	2.5 m Pan	5 days	1.440€ (PanA and PanF)	4,80€	*Stereo pair, tasking*
SPOT-6/7	Pan, R,G,B,NIR	Pan: 1.5 m, MS: 6 m	1 day	1.140€ (Pan and 4MS)	3,80€	*Stereo pair, tasking*
FORMOSAT-2	Pan, R,G,B,NIR	Pan: 2 m, MS: 8 m	3 days	990€ (Pan and 4MS)	3,30€	
RAPIDEYE	R,G,B,Red.,NIR	MS: 5 m	1 day	285€ (5xMS)	0,90€	*with red edge*
IKONOS	Pan, R,G,B,NIR	Pan: 0,8 m, MS: 3,2 m	3 days	3.900€ (Pan and 4MS)	13,00€	
QUICKBIRD	Pan, R,G,B,NIR	Pan: 0,65 m; MS: 2,5 m	3 days	4.500€ (Pan and 4MS)	15,00€	
PLEIADES	Pan, R,G,B,NIR	Pan: 0,5 m; MS: 2 m	1 day	3.600€ (Pan and 4MS)	12,00€	
GEOEYE-1	Pan, R,G,B,NIR	Pan: 0,5 m; MS: 2 m	2–8 days	4.800€ (Pan and 4MS)	16,00€	
WORLDVIEW-2	Pan, 8xMS	Pan: 0,46 m; MS: 1,84 m	1,1 days	3.600€ (Pan)	12,00€	
				5.400€ (Pan and 8MS)	18,00€	
WORLDVIEW-3	Pan, 8xMS, 8xSWIR, 12xCAVIS	Pan: 0,31 m; MS: 1,24 m, SWIR: 3,70 m, CAVIS: 30,00 m	<1 day	6.600€ (Pan)	22,00€	*Highest spatial & spectral resolution*
				9.600€ (Pan and 8MS)	32,00€	
				>12.000€ (All bands)	>40,00€	
ALOS	*PALSAR*	10–100 m	46 days	600€ (FBS fine)	0,12€	*Low temporal resolution*
	SAR-L					
COSMO-SkyMed	*Spotlight-2*	1 m	1 day	4700€ (10 km × 10 km)	47,00€	
TerraSAR-X	HR SpotLight	1 m	1 day	3.000€ (10 km × 5 km)	60,00€	

which were employed in recent change detection studies, is reported along with the major data specifications and cost. Apart from their spatial, spectral, and temporal resolution, their cost is referring to archive data (apart from the Cartosat-1 case) and is associated with the specific product/mode which offers the highest spatial resolution. The cost refers to list prices (e-geos 2013; GeoStore 2013) and has been estimated for the minimum ("per scene") order and per square kilometer (km^2) in order to ease the comparison. It is obvious that when moving from the medium- and high-spatial-resolution products to the very high-resolution ones, the cost per square kilometer increases significantly *i.e.*, from about 1€ per km^2 to about 20€. The high-spatial-resolution SAR satellite sensors are, also, offering costly products, similar with or higher than the optical ones. In addition, it should be noted that as we are moving from smaller to larger spatial scales, the number of images required to cover the same area increases significantly. Therefore, the cost for delivering change detection geospatial products increases exponentially as we are moving from regional land cover/use or urban growth studies to local building change detection and cadastral map updating.

In Table 10.3, recent change detection approaches are classified according to the type of the remote sensing data used in each recent study. Medium- to high-resolution optical data, radar data, and multimodal data (Fig. 10.1) are holding the biggest share among the recent change detection research activity. However, 3D data (satellite or airborne) and vector data from existing geodatabases are gaining increasing attention for spatiotemporal monitoring in local scales. In region scales, the research activity, as has been already mentioned, has been empowered from the increasing US and EU open data policies. Moreover, new open products which include basic but necessary preprocessing procedures will boost more research and development for quantifying global and regional transitions given the changing state of global/regional climate, biodiversity, food, and other critical environmental/ecosystem issues. Web-enabled Landsat data is an example, where large volumes of preprocessed Landsat 7 Enhanced Thematic Mapper Plus data are operationally offered for easing the mapping procedure of land-cover extent and change (Hansen et al. 2014).

10.4 Data Preprocessing

Certain factors, such as the radiometric calibration and normalization between multitemporal datasets, the quality of atmospheric corrections, the quality of data registration, the complexity of the landscape and topography under investigation, the analyst's skill and experience, and last but not least, the selected change detection algorithm, are directly associated with quality of the change detection product. The initial preprocessing stage, which current efforts try to standardize (Yang and Lo 2000; Chander et al. 2009; Hansen et al. 2014), addresses important issues regarding the radiometric, atmospheric, and geometric corrections in the available datasets transforming them from raw to geospatial ready-for-analysis data. However, there

Table 10.3 Remote sensing data and recent change detection and monitoring research studies

Optical satellite data		3D data	Vector data	Radar data	Multimodal data
Medium to high resolution (LANDSAT, *etc.*)	Very high resolution (IKONOS, *etc.*)	ALS, LiDAR, DEM, DSM,	Geodatabase, Cadastral, *etc.*	Satellite, airborne	Optical, radar, DSM, *etc.*
Bagan and Yamagata (2012)	Bouziani et al. (2010)	Boehm et al. (2013)	Poulain et al. (2011)	Ahmad and Amin (2013)	Berger et al. (2013)
Deng et al. (2009a)	Doxani et al. (2012)	Champion et al. (2010)	Gonzalez-Aguilera et al. (2013)	Aiazzi et al. (2013)	Bouziani et al. (2010)
Du et al. (2012)	Du et al. (2013)	Hebel et al. (2013)	Bouziani et al. (2010)	Bovolo et al. (2013)	Deng et al. (2009b)
Hansen et al. (2014)	Falco et al. (2013)	James et al. (2012)	Taneja et al. (2013)	Celik and Ma (2011)	Desclee et al. (2013)
He et al. (2011a)	Hao et al. (2014)	Sesnie et al. (2008)		Chatelain et al. (2008)	Leinenkugel et al. (2011)
Irons and Loveland (2013)	Im et al. (2007)	Tian et al. (2013)		Del Frate et al. (2008)	Longbotham et al. (2012)
Michishita et al. (2012)	Im et al. (2008)			Giustarini et al. (2013)	Lu et al. (2011a)
Shafizadeh-Moghadam and Helbich (2013)	Kit and Lüdeke (2013)			Gong et al. (2012)	Lu et al. (2008)
Schneider (2012)				Ma et al. (2012)	Poulain et al. (2011)
Sexton et al. (2013)	Volpi et al. (2013)			Marino et al. (2013)	Taubenböck et al. (2012)
Taubenböck et al. (2012) and Tian et al. (2013)	Liu et al. (2012)			Moser and Serpico (2009)	Wulder et al. (2008)
Villa et al. (2012)				Pratola et al. (2013)	
Xian and Homer (2010)				Wang et al. (2013)	
Zhang et al. (2013)				Yousif and Ban (2013)	

are still a number of challenges that should be addressed (Villa et al. 2012) in order to exploit raw big remote sensing data and transform them to big geospatial reflectance surfaces. The most important is automation. In the following two subsections, the main preprocessing procedures are briefly described and discussed. It should be noted that for Landsat datasets, certain protocols have been proposed

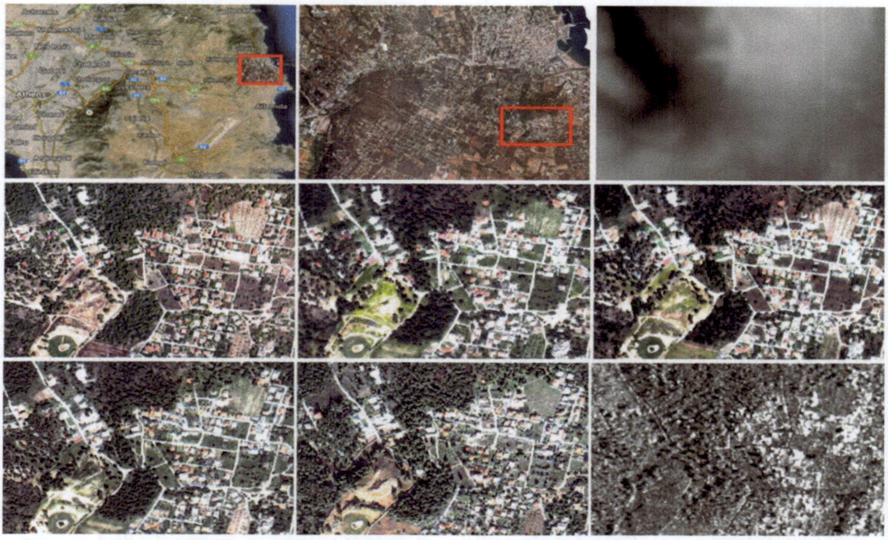

Fig. 10.1 A multimodal, multitemporal remote sensing dataset covering a 25 km^2 region in the East Prefecture of Attica, Greece. The corresponding DEM is shown in the *upper right* image. *Middle row*: An aerial orthomosaic acquired in 2010 (*left*), a WorldView-2 image acquired in 2011 (*middle*) and a WorldView-2 image acquired in 2010 (*right*). *Bottom row*: A QuickBird image acquired in 2009 (*left*), a QuickBird image acquired in 2007 (*middle*) and a TerraSAR-X image acquired in 2013 (*right*)

and widely adopted (Han et al. 2007; Vicente-Serrano et al. 2008) including *(i)* geometric correction, *(ii)* calibration of the satellite signal to obtain "top of atmosphere" radiance, *(iii)* atmospheric correction to estimate surface reflectance, *(iv)* topographic correction, and *(v)* relative radiometric normalization between images obtained at different dates. The latter is not required in cases where, e.g., an absolute physical correction model has been employed. The radiometric processing should be the initial one; however, this is not always the case, since, for example, the former Landsat datasets in Europe were available already and geometrically corrected (*e.g.,* level 1 system corrected from the European Space Agency).

10.4.1 Radiometric and Atmospheric Correction and Calibration

The main goal of radiometric and atmospheric corrections is to model the various sources of noise which affect the information captured by the sensor, making it difficult to differentiate the surface signal from any type of noise. Despite the efforts that are persistently made to calibrate satellite sensors towards correcting lifetime radiometric trends and minimize the effect from atmospheric noise, certain studies have shown that the application of accurate sensor calibrations and complex atmospheric corrections does not guarantee the multitemporal homogeneity of (*e.g.*, Landsat) datasets since complete atmospheric properties are difficult to quantify and simplifications are commonly assumed (Han et al. 2007). Therefore, a cross-calibration between the data stack and time series can address the problem.

Given a remote sensing optical dataset, the first step is to convert the capture radiance, the raw digital numbers to the "top of atmosphere" values (Chander et al. 2009; Villa et al. 2012, and the references therein). Then the second step is to model the upward and downward irradiance which is constrained by the gases absorption and the water molecules and aerosols scattering. Complex radiative transference models simulate the atmosphere and light interactions between the sun-to-terrain and terrain-to-sensor trajectories. Although, such an atmospheric correction can account for signal attenuation and restore in some extent the intercomparability of satellite images taken on different dates, "top of atmosphere" values are widely used directly for inventory and ecosystem studies or in procedures that are based on post-classification change detection approaches. However, recent studies indicate that cross-calibration and atmospheric corrections are required prior to relative normalization since certain remote sensing products and accurate biophysical parameters like vegetation indices cannot be calculated (Vicente-Serrano et al. 2008).

The third step is to model the modified illumination conditions due to the scene topography. In order to simplify this extremely complex setting, in practice one concentrates on the shaded areas which deliver less than expected reflectance and on the sunny areas which deliver more than expected. Then, usually, we assume a Lambertian terrain behavior or model non-Lambertian effects. Last but not least, a relative radiometric normalization should be performed between the images of the time series/dataset, in case where an absolute physical correction model was not employed. The normalization process is based on a linear comparison between the images which have been acquired on different dates. To this end, linear regression or other automated techniques like the pseudo-invariant feature regression has given promising results (Vicente-Serrano et al. 2008) while indicating that the relative radiometric normalization is an absolutely essential step to ensure high levels of homogeneity between the images of the dataset.

10.4.2 Geometric Corrections and Data Registration

Once the radiometric and atmospheric calibration has been performed, the next step is to register, co-register, and geo-reference the available data. Early studies (Dai and Khorram 1998; Roy 2000; Bovolo et al. 2009) have underlined the important problems which occurred from data misregistration and how significantly the change detection product is affected. Therefore, in order to develop operational detection systems, the registration problem must be addressed with an optimal way (Klaric et al. 2013). In particular, this is a common challenge in most computer vision, medical imaging, remote sensing, and robotics applications, and this is the reason why image registration, segmentation, and object detection hold the biggest share in modern image analysis and computer vision research and development (Sotiras et al. 2013).

Speaking briefly, the image registration task involves three main components: a transformation model, an objective function, and an optimization method. The success of the procedure depends naturally on the transformation model and the objective function. The dependency on the optimization process follows from the fact that image registration is inherently an ill-posed problem. Actually, in almost all realistic scenarios and computer vision applications, the registration is ill-posed according to Hadamard's definition of well-posed problems. Therefore, devising each component of the registration algorithm in such way that the requirements (regarding accuracy, automation, speed, *etc.*) are met is a demanding and challenging process (Eastman et al. 2007; Le Moigne et al. 2011; Sotiras et al. 2013).

The intensive research on invariant feature descriptors (Lowe 2004) empowered the automation in the feature detection (points, lines, regions, templates, *etc.*) procedure. Along with the model fitting approaches, through iterative non-deterministic algorithms, an optimal set of the selected mathematical model parameters (*i.e.*, transformation, deformation, *etc.*) is detected excluding outliers. Area-based methods, mutual information methods, and descriptor-based algorithms restore data deformations and through a resampling data are warped to the reference. Furthermore, since the effective modeling requires rich spatial, spectral, and temporal observations over the structured environment recent approaches fuse data from various sensors, *i.e.*, multimodal data (Fig. 10.1). The various sensors include frame and push-broom cameras and multispectral, hyperspectral, and thermal cameras, while the various platforms include satellite, airborne, UAV, and ground systems.

In multimodal data registration (De Nigris et al. 2012; He et al. 2011b), mutual information techniques have become a standard reference, mainly in medical imaging (Legg et al. 2013; Wachinger and Navab 2012; Sotiras et al. 2013). However, being an area-based technique, the mutual information process possesses natural limitations. To address them, a combination with other, preferably feature-based, methods have gain high robustness and reliability. To speed up the computation, scale space representations (Tzotsos et al. 2014) are employed along with fast optimization algorithms. However, when data have significant rotation and/or scaling differences, these methods either fail or become extremely time expensive. Future development on addressing the multimodal data challenges may concentrate more on feature-based methods, where appropriate invariant and modality-insensitive features (Heinrich et al. 2012) can provide the reliable and adequate volume of features for a generic and automated multimodal data registration.

To sum up, the described radiometric and geometric corrections between all the available data of a given time series transform raw data to valuable "ready-for-analysis" geospatial datasets and ensure an optimal exploitation from the following, in the processing chain, core change detection algorithms.

10.5 Unsupervised Change Detection Methods

Unsupervised approaches are based on automated computational frameworks that usually produce binary maps indicating whether a change has occurred or not. Therefore, standard unsupervised change detection techniques are not usually based on a detailed analysis of the concept of change but rather compare two or more images by assuming that their radiometric properties are similar, excluding real change detection phenomenon (Bruzzone and Bovolo 2013). However, this assumption in realist scenarios is not satisfied, especially, in local scales. In particular, the captured complexity of terrain objects, with different spectral behaviors at different dates and environmental conditions, is significant especially in very high-resolution data. That is the main reason why although unsupervised change detection methods have validated so far, their effectiveness on medium- to high-resolution data and usually under pixel-based image analysis, when the spatial resolution reaches submeter accuracies, they become less accurate (Hussain et al. 2013).

Unsupervised approaches have accumulated a significant amount of research efforts since i) on the one hand, they are more attractive from an operational point of view, allowing automation without the need for manual collection of reference data/samples and ii) on the other hand, they can possible address the aforementioned challenges and move towards a semantic change labeling by identifying the exact land-cover transition.

In Table 10.4, a summary of the recent unsupervised change detection studies is presented. Recent methods are classified according to the core technique on which

Table 10.4 Summary of recent change detection studies classified according to their unsupervised or supervised nature and the main technique that they were based on

	Methods	
Employed techniques	Unsupervised	Supervised
Direct comparison, transformations, similarity (ratios, kernels, change vector analysis, *etc.*)	Bovolo et al. (2012), Canty and Nielsen (2008), Celik (2009), Chen et al. (2011), Dalla Mura et al. (2008), Renza et al. (2013), Demir et al. (2013), Gueguen et al. (2011), Marchesi and Bruzzone (2009), Marpu et al. (2011), and Volpi et al. (2012)	Brunner et al. (2010), Deng et al. (2008), and Falco et al. (2013)
Multiscale analysis (wavelets, *etc.*)	Bovolo et al. (2013), Celik and Ma (2010), Celik and Ma (2011), Dalla Mura et al. (2008), and Moser et al. (2011)	Bovolo et al. (2009)
Fuzzy theory	Ling et al. (2011), Luo and Li (2011), and Robin et al. (2010)	

(continued)

Table 10.4 (continued)

Clustering, Bayesian classifier	Aiazzi et al. (2013), Celik (2010), Ghosh et al. (2011), and Salmon et al. (2011)	
Spectral mixture analysis (Gaussian, *etc.*), unmixing	Yetgin (2012)	Michishita et al. (2012)
Active contours, level sets	Bazi et al. (2010) and Hao et al. (2014)	Celik and Ma (2011)
Support vector machines, neural networks, learning	Bovolo et al. (2008)	Bovolo et al. (2010), Chini et al. (2008), Camps-Valls et al. (2008), Habib et al. (2009), Pacifici and Del Frate (2010), Demir et al. (2012), Pagot and Pesaresi (2008), Taneja et al. (2013), and Volpi et al. (2013)
MRFs	Ghosh et al. (2013), Moser and Serpico (2009), Moser et al. (2011), and Wang et al. (2013)	Fernandez-Prieto and Marconcini (2011)
Data fusion	Du et al. (2012), Moser and Serpico (2009), Ma et al. (2012), Gong et al. (2012), and Du et al. (2013)	
Post-classification comparison		Del Frate et al. (2008), Dewan and Yamaguchi (2009), Abd El-Kawy et al. (2011), Knudby et al. (2010), Sexton et al. (2013)
	Methods	
Employed techniques	Unsupervised	Supervised
Object-based	Bouziani et al. (2010)	Berberoglu and Akin (2009), Brunner et al. (2010), Doxani et al. (2012), Gamanya et al. (2009), Hebel et al. (2013), Huo et al. (2010), Lu et al. (2011b), Xian and Homer (2010), and Zhou et al. (2009)
Data mining		Boulila et al. (2011), Dos Santos Silva et al. (2008), Schneider (2012), and Vieira et al. (2012)

they were mainly based on. The majority of recent studies is based on standard direct comparisons, data transformations, data fusion, multiscale analysis, and clustering. Most of the recent unsupervised methods are, also, pixel-based approaches and focus on the pixel-by-pixel analysis of the multispectral multitemporal data.

More specifically, they calculate after a certain computation (like a transformation, a spectral analysis, *etc.*) the magnitude of change vectors and apply a thresholding technique in order to detect possible changes.

An important number of approaches are based on ratios, kernels, change vector analysis, and indices (Bovolo et al. 2012; Canty and Nielsen 2008; Celik 2009; Chen et al. 2011; Dalla Mura et al. 2008; Renza et al. 2013; Demir et al. 2013; Gueguen et al. 2011; Marchesi and Bruzzone 2009; Marpu et al. 2011; Volpi et al. 2012). Other efforts are based on multiscale analysis like wavelets (Bovolo et al. 2013; Celik and Ma 2010, 2011; Dalla Mura et al. 2008; Moser et al. 2011), fuzzy theory (Ling et al. 2011; Luo and Li 2011; Robin et al. 2010), clustering and MRFs (Aiazzi et al. 2013; Celik 2010; Ghosh et al. 2011, 2013; Salmon et al. 2011; Moser and Serpico 2009; Moser et al. 2011; Wang et al. 2013).

Spectral mixture analysis (Yetgin 2012), level sets (Bazi et al. 2010; Hao et al. 2014), and data fusion approaches (Du et al. 2012, 2013; Moser and Serpico 2009; Ma et al. 2012; Gong et al. 2012) are holding an important share also. Moreover, and despite the fact that their core employed algorithms are supervised, recent proposed automated studies are based on object-based techniques (Bouziani et al. 2010), semi-supervised support vectors (Bovolo et al. 2008), and neural networks (Pacifici and Del Frate 2010).

In addition, among the recent unsupervised techniques, a clear computational advantage possess the ones who can address the dependence between spatially adjacent image neighbors either by standard texture or morphological measures or either by clustering, Markov random fields, Bayesian networks, and context-sensitive analysis. Such frameworks (Celik 2009, 2010; Ghosh et al. 2013; Volpi et al. 2012; Bruzzone and Bovolo 2013) can cope more efficiently with the complexity pictured in very high-resolution data.

Promising experimentalresults after the application of an unsupervised change detection procedure, which is based on the iterative reweighting multivariate alteration detection (IR-MAD) algorithm (Nielsen 2007; Canty and Nielsen 2008), are presented in Figs. 10.2, 10.3, and 10.4. Based on the invariant properties of the standard MAD transform where we assume that the orthogonal differences contain the maximum information in all spectral bands, an iterative reweighting procedure involving no-change probabilities can account for the efficient detection of changes. In the upper row of Fig. 10.2, the QuickBird image acquired in 2007 is shown, while the corresponding QuickBird image acquired in 2009 is presented in the middle row. The detected changes after the application of the IR-MAD and post-processing morphological algorithms are shown in the bottom. All changes represent the new buildings that were constructed in the region after 2007. The detected changes/buildings are overlaid in the 2009 image and shown with a red color. The ground truth data are shown with the same manner in green.

In Fig. 10.3, the IR-MAD output and the corresponding binary image after a thresholding are shown in the upper row. The detected changes (new buildings) after the application of a morphological post-processing procedure and the corresponding ground truth data are shown in the bottom. All the changes (all new buildings) have been successfully detected by the unsupervised procedure. The quantitative evaluation reported a low detection completeness of around 60 % and a high detection correctness of 95 %. This can be, also, observed in Fig. 10.4 where the

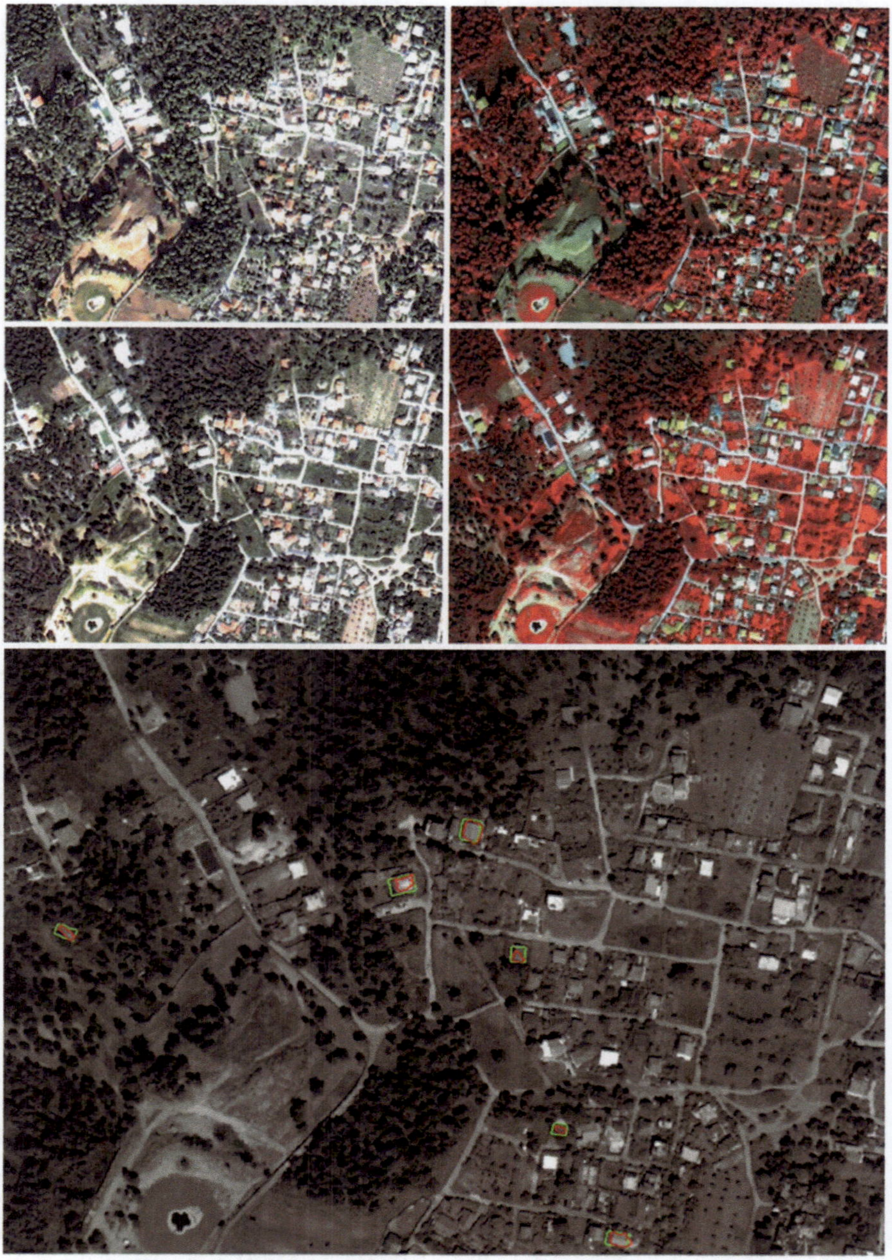

Fig. 10.2 Unsupervised change detection in multitemporal high resolution data. *Upper row*: The raw QuickBird image, acquired in 2007, in RGB321 (*left*) and R432 (*right*). *Middle row*: The raw QuickBird image, acquired in 2009, in RGB321 (*left*) and R432 (*right*). *Bottom row*: The detected changes (they are all new buildings), overlaid in the 2009 image, are shown with a *red color*. Ground truth data are shown in *green*

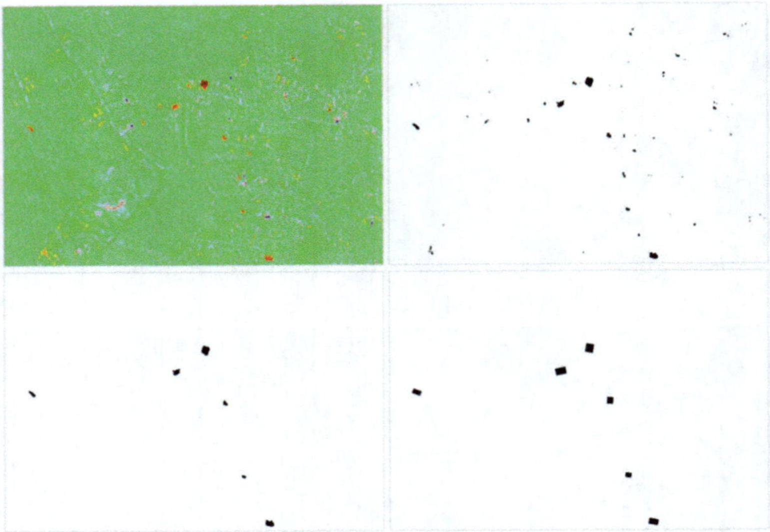

Fig. 10.3 The detected under an unsupervised manner changes (buildings) and the corresponding ground truth data. *Upper row*: A map with the possible changes after the application of the regularized iteratively reweighted MAD algorithm (*left*) and after thresholding (*right*). *Bottom row*: The detected changes (buildings) after the application of morphological post-processing (*left*) and the ground truth (*right*). All the changes (new buildings) have been successfully detected. The quantitative evaluation reported a low detection completeness of around 60 % and a much higher detection correctness of 95 %

detected changes have been associated with the corresponding DEM. The detected new buildings in 3D are shown in the upper part of Fig. 10.4, while the 3D buildings from the ground truth data are shown in the bottom.

10.6 Supervised Change Detection Methods

The supervised classification approaches traditionally are based on the detection of changes from a post-classification process (which is usually another classification). This process enables, also, the detection of actual class transitions instead of a binary "change or not change" product. However, errors from each step and each individual classification are propagating and are summed up at the end product. Moreover, collecting reliable, dense training sample sets can be difficult and time-consuming for certain cases (*e.g.*, historical data) or even unrealistic if one has to deal with extensive dense time series and multimodal data. In practice, however, the post-classification approach is, nowadays, the most standard one especially for global and regional scales, for land-cover, land-use, and urbanization monitoring.

In more local scales and for very high-resolution data, the standard supervised approach is an object-oriented one under an object-based image analysis framework

Fig. 10.4 The detected changes (new buildings) in 3D after the application of the unsupervised change detection procedure on QuickBird 2007 and 2009 satellite data. The detected new building in 3D are shown in the *upper* part, while the 3D buildings from the ground truth data are shown in the *bottom*. After a close inspection one can observe the low completeness and high correctness detection rates of the unsupervised change detection algorithm

(Blaschke 2010). Multilevel segmentation and supervised classification are the main key process there (Tzotsos et al. 2011, 2014). Recent object-based change detection approaches (Table 10.4) include scale space filtering and multivariate alteration detection (Doxani et al. 2012), the combination with multi-view airborne laser scanning data (Hebel et al. 2013), the detection of impervious surfaces (Xian and Homer 2010), shaded areas (Zhou et al. 2009), landslides (Lu et al. 2011b), and building damage assessment after earthquakes (Brunner et al. 2010). Another promising combination is to employ data mining techniques under an object-based framework in order to address big datasets and dense, long-term time series (Schneider 2012).

To this end, algorithms focusing on knowledge discovery in databases aim at extracting/mining nontrivial, implicit information from unstructured datasets. In particular, for geospatial datasets, data mining techniques are exploiting spatial and nonspatial properties in order to discover the desired knowledge/data. Dos Santos Silva et al. (2008) proposed a data mining framework which associates each change pattern to one predefined type of change by employing a decision-tree classifier to describe shapes found in land-use maps. Boulila et al. (2011) employed fuzzy sets and a data mining procedure to build predictions and decisions. Based on the imperfections related to the spatiotemporal mining process, they proposed an approach towards a more accurate and reliable information extraction of the spatiotemporal land-cover changes. Vieira et al. (2012) introduced a joint object-based data mining framework during which instead of the standard supervised classification step, a data mining algorithm was employed to generate decision trees from certain training sets. Schneider (2012) proposed an approach that exploits multi-seasonal information in dense time stacks of Landsat imagery comparing the performance of maximum likelihood, boosted decision trees, and support vector machines. Experimental results indicated only minor differences in the overall detection accuracy between boosted decision trees and support vector machines, while for band combinations across the entire dataset, both classifiers achieved similar accuracy and success rates.

This observation is in accordance with similar recent studies (Table 10.4) which employ powerful machine learning classifiers (Bovolo et al. 2010; Chini et al. 2008; Camps-Valls et al. 2008; Habib et al. 2009; Pacifici and Del Frate 2010; Demir et al. 2012; Pagot and Pesaresi 2008; Taneja et al. 2013; Volpi et al. 2013) for supervised change detection and indicate why they are so popular for remote sensing classification and change detection problems. However, machine learning algorithms are, usually, time-consuming and efforts towards a more computational efficient design and algorithmic optimization are required (Habib et al. 2009). Moreover, in local scales and very high-resolution data, including 3D or vector data, there is a lot of room for research and development in order to exploit the entire multimodal datasets. In particular, an important outcome from the recent 2012 multimodal remote sensing data contest (Berger et al. 2013) indicates that none of the submitted algorithms actually exploited in full synergy the entire available dataset, which included very high-resolution multispectral images (with a 50 cm spatial resolution for the panchromatic channel), very high-resolution radar data (TerraSAR-X), and LiDAR 3D data from the city of San Francisco, USA.

Therefore, in local scales, but not only, novel sophisticated, generic solutions should exploit the recent advances in 2D and 3D building extraction, reconstruction, and 3D city modeling which have gain a lot of attention during the last decade due to emerging new engineering applications including augmented reality, virtual tourism, location-based services, navigation, wireless telecommunications, disaster management, *etc.* In a similar manner like the post-classification change detection, monitoring the structured environment, both in 2D and 3D, can be based on the recent advancements on building extraction and reconstruction by, for instance, a similar direct comparison between two different dates. In the following subsection, recent building detection and modeling methods are briefly reviewed.

10.7 Computational Methods for 2D and 3D Building Extraction and Modeling

The accurate extraction and recognition of man-made objects from remote sensing data has been an important topic in remote sensing, photogrammetry, and computer vision for more than two decades. Urban object extraction is, still, an active research field, with the focus shifting to object detailed representation, the use of data from multiple sensors, and the design of novel generic algorithms.

Recent quantitative results from the ISPRS (WGIII/4) benchmark on urban object detection and 3D building reconstruction (Rottensteiner et al. 2013) indicated that, in 2D, buildings can be recognized and separated from the other terrain objects; however, there is room for improvement towards the detection of small building structures and the precise delineation of building boundaries.

In 3D, none of the methods was able to fully exploit the spatial accuracy of the available datasets. Therefore, although for visualization purposes 3D building reconstruction may be considered as a solved problem, for geospatial applications, and when geometrically and topologically accurate building models are required, novel efficient algorithms are, also, required. Moreover, regarding other urban object like trees, there is a lot of room, also, for research and development towards their efficient extraction and discrimination in complex urban regions.

In Table 10.5, a summary of recent building and road network extraction and reconstruction approaches are presented. They are classified in three categories,

Table 10.5 Summary of recent building and road network extraction and reconstruction approaches

2D building detection/extraction	Road network detection	3D building/city extraction and reconstruction
dos Santos Galvanin and Porfírio Dal Poz (2012)	Chaudhuri et al. (2012)	Crispell et al. (2012)
Benedek et al. (2010)	Das et al. (2011)	
Yang et al. (2013)	Gilles and Meyer (2010)	Garcia-Dorado et al. (2013)
Champion et al. (2010)	Poullis and You (2010)	Haala and Kada (2010)
	Unsalan and Sirmacek (2012)	Hane et al. (2013)
Katartzis and Sahli (2008)		Heo Joon et al. (2013)
Rutzinger et al. (2009)		
		Izadi and Saeedi (2012)
		Karantzalos and Paragios (2010)
Senaras et al. (2013)		Lafarge et al. (2010)
Karantzalos and Argialas (2009)		Loch-Dehbi and Plümer (2011)
Karantzalos and Paragios (2009)		Matei et al. (2008)
Senaras et al. (2013)		Rottensteiner et al. (2013)
Sirmacek and Unsalan (2011)		Rutzinger et al. (2009)
Stankov and He (2013)		Sampath and Jie Shan (2010)
Wegner et al. (2011)		Shaohui Sun and Salvaggio (2013)
Huang and Zhang (2012)		Sirmacek et al. (2012)
Zhou et al. (2009)		Sportouche et al. (2011)
		Tack et al. (2012)
		Taneja et al. (2013)
		Turlapaty et al. (2012)
		Zebedin et al. (2008)

i.e., 2D building detection/extraction, 3D building extraction/reconstruction, and road network detection. Buildings among the other man-made object dominate the research interest due to the aforementioned emerging applications that their efficient modeling can guarantee. In general, advanced methods are much likely to have a model-based structure and take into consideration the available intrinsic information such as color, texture, shape, and size and topological information as location and neighborhood. Novel expressive ways for the efficient modeling of urban terrain objects both in 2D and 3D have, already, received significant attention from the research community. From the standard generic, parametric, polyhedral, and structural models, novel ones have been, recently, proposed like the statistical ones, the geometric shape priors, and the procedural modeling with L-system grammar or other shape grammars (Rousson and Paragios 2008; Matei et al. 2008; Zebedin et al. 2008; Poullis and You 2010; Karantzalos and Paragios 2010; Simon et al. 2010). Furthermore, focusing on automation and efficiency, certain optimization algorithm have been developed for the model-based object extraction and reconstruction like discrete optimization algorithms, random Markov fields, and Markov chain Monte Carlo (Szeliski et al. 2008).

Focusing on 2D building boundaries detection, various techniques have been proposed (Champion et al. 2010; Katartzis and Sahli 2008; Senaras et al. 2013; Karantzalos and Argialas 2009; Stankov and He 2013; Wegner et al. 2011; Huang and Zhang 2012; Zhou et al. 2009), including unsupervised, semi-supervised, and supervised ones.

Even if the end product is in 2D, certain studies are based on 3D data (*e.g.*, DSM, LiDAR) (dos Santos Galvanin and Porfírio Dal Poz 2012; Yang et al. 2013; Rutzinger et al. 2009; Sampath and Shah 2010). In particular, buildings can be detected by calculating the difference between objects and terrain height. In case other data are, also, available, data fusion and classification approaches are employed. Other approaches are focusing on processing very high-resolution satellite data and certain of those have proposed algorithms for building detection from just a single aerial or satellite panchromatic image (Benedek et al. 2010; Karantzalos and Paragios 2009; Katartzis and Sahli 2008; Wegner et al. 2011; Huang and Zhang 2012).

The reported qualitative and quantitative validation indicates that the automated detection is hindered by certain factors. The major difficulty is to address scene complexity, as most urban scenes contain, usually, very rich information and various cues. These cues, which are mainly other artificial surfaces and man-made objects, possess important geometric and radiometric similarities with buildings. In addition, addressing occlusions, shadows, different perspectives and data quality issues constrain significantly the operational performance of the developed automated algorithms.

In 3D, a number of methods are based only on a digital surface model or a set of point clouds (Lafarge et al. 2010; Rutzinger et al. 2009; Sampath and Jie Shan 2010; Shaohui Sun and Salvaggio 2013; Sirmacek et al. 2012; Heo Joon et al. 2013). Other ones are exploiting multimodal data like optical and 3D data (Karantzalos and Paragios 2010) or optical and SAR data (Sportouche et al. 2011). Even in 3D there are efforts that are based on a single optical satellite image (Izadi and Saeedi

2012) or a single SAR one (Ferro et al. 2013). Image-based 3D reconstruction has been, also, demonstrated from user-contributed photos (Irschara et al. 2012) and multiangular optical images (Turlapaty et al. 2012).

Experimental results demonstrating the performance of supervised classification algorithms combined with post-classification procedures for building extraction from high-resolution satellite data are shown in Figs. 10.5 and 10.6.

Standard pixel-based classification algorithms like the minimum distance, maximum likelihood, and SVMs deliver detection outcomes with a low correctness rate. In particular, in the upper left part of Fig. 10.5, the raw Pleiades image acquired in 2013 is shown. The result from the minimum distance algorithm, showing only classes related to buildings, is shown in the upper right part of the figure. The quantitative evaluation reported a low detection overall quality of 62 % for the minimum distance algorithm. With the same ground samples, the maximum likelihood algorithm reported an overall detection rate of 67 % and the result is shown in the middle row (left). The SVM classifier scores higher with an overall detection quality of 74 % (middle right).

After post-classification procedures, including mathematical morphology, object radiometric and geometric properties calculation, and spatial relation analysis, the result from the supervised classification has been refined and its correctness rate is significantly improved. The detected buildings, based on the SVM output, which have been recognized and labeled by the algorithm, are shown in 2D with different colors in the bottom row of Fig. 10.5 (left). The detected buildings overlaid on the raw Pleiades image are, also, presented in the bottom right of Fig. 10.5. Moreover, the low detection rate can be observed in Fig. 10.6 where the detected buildings are presented. In particular, the detected buildings are shown in 3D, in the top of Fig. 10.6, while all scene buildings are shown in the bottom as they have been extracted from the ground truth data.

10.8 Conclusion and Future Directions

Computational change detection is a mature field that has been extensively studied from the geography, geoscience, and computer vision scientific communities during the past decades. An important amount of research and development has been devoted to comprehensive problem formulation, generic and standardize procedures, various applications, and validation for real and critical earth observation challenges.

In this review, we have made an effort to provide a comprehensive survey of the recent developments in the field of 2D and 3D change detection approaches in urban environments. Our approach was structured around the key change detection components, *i.e., (i)* the properties of the change detection targets and end products; *(ii)* the characteristics of the remote sensing data; *(iii)* the initial radiometric, atmospheric, and geometric corrections; *(iv)* the unsupervised methodologies; *(v)* the supervised methodologies; and *(vi)* the building extraction and reconstruction algorithms.

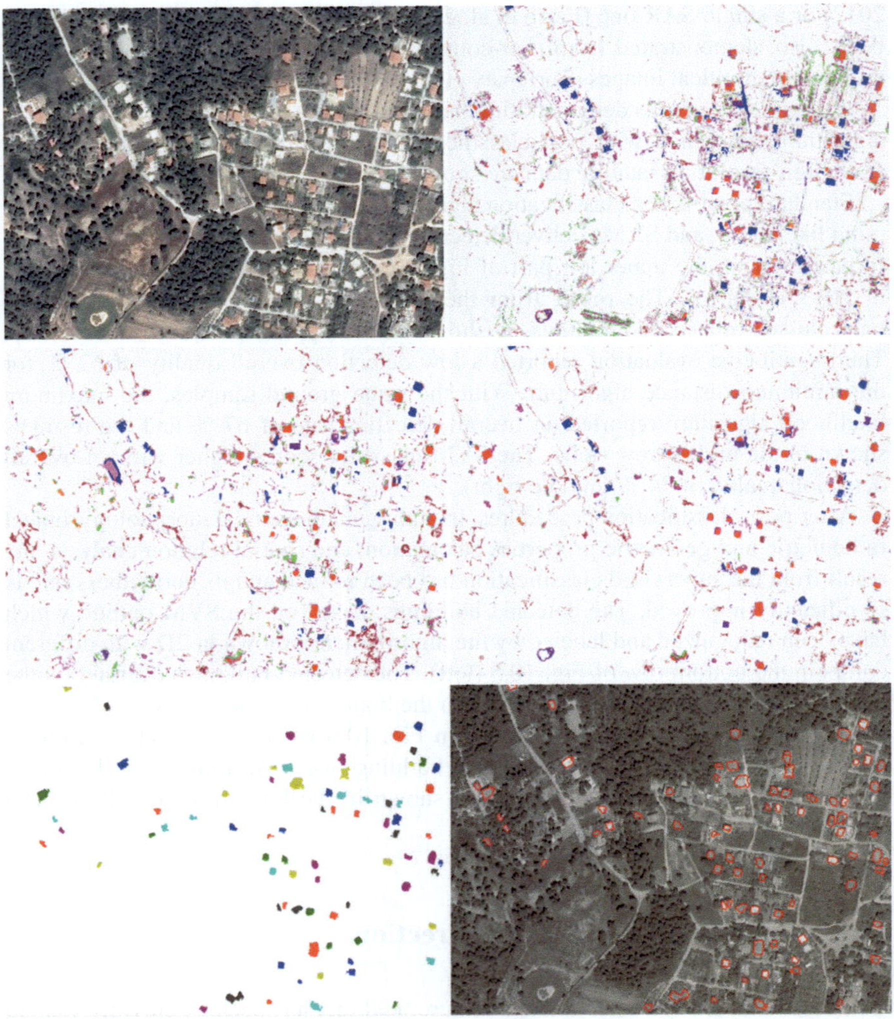

Fig. 10.5 Supervised 2D building detection based on data classification algorithms. *Upper row*: A Pleiades image acquired in 2013 (*left*) and the result from a standard minimum distance classification algorithm (showing only classes related to buildings). The quantitative evaluation reported a low detection overall quality of 62 %. *Middle row*: The result from a standard maximum likelihood classification algorithm with an overall detection rate of 67 % (*left*). A SVM classifier scores higher with an overall detection quality of 74 % (*right*). *Bottom row*: The detected buildings, after post-classification processing in the SVM output, are labeled and shown with different colors (*left*). The detected buildings overlaid on the raw Pleiades image (*right*)

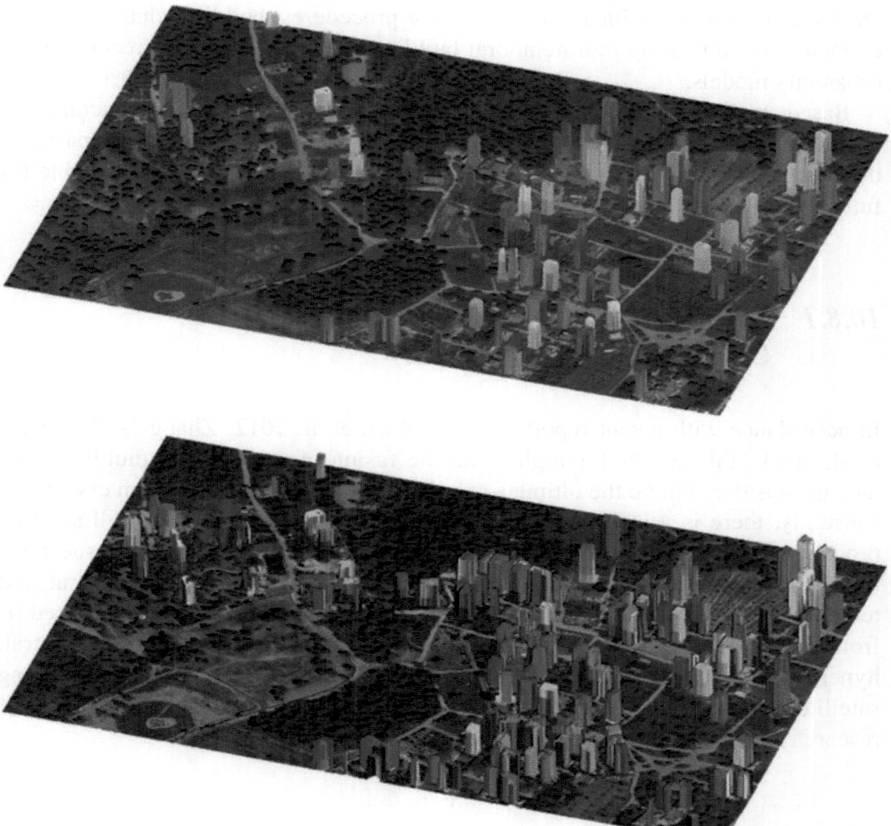

Fig. 10.6 The detected building in 3D after the application of a supervised SVM classifier and post-processing procedures on the high spatial resolution Pleiades data (*top*). Scene buildings in 3D as extracted from the ground truth data (*bottom*)

The aim was to focus our presentation on giving an account of recent approaches that have not been covered in previous surveys, and therefore, recent advances during the last 6 years have been reviewed. In addition, the change detection approaches were classified according to the monitoring targets (Table 10.1) and according to the remote sensing data that were design to process (Table 10.3). The unsupervised and supervised methods were classified according to their core algorithm that they were, mainly, based on (Table 10.4). Moreover, a summary of the currently available satellite remote sensing sensors, which were employed in recent studies, and their major specifications and cost are given in Table 10.2. Recent approaches focusing on 2D and 3D building extraction and modeling are given in Table 10.5, providing important computational frameworks which can be directly or partially adopted for addressing more efficiently the change detection problem. In particular, in a similar way with the change detection approaches that are based on post-classification comparison procedures, building changes can be extracted by comparing multitemporal building detection maps and reconstructed urban/city models.

Based on the current status and state of the art, the validation outcomes of relevant studies, and the special challenges of each detection component separately, the present study highlights certain issues and insights that may be applicable for future research and development.

10.8.1 Need to Design Novel Multimodal Computational Frameworks

In accordance with recent reports (Longbotham et al. 2012; Zhang 2012; Berger et al. 2013), this survey highlights that the fusion of multimodal, multitemporal data is considered to be the ultimate solution for optimized information extraction. Currently, there is a lack in single, generic frameworks that can in full synergy process and exploit all available geospatial data. This is a rather crucial issue since the effective and accurate detection and modeling requires rich spatial, spectral, and temporal (remote or not) observations over the structured environment acquired *(i)* from various sensors, including frame and push-broom cameras and multispectral, hyperspectral, thermal, and radar sensors, and *(ii)* from various platforms, including satellite, airborne, UAV, and ground systems. This is not a trivial task and a lot of research and development is, thus, required.

10.8.2 Need for Efficient Unsupervised Techniques Able to Identify "From-To" Change Trajectories

Unsupervised and supervised approaches are holding the same share of research interest. In particular, the unsupervised ones in many cases achieve the same overall detection accuracy levels as the supervised ones do (*e.g.*, Longbotham et al. 2012). This is a really promising fact given the possible capability of (near) real-time response to urgent and timely crucial change detection tasks, without training samples available. In dense time series and big geospatial data analysis, this seems, also, the only possible direction. However, most applications require end products which report on the detailed land-cover/land-use "from-to" change trajectories instead of a binary "change or not" map (Lu et al. 2011c; Bruzzone and Bovolo 2013). The need for incorporating spatial context and relationships into the detection procedure and introduce automated algorithms able to detect changes with a semantic meaning is underlined from the present study.

10.8.3 The Importance of Open Data Policies

Furthermore, this survey exhibits the importance of open data policies. This is, mainly, due to the fact that the extensive recent research activity in regional scales has been boosted by the currently increasing US and EU open data policies and mostly by the opening of the United States Geological Survey's Landsat data archive (Woodcock et al. 2008; Wulder et al. 2012) including current and future missions. Even not in a raw or quality-controlled format and not in a formal open data framework, there is an increasing availability of Google Earth/Street View, Microsoft Bing Maps/Streetside data which can also ease certain applications and studies. All these open data and open source (regarding software) initiatives and polices ensure the availability of big geospatial data and the availability of remote sensing datasets spanning densely over longer periods which, moreover, can enable further research towards quantifying global and regional transitions given the changing state of the urban environment, global and regional climate, biodiversity, food, and other critical environmental/ecosystem issues.

10.8.4 The Importance of Automation

The aforementioned availability of open big geospatial data impose as never before the need for automation. Despite the important advances and the available image processing technologies, powered mainly from the computer vision community, still, the skills and experience of an analyst are very important for the success of a classification/post-classification procedure (Weng 2011; Lu et al. 2011c), requiring human intervention which is labor consuming and subjective. Therefore, introducing generic, automated computational methods in every change detection component is of fundamental importance.

10.8.5 The Importance of Innovative Basic Research in the Core of the Change Detection Mechanism

Recent state-of-the-art change detection, classification, and modeling methodologies are not reaching high (>80 %) levels of accuracy and success rates when complex and/or extensive regions and/or local scales and/or relative small urban objects and/or dense time series have been explored in the urban environment (Wilkinson 2005; Longbotham et al. 2012; Berger et al. 2013; Rottensteiner et al. 2013). Thus, there is a strong need for designing new core classification, change detection, and modeling approaches being able to properly handle the high amount of spatial, spectral, and temporal information from the new generation sensors, being able to search effectively through huge archives of remote sensing datasets.

10.8.6 The Importance of Operational Data Preprocessing

Most standard remote sensing algorithms and techniques (classifications, indices, biophysical parameters, model inversions, object detection, *etc.*) assume cloud-free data, already radiometric, atmospheric, and geometric corrected. However, this is not an operationally solved problem yet. The production of a European cloud-free mosaic, two times per year, was not 100 % feasible despite the availability of three different satellite sensors and a considerable flexibility in the date windows around every region (Hoersch and Amans 2012). Moreover, in accordance with recent relevant studies (Vicente-Serrano et al. 2008), this survey underlines the fact that it is essential to accurately ensure the homogeneity of multitemporal datasets through operational radiometric and geometric data corrections including sensor calibration, cross-calibration, atmospheric, geometric, and topographic corrections and relative radiometric normalization using objective statistical techniques. Being able to address for the same invariant terrain object, the pictured different spectral signatures in time series data, being able to construct operationally cloud-free reflectance surfaces (Villa et al. 2012), will further boost the effectiveness and applicability of remote sensing methods in emerging urban environmental applications.

To sum up, the significant research interest on urban change detection and modeling is driven from real, critical, and current environmental and engineering problems, which pose emerging technological questions and challenges. Recent advances on the domain indicate that remote sensing and computer vision state-of-the-art approaches can be fused and further expanded towards the fruitful and comprehensive exploitation of open, big geospatial data.

Acknowledgment This research has been co-financed by the European Union (European Social Fund – ESF) and Greek national funds through the Operational Program "Education and Lifelong Learning" of the National Strategic Reference Framework (NSRF) – Research Funding Program: THALES: Reinforcement of the interdisciplinary and/or interinstitutional research and innovation.

References

Abd El-Kawy OR, Rød JK, Ismail HA, Suliman AS (2011) Land use and land cover change detection in the western Nile delta of Egypt using remote sensing data. Appl Geogr 31(2):483–494, ISSN 0143-6228. http://dx.doi.org/10.1016/j.apgeog.2010.10.012. Keywords: Change detection, Land use, Land cover, Post-classification comparison, Western Nile delta

Ahmad F, Amin MG (2013) Through-the-wall human motion indication using sparsity-driven change detection. IEEE Trans Geosci Remote Sens 51(2):881–890. doi:10.1109/TGRS.2012.2203310

Aiazzi B, Alparone L, Baronti S, Garzelli A, Zoppetti C (2013) Nonparametric change detection in multitemporal SAR images based on mean-shift clustering. IEEE Trans Geosci Remote Sens 51(4):2022–2031. doi:10.1109/TGRS.2013.2238946

Bagan H, Yamagata Y (2012) Landsat analysis of urban growth: how Tokyo became the world's largest megacity during the last 40 years. Remote Sens Environ 127:210–222, ISSN 0034-4257. http://dx.doi.org/10.1016/j.rse.2012.09.011

Bazi Y, Melgani F, Al-Sharari HD (2010) Unsupervised change detection in multispectral remotely sensed imagery with level set methods. IEEE Trans Geosci Remote Sens 48(8):3178–3187

Benedek C, Descombes X, Zerubia J (2010) Building detection in a single remotely sensed image with a point process of rectangles. In: 20th international conference on Pattern Recognition (ICPR), 23–26 Aug 2010, pp 1417–1420. doi:E10.1109/ICPR.2010.350

Benedek C, Descombes X, Zerubia J (2012) Building development monitoring in multitemporal remotely sensed image pairs with stochastic birth-death dynamics. IEEE Trans Pattern Anal Mach Intell 34(1):33–50. doi:10.1109/TPAMI.2011.94

Berberoglu S, Akin A (2009) Assessing different remote sensing techniques to detect land use/cover changes in the eastern Mediterranean. Int J Appl Earth Obs Geoinf 11(1):46–53, ISSN 0303-2434. http://dx.doi.org/10.1016/j.jag.2008.06.002

Berger C, Voltersen M, Eckardt R, Eberle J, Heyer T, Salepci N, Hese S, Schmullius C, Tao J, Auer S, Bamler R, Ewald K, Gartley M, Jacobson J, Buswell A, Du Q, Pacifici F (2013) Multi-modal and multi-temporal data fusion: outcome of the 2012 GRSS data fusion contest. IEEE J Sel Top Appl Earth Obs Remote Sens 6(3):1324–1340. doi:10.1109/JSTARS.2013.2245860

Blaschke T (2010) Object based image analysis for remote sensing. ISPRS J Photogramm Remote Sens 65(1):2–16

Boehm H-DV, Liesenberg V, Limin SH (2013) Multi-temporal airborne LiDAR-survey and field measurements of tropical peat swamp forest to monitor changes. IEEE J Sel Top Appl Earth Obs Remote Sens 6(3):1524–1530. doi:10.1109/JSTARS.2013.2258895

Boulila W, Farah IR, Saheb Ettabaa K, Solaiman B, Ben Ghézala H (2011) A data mining based approach to predict spatiotemporal changes in satellite images. Int J Appl Earth Obs Geoinf 13(3):386–395, ISSN 0303-2434. http://dx.doi.org/10.1016/j.jag.2011.01.008

Bouziani M, Goïta K, He D-C (2010) Automatic change detection of buildings in urban environment from very high spatial resolution images using existing geodatabase and prior knowledge. ISPRS J Photogramm Remote Sens 65(1):143–153, ISSN 0924-2716. http://dx.doi.org/10.1016/j.isprsjprs.2009.10.002

Bovolo F, Bruzzone L, Marconcini M (2008) A novel approach to unsupervised change detection based on a semisupervised SVM and a similarity measure. IEEE Trans Geosci Remote Sens 46(7):2070–2082

Bovolo F, Bruzzone L, Marchesi S (2009) Analysis and adaptive estimation of the registration noise distribution in multitemporal VHR images. IEEE Trans Geosci Remote Sens 47(8):2658–2671

Bovolo F, Camps-Valls G, Bruzzone L (2010) A support vector domain method for change detection in multitemporal images. Pattern Recognit Lett 31(10):1148–1154

Bovolo F, Marchesi S, Bruzzone L (2012) A framework for automatic and unsupervised detection of multiple changes in multitemporal images. IEEE Trans Geosci Remote Sens 50(6):2196–2212. doi:10.1109/TGRS.2011.2171493

Bovolo F, Marin C, Bruzzone L (2013) A hierarchical approach to change detection in very high resolution SAR images for surveillance applications. IEEE Trans Geosci Remote Sens 51(4):2042–2054. doi:10.1109/TGRS.2012.2223219

Brunner D, Lemoine G, Bruzzone L (2010) Earthquake damage assessment of buildings using VHR optical and SAR imagery. IEEE Trans Geosci Remote Sens 48(5):2403–2420. doi:10.1109/TGRS.2009.2038274

Bruzzone L, Bovolo F (2013) A novel framework for the design of change-detection systems for very-high-resolution remote sensing images. Proc IEEE 101(3):609–630. doi:10.1109/JPROC.2012.2197169

Camps-Valls G, Gómez-Chova L, Muñoz-Marí J, Rojo-Álvarez JL, Martínez-Ramón M (2008) Kernel-based framework for multi-temporal and multi-source remote sensing data classification and change detection. IEEE Trans Geosci Remote Sens 46(6):1822–1835

Canty MJ, Nielsen AA (2008) Automatic radiometric normalization of multitemporal satellite imagery with the iteratively re-weighted MAD transformation. Remote Sens Environ 112:1025–1036

Celik T (2009) Unsupervised change detection in satellite images using principal component analysis and k -means clustering. IEEE Geosci Remote Sens Lett 6(4):772–776

Celik T (2010) A Bayesian approach to unsupervised multiscale change detection in synthetic aperture radar images. Signal Process 90(5):1471–1485, ISSN 0165-1684. http://dx.doi.org/10.1016/j.sigpro.2009.10.018

Celik T, Ma K-K (2010) Unsupervised change detection for satellite images using dual-tree complex wavelet transform. IEEE Trans Geosci Remote Sens 48(3):1199–1210. doi:10.1109/TGRS.2009.2029095

Celik T, Ma K-K (2011) Multitemporal image change detection using undecimated discrete wavelet transform and active contours. IEEE Trans Geosci Remote Sens 49(2):706–716. doi:10.1109/TGRS.2010.2066979

Champion N, Boldo D, Pierrot-Deseilligny M, Stamon G (2010) 2D building change detection from high resolution satellite imagery: a two-step hierarchical method based on 3D invariant primitives. Pattern Recognit Lett 31(10):1138–1147, ISSN 0167-8655. http://dx.doi.org/10.1016/j.patrec.2009.10.012

Chander G, Markham BL, Helder DL (2009) Summary of current radiometric calibration coefficients for Landsat MSS, TM, ETM+, and EO-1 ALI sensors. Remote Sens Environ 113:893

Chatelain F, Tourneret J-Y, Inglada J (2008) Change detection in multisensor SAR images using bivariate gamma distributions. IEEE Trans Image Process 17(3):249–258. doi:10.1109/TIP.2008.916047

Chaudhuri D, Kushwaha NK, Samal A (2012) Semi-automated road detection from high resolution satellite images by directional morphological enhancement and segmentation techniques. IEEE J Sel Top Appl Earth Obs Remote Sens 5(5):1538–1544. doi:10.1109/JSTARS.2012.2199085

Chen J, Chen X, Cui X, Chen J (2011) Change vector analysis in posterior probability space: a new method for land cover change detection. IEEE Geosci Remote Sens Lett 8(2):317–321

Chen X, Chen J, Shi Y, Yamaguchi Y (2012) An automated approach for updating land cover maps based on integrated change detection and classification methods. ISPRS J Photogramm Remote Sens 71:86–95, ISSN 0924-2716. http://dx.doi.org/10.1016/j.isprsjprs.2012.05.006

Chen J, Lu M, Chen X, Chen J, Chen L (2013) A spectral gradient difference based approach for land cover change detection. ISPRS J Photogramm Remote Sens 85:1–12, ISSN 0924-2716. http://dx.doi.org/10.1016/j.isprsjprs.2013.07.009

Chini M, Pacifici F, Emery WJ, Pierdicca N, Del Frate F (2008) Comparing statistical and neural network methods applied to very high resolution satellite images showing changes in man-made structures at Rocky flats. IEEE Trans Geosci Remote Sens 46(6):1812–1821

Craglia M, de Bie K, Jackson D, Pesaresi M, Remetey-Fülöpp G, Wang C, Annoni A, Bian L, Campbell F, Ehlers M, van Genderen J, Goodchild M, Guo H, Lewis A, Simpson R, Skidmore A, Woodgate P (2012) Digital earth 2020: towards the vision for the next decade. Int J Digit Earth 5(1):4–21. doi:10.1080/17538947.2011.638500

Crispell D, Mundy J, Taubin G (2012) A variable-resolution probabilistic three-dimensional model for change detection. IEEE Trans Geosci Remote Sens 50(2):489–500

Dai X, Khorram S (1998) The effects of image misregistration on the accuracy of remotely sensed change detection. IEEE Trans Geosci Remote Sens 36(5):1566–1577

Dalla Mura M, Benediktsson JA, Bovolo F, Bruzzone L (2008) An unsupervised technique based on morphological filters for change detection in very high resolution images. IEEE Geosci Remote Sens Lett 5(3):433–437

Das S, Mirnalinee TT, Varghese K (2011) Use of salient features for the design of a multistage framework to extract roads from high-resolution multispectral satellite images. IEEE Trans Geosci Remote Sens 49(10):3906–3931. doi:10.1109/TGRS.2011.2136381

De Nigris D, Collins DL, Arbel T (2012) Multi-modal image registration based on gradient orientations of minimal uncertainty. IEEE Trans Med Image 31(12):2343–2354

Del Frate F, Pacifici F, Solimini D (2008) Monitoring urban land cover in Rome, Italy, and its changes by single-polarization multitemporal SAR images. IEEE J Sel Top Appl Earth Obs Remote Sens 1(2):87–97. doi:10.1109/JSTARS.2008.2002221

Demir B, Bovolo F, Bruzzone L (2012) Detection of land-cover transitions in multitemporal remote sensing images with active learning based compound classification. IEEE Trans Geosci Remote Sens 50(5):1930–1941

Demir B, Bovolo F, Bruzzone L (2013) Updating land-cover maps by classification of image time series: a novel change-detection-driven transfer learning approach. IEEE Trans Geosci Remote Sens 51:300–312. doi:10.1109/TGRS.2012.2195727

Deng JS, Wang K, Deng YH, Qi GJ (2008) PCA-based land-use change detection and analysis using multitemporal and multisensor satellite data. Int J Remote Sens 29:4823–4838. doi:10.1080/01431160801950162

Deng JS, Wang K, Hong Y, Qi JG (2009a) Spatio-temporal dynamics and evolution of land use change and landscape pattern in response to rapid urbanization. Landsc Urban Plan 92(3–4):187–198, ISSN 0169-2046. http://dx.doi.org/10.1016/j.landurbplan.2009.05.001

Deng J-S, Wang K, Li J, Deng Y-H (2009b) Urban land use change detection using multisensor satellite images. Pedosphere 19(1):96–103, ISSN 1002-0160. http://dx.doi.org/10.1016/S1002-0160(08)60088-0

Desclee B, Simonetti D, Mayaux P, Achard A (2013) Multi-sensor monitoring system for forest cover change assessment in central Africa. IEEE J Sel Top Appl Earth Obs Remote Sens 6(1):110–120. doi:10.1109/JSTARS.2013.2240263

Dewan AM, Yamaguchi Y (2009) Land use and land cover change in Greater Dhaka, Bangladesh: using remote sensing to promote sustainable urbanization. Appl Geogr 29(3):390–401, ISSN 0143–6228. http://dx.doi.org/10.1016/j.apgeog.2008.12.005

Dong L, Shan J (2013) A comprehensive review of earthquake-induced building damage detection with remote sensing techniques. ISPRS J Photogramm Remote Sens 84:85–99, ISSN 0924-2716. http://dx.doi.org/10.1016/j.isprsjprs.2013.06.011

Dos Santos Silva MP, Câmara G, Escada MIS, de Souza RCM (2008) Remote sensing image mining: detecting agents of land-use change in tropical forest areas. Int J Remote Sens 29: 4803–4822

dos Santos Galvanin EA, Porfirio Dal Poz A (2012) Extraction of building roof contours from LiDAR data using a Markov-random-field-based approach. IEEE Trans Geosci Remote Sens 50(3):981–987. doi:10.1109/TGRS.2011.2163823

Doxani G, Karantzalos K, Tsakiri-Strati M (2012) Monitoring urban changes based on scale-space filtering and object-oriented classification. Int J Appl Earth Obs Geoinf 15:38–48, ISSN 0303-2434. http://dx.doi.org/10.1016/j.jag.2011.07.002

Du P, Liu S, Gamba P, Tan K, Xia J (2012) Fusion of difference images for change detection over urban areas. IEEE J Sel Top Appl Earth Obs Remote Sens 5(4):1076–1086. doi:10.1109/JSTARS.2012.2200879

Du P, Liu S, Xia J, Zhao Y (2013) Information fusion techniques for change detection from multi-temporal remote sensing images. Inf Fusion 14(1):19–27, ISSN 1566-2535. http://dx.doi.org/10.1016/j.inffus.2012.05.003

Eastman RD, Le-Moigne J, Netanyahu NS (2007) Research issues in image registration for remote sensing. In: Proceedings of the IEEE conference on computer vision and pattern recognition, CVPR'07, 17–22 June 2007, pp 1–8

e-geos (2013) Price list. SD-COS 13-010, 33 p. http://www.e-geos.it/products/pdf/prices.pdf. Accessed 16 Nov 2013

Falco N, Mura MD, Bovolo F, Benediktsson JA, Bruzzone L (2013) Change detection in VHR images based on morphological attribute profiles. IEEE Geosci Remote Sens Lett 10(3):636–640. doi:10.1109/LGRS.2012.2222340

Fernandez-Prieto D, Marconcini M (2011) A novel partially supervised approach to targeted change detection. IEEE Trans Geosci Remote Sens 49(12):5016–5038

Ferro A, Brunner D, Bruzzone L (2013) Automatic detection and reconstruction of building radar footprints from single VHR SAR images. IEEE Trans Geosci Remote Sens 51:935–952. doi:http://dx.doi.org/10.1109/TGRS.2012.2205156

Gamanya R, De Maeyer P, De Dapper M (2009) Object-oriented change detection for the city of Harare, Zimbabwe. Expert Syst Appl 36(1):571–588, ISSN 0957-4174. http://dx.doi.org/10.1016/j.eswa.2007.09.067

Garcia-Dorado I, Demir I, Aliaga DG (2013) Automatic urban modeling using volumetric reconstruction with surface graph cuts. Comput Graph 37(7):896–910. doi:http://dx.doi.org/10.1016/j.cag.2013.07.003

GeoStore (2013) Astrium. http://www.astrium-geo.com/geostore

Ghosh A, Mishra NS, Ghosh S (2011) Fuzzy clustering algorithms for unsupervised change detection in remote sensing images. Inf Sci 181(4):699–715, ISSN 0020-0255. http://dx.doi.org/10.1016/j.ins.2010.10.016

Ghosh A, Subudhi BN, Bruzzone L (2013) Integration of Gibbs Markov random field and Hopfield-type neural networks for unsupervised change detection in remotely sensed multitemporal images. IEEE Trans Image Process 22(8):3087–3096. doi:10.1109/TIP.2013.2259833

Gilles J, Meyer Y (2010) Properties of BV-G structures + textures decomposition models. Application to road detection in satellite images. IEEE Trans Image Process 19(11):2793–2800. doi:10.1109/TIP.2010.2049946

Giustarini L, Hostache R, Matgen P, Schumann GJ-P, Bates PD, Mason DC (2013) A change detection approach to flood mapping in urban areas using TerraSAR-X. IEEE Trans Geosci Remote Sens 51(4):2417–2430. doi:10.1109/TGRS.2012.2210901

Gong M, Zhou Z, Ma J (2012) Change detection in synthetic aperture radar images based on image fusion and fuzzy clustering. IEEE Trans Image Process 21(4):2141–2151. doi:10.1109/TIP.2011.2170702

Gonzalez-Aguilera D, Crespo-Matellan E, Hernandez-Lopez D, Rodriguez-Gonzalvez P (2013) Automated urban analysis based on LiDAR-derived building models. IEEE Trans Geosci Remote Sens 51(3):1844–1851

Gueguen L, Soille P, Pesaresi M (2011) Change detection based on information measure. IEEE Trans Geosci Remote Sens 49(11):4503–4515

Haala N, Kada M (2010) An update on automatic 3D building reconstruction. ISPRS J Photogramm Remote Sens 65(6):570–580. doi:http://dx.doi.org/10.1016/j.isprsjprs.2010.09.006

Habib T, Inglada J, Mercier G, Chanussot J (2009) Support vector reduction in SVM algorithm for abrupt change detection in remote sensing. IEEE Geosci Remote Sens Lett 6(3):606–610

Han T, Wulder MA, White JC, Coops NC, Alvarez MF, Butson C (2007) An efficient protocol to process Landsat images for change detection with tasselled cap transformation. IEEE Geosci Remote Sens Lett 4:147–151

Hane C, Zach C, Cohen A, Angst R, Pollefeys M (2013) Joint 3D scene reconstruction and class segmentation. In: 2013 IEEE conference on Computer Vision and Pattern Recognition (CVPR), 23–28 June 2013, pp 97–104. doi:10.1109/CVPR.2013.20

Hansen MC, Loveland TR (2012) A review of large area monitoring of land cover change using Landsat data. Remote Sens Environ 122:66–74, ISSN 0034-4257. http://dx.doi.org/10.1016/j.rse.2011.08.024

Hansen MC, Egorov A, Potapov PV, Stehman SV, Tyukavina A, Turubanova SA, Roy DP, Goetz SJ, Loveland TR, Ju J, Kommareddy A, Kovalskyy V, Forsyth C, Bents T (2014) Monitoring conterminous United States (CONUS) land cover change with Web-Enabled Landsat Data (WELD). Remote Sens Environ 140:466–484, ISSN 0034-4257. http://dx.doi.org/10.1016/j.rse.2013.08.014

Hao M, Shi W, Zhang H, Li C (2014) Unsupervised change detection with expectation-maximization-based level set. IEEE Geosci Remote Sens Lett 11(1):210–214. doi:10.1109/LGRS.2013.2252879

He C, Wei A, Shi P, Zhang Q, Zhao Y (2011a) Detecting land-use/land-cover change in rural–urban fringe areas using extended change-vector analysis. Int J Appl Earth Obs Geoinf 13(4):572–585, ISSN 0303-2434. http://dx.doi.org/10.1016/j.jag.2011.03.002

He R, Xiong N, Yang LT, Park JH (2011b) Using multi-modal semantic association rules to fuse keywords and visual features automatically for Web image retrieval. Inf Fusion 12(3):223–230, ISSN 1566-2535. http://dx.doi.org/10.1016/j.inffus.2010.02.001

Hebel M, Arens M, Stilla U (2013) Change detection in urban areas by object-based analysis and on-the-fly comparison of multi-view ALS data. ISPRS J Photogramm Remote Sens 86:52–64, ISSN 0924-2716

Heinrich MP, Jenkinson M, Bhushan M, Matin T, Gleeson FV, Brady SM, Schnabel JA (2012) MIND: modality independent neighbourhood descriptor for multi-modal deformable registration. Med Image Anal 16(7):1423–1435, ISSN 1361-8415. http://dx.doi.org/10.1016/j.media.2012.05.008

Heo J, Jeong S, Park H-K, Jung JH, Han S, Hong S, Sohn H-G (2013) Productive high complexity 3D city modeling with point clouds collected from terrestrial LiDAR. Comput Environ Urban Syst 41:26–38

Hoersch B, Amans V (2012) GMES space component data access portfolio: data warehouse 2011–2014. European Space Agency, Frascati. http://gmesdata.esa.int/web/gsc/about-gscdata-access

Hu J, Zhang Y (2013) Seasonal change of land-use/land-cover (LULC) detection using MODIS data in rapid urbanization regions: a case study of the pearl river delta region (China). IEEE J Sel Top Appl Earth Obs Remote Sens 6(4):1913–1920

Huang X, Zhang L (2012) Morphological building/shadow index for building extraction from high-resolution imagery over urban areas. IEEE J Sel Top Appl Earth Obs Remote Sens 5(1): 161–172

Huo C, Zhou Z, Lu H, Chen K, Pan C (2010) Fast object-level change detection for VHR images. IEEE Geosci Remote Sens Lett 7(1):118–122

Hussain M, Chen D, Cheng A, Wei H, Stanley D (2013) Change detection from remotely sensed images: from pixel-based to object-based approaches. ISPRS J Photogramm Remote Sens 80:91–106, ISSN 0924-2716. http://dx.doi.org/10.1016/j.isprsjprs.2013.03.006

Im J, Rhee J, Jensen JR, Hodgson ME (2007) An automated binary change detection model using a calibration approach. Remote Sens Environ 106(1):89–105, ISSN 0034-4257. http://dx.doi.org/10.1016/j.rse.2006.07.019

Im J, Jensen JR, Hodgson ME (2008) Optimizing the binary discriminant function in change detection applications. Remote Sens Environ 112(6):2761–2776, ISSN 0034-4257. http://dx.doi.org/10.1016/j.rse.2008.01.007

Irons JR, Loveland TR (2013) Eighth Landsat satellite becomes operational. Photogramm Eng Remote Sens 79:398–401

Irschara A, Rumpler M, Meixner P, Pock T, Bischof H (2012) Efficient and globally optimal multi view dense matching for aerial images. In: ISPRS annals of the photogrammetry, remote sensing and spatial information sciences, Melbourne

Izadi M, Saeedi P (2012) Three-dimensional polygonal building model estimation from single satellite images. IEEE Trans Geosci Remote Sens 50(6):2254–2272. doi:10.1109/TGRS.2011.2172995

James LA, Hodgson ME, Ghoshal S, Latiolais MM (2012) Geomorphic change detection using historic maps and DEM differencing: the temporal dimension of geospatial analysis. Geomorphology 137(1):181–198, ISSN 0169-555X. http://dx.doi.org/10.1016/j.geomorph.2010.10.039

Karantzalos K, Argialas D (2009) A region-based level set segmentation for automatic detection of man-made objects from aerial and satellite images. Photogramm Eng Remote Sens 75(6): 667–678

Karantzalos K, Paragios N (2009) Recognition-driven 2D competing priors towards automatic and accurate building detection. IEEE Trans Geosci Remote Sens 47(1):133–144

Karantzalos K, Paragios N (2010) Large-scale building reconstruction through information fusion and 3D priors. IEEE Trans Geosci Remote Sens 48(5):2283–2296

Katartzis A, Sahli H (2008) A stochastic framework for the identification of building rooftops using a single remote sensing image. IEEE Trans Geosci Remote Sens 46(1):259–271. doi:10.1109/TGRS.2007.904953

Kit O, Lüdeke M (2013) Automated detection of slum area change in Hyderabad, India using multitemporal satellite imagery. ISPRS J Photogramm Remote Sens 83:130–137, ISSN 0924-2716. http://dx.doi.org/10.1016/j.isprsjprs.2013.06.009

Klaric MN, Claywell BC, Scott GJ, Hudson NJ, Sjahputera O, Li Y, Barratt ST, Keller JM, Davis CH (2013) GeoCDX: an automated change detection and exploitation system for high-resolution satellite imagery. IEEE Trans Geosci Remote Sens 51(4):2067–2086

Klonus S, Tomowski D, Ehlers M, Reinartz P, Michel U (2012) Combined edge segment texture analysis for the detection of damaged buildings in crisis areas. IEEE J Sel Top Appl Earth Obs Remote Sens 5(4):1118–1128

Knudby A, Newman C, Shaghude Y, Muhando C (2010) Simple and effective monitoring of historic changes in nearshore environments using the free archive of Landsat imagery. Int J Appl Earth Obs Geoinf 12(Suppl 1):S116–S122, ISSN 0303-2434. http://dx.doi.org/10.1016/j.jag.2009.09.002

Lafarge F, Descombes X, Zerubia J, Pierrot-Deseilligny M (2010) Structural approach for building reconstruction from a single DSM. IEEE Trans Pattern Anal Mach Intell 32(1):135–147. doi:10.1109/TPAMI.2008.281

Le Moigne J, Netanyahu NS, Eastman RD (2011) Image registration for remote sensing. Cambridge University Press, Cambridge, 497p. ISBN 9780521516112

Legg PA, Rosin PL, Marshall D, Morgan JE (2013) Improving accuracy and efficiency of mutual information for multi-modal retinal image registration using adaptive probability density estimation. Comput Med Image Graph 37:597–606

Leinenkugel P, Esch T, Kuenzer C (2011) Settlement detection and impervious surface estimation in the Mekong Delta using optical and SAR remote sensing data. Remote Sens Environ 115(12):3007–3019, ISSN 0034-4257. http://dx.doi.org/10.1016/j.rse.2011.06.004

Ling F, Li W, Du Y, Li X (2011) Land cover change mapping at the subpixel scale with different spatial-resolution remotely sensed imagery. IEEE Geosci Remote Sens Lett 8(1): 182–186

Liu Z-g, Dezert J, Mercier G, Pan Q (2012) Dynamic evidential reasoning for change detection in remote sensing images. IEEE Trans Geosci Remote Sens 50(5):1955–1967. doi:10.1109/TGRS.2011.2169075

Loch-Dehbi S, Plümer L (2011) Automatic reasoning for geometric constraints in 3D city models with uncertain observations. ISPRS J Photogramm Remote Sens 66(2):177–187. doi:http://dx.doi.org/10.1016/j.isprsjprs.2010.12.003

Longbotham N, Pacifici F, Glenn T, Zare A, Volpi M, Tuia D, Christophe E, Michel J, Inglada J, Chanussot J, Du Q (2012) Multi-modal change detection, application to the detection of flooded areas: outcome of the 2009–2010 data fusion contest. IEEE J Sel Top Appl Earth Obs Remote Sens 5(1):331–342. doi:10.1109/JSTARS.2011.2179638

Lowe DG (2004) Distinctive image features from scale-invariant keypoints. Int J Comput Vis 60(2):91–110

Lu D, Mausel P, Brondízio E, Moran E (2004) Change detection techniques. Int J Remote Sens 25(12):2365–2407

Lu D, Tian H, Zhou G, Ge H (2008) Regional mapping of human settlements in southeastern China with multisensor remotely sensed data. Remote Sens Environ 112(9):3668–3679

Lu D, Moran E, Hetrick S (2011a) Detection of impervious surface change with multitemporal Landsat images in an urban–rural frontier. ISPRS J Photogramm Remote Sens 66(3):298–306, ISSN 0924-2716. http://dx.doi.org/10.1016/j.isprsjprs.2010.10.010

Lu P, Stumpf A, Kerle N, Casagli N (2011b) Object-oriented change detection for landslide rapid mapping. IEEE Geosci Remote Sens Lett 8(4):701–705

Lu D, Moran E, Hetrick S, Li G (2011c) Land-use and land-cover change detection. In: Weng Q (ed) Advances in environmental remote sensing sensors, algorithms, and applications. CRC Press/Taylor & Francis Group, New York, pp 273–290

Luo W, Li H (2011) Soft-change detection in optical satellite images. IEEE Geosci Remote Sens Lett 8(5):879–883

Ma J, Gong M, Zhou Z (2012) Wavelet fusion on ratio images for change detection in SAR images. IEEE Geosci Remote Sens Lett 9(6):1122–1126. doi:10.1109/LGRS.2012.2191387

Marchesi S, Bruzzone L (2009) ICA and kernel ICA for change detection in multispectral remote sensing images. IEEE Int Geosci Remote Sens Symp (IGARSS) 2:II-980–II-983

Marino A, Cloude SR, Lopez-Sanchez JM (2013) A new polarimetric change detector in radar imagery. IEEE Trans Geosci Remote Sens 51(5):2986–3000. doi:10.1109/TGRS.2012.2211883

Marpu PR, Gamba P, Canty MJ (2011) Improving change detection results of IR-MAD by eliminating strong changes. IEEE Geosci Remote Sens Lett 8(4):799–803

Matei BC, Sawhney H, Samarasekera S, Kim J, Kumar R (2008) Building segmentation for densely built urban regions using aerial LiDAR data. In: Proceeding of the IEEE conference on computer vision and pattern recognition. IEEE, Piscataway, pp 1–8

Michishita R, Jiang Z, Xu B (2012) Monitoring two decades of urbanization in the Poyang Lake area, China through spectral unmixing. Remote Sens Environ 117:3–18, ISSN 0034-4257. http://dx.doi.org/10.1016/j.rse.2011.06.021

Shafizadeh-Moghadam H, Helbich M (2013) Spatiotemporal urbanization processes in the megacity of Mumbai, India: a Markov chains-cellular automata urban growth model. Appl Geogr 40:140–149, ISSN 0143-6228. http://dx.doi.org/10.1016/j.apgeog.2013.01.009

Moser G, Serpico SB (2009) Unsupervised change detection from multichannel SAR data by Markovian data fusion. IEEE Trans Geosci Remote Sens 47(7):2114–2128. doi:10.1109/TGRS.2009.2012407

Moser G, Angiati E, Serpico SB (2011) Multiscale unsupervised change detection on optical images by Markov random fields and wavelets. IEEE Geosci Remote Sens Lett 8(4):725–729

Nielsen AA (2007) The regularized iteratively reweighted MAD method for change detection in multi- and hyperspectral data. IEEE Trans Image Process 16(2):463–478

Pacifici F, Del Frate F (2010) Automatic change detection in very high resolution images with pulse-coupled neural networks. IEEE Geosci Remote Sens Lett 7(1):58–62. doi:10.1109/LGRS.2009.2021780

Pagot E, Pesaresi M (2008) Systematic study of the urban postconflict change classification performance using spectral and structural features in a support vector machine. IEEE Trans Geosci Remote Sens 1(2):120–128

Phelps J, Webb EL, Adams WM (2013) Biodiversity co-benefits of policies to reduce forest-carbon emissions. Nat Clim Chang 2:497–503. doi:10.1038/nclimate1462

Poulain V, Inglada J, Spigai M, Tourneret J-Y, Marthon P (2011) High-resolution optical and SAR image fusion for building database updating. IEEE Trans Geosci Remote Sens 49(8):2900–2910. doi:10.1109/TGRS.2011.2113351

Poullis C, You S (2010) Delineation and geometric modeling of road networks. ISPRS J Photogramm Remote Sens 65(2):165–181, ISSN 0924-2716. http://dx.doi.org/10.1016/j.isprsjprs.2009.10.004

Pratola C, Del Frate F, Schiavon G, Solimini D (2013) Toward fully automatic detection of changes in suburban areas from VHR SAR images by combining multiple neural-network models. IEEE Trans Geosci Remote Sens 51(4):2055–2066. doi:10.1109/TGRS.2012.2236846

Radke AJ, Andra S, Al-Kofahi O, Roysam B (2005) Image change detection algorithms: a systematic survey. IEEE Trans Image Process 14(3):294–307

Renza D, Martinez E, Arquero A (2013) A new approach to change detection in multispectral images by means of ERGAS index. IEEE Geosci Remote Sens Lett 10(1):76–80. doi:10.1109/LGRS.2012.2193372

Robin A, Moisan L, Le Hegarat-Mascle S (2010) An a-contrario approach for subpixel change detection in satellite imagery. IEEE Trans Pattern Anal Mach Intell 32(11):1977–1993. doi:10.1109/TPAMI.2010.37

Rottensteiner F, Sohn G, Gerke M, Wegner JD, Breitkopf U, Jung J (2013) Results of the ISPRS benchmark on urban object detection and 3D building reconstruction. ISPRS J Photogramm Remote Sens. http://dx.doi.org/10.1016/j.isprsjprs.2013.10.004

Rousson M, Paragios N (2008) Prior knowledge, level set representations and visual grouping. Int J Comput Vis 76(3):231–243
Roy DP (2000) The impact of misregistration upon composited wide field of view satellite data and implications for change detection. IEEE Trans Geosci Remote Sens 38(4):2017–2032
Rutzinger M, Rottensteiner F, Pfeifer N (2009) A comparison of evaluation techniques for building extraction from airborne laser scanning. IEEE J Sel Top Appl Earth Obs Remote Sens 2(1):11–20. doi:10.1109/JSTARS.2009.2012488
Salmon BP, Olivier JC, Wessels KJ, Kleynhans W, Van den Bergh F, Steenkamp KC (2011) Unsupervised land cover change detection: meaningful sequential time series analysis. IEEE J Sel Top Appl Earth Obs Remote Sens 4(2):327–335
Sampath A, Shan J (2010) Segmentation and reconstruction of polyhedral building roofs from aerial lidar point clouds. IEEE Trans Geosci Remote Sens 48(3):1554–1567. doi:10.1109/TGRS.2009.2030180
Schneider A (2012) Monitoring land cover change in urban and peri-urban areas using dense time stacks of Landsat satellite data and a data mining approach. Remote Sens Environ 124:689–704, ISSN 0034-4257. http://dx.doi.org/10.1016/j.rse.2012.06.006
Senaras C, Ozay M, Yarman Vural FT (2013) Building detection with decision fusion. IEEE J Sel Top Appl Earth Obs Remote Sens 6(3):1295–1304. doi:10.1109/JSTARS.2013.2249498
Sesnie SE, Gessler PE, Finegan B, Thessler S (2008) Integrating Landsat TM and SRTM-DEM derived variables with decision trees for habitat classification and change detection in complex neotropical environments. Remote Sens Environ 112(5):2145–2159, ISSN 0034-4257. http://dx.doi.org/10.1016/j.rse.2007.08.025
Sexton JO, Urban DL, Donohue MJ, Song C (2013) Long-term land cover dynamics by multi-temporal classification across the Landsat-5 record. Remote Sens Environ 128:246–258, ISSN 0034-4257. http://dx.doi.org/10.1016/j.rse.2012.10.010
Simon L, Teboul O, Koutsourakis P, Paragios N (2010) Random exploration of the procedural space for single-view 3D modeling of buildings. Int J Comput Vis 93(2):253–271
Singh A (1989) Digital change detection techniques using remotely-sensed data. Int J Remote Sens 10(6):989–1003
Sirmacek B, Taubenbock H, Reinartz P, Ehlers M (2012) Performance evaluation for 3-D city model generation of six different DSMs from air- and spaceborne sensors. IEEE J Sel Top Appl Earth Obs Remote Sens 5(1):59–70. doi:10.1109/JSTARS.2011.2178399
Sirmacek B, Unsalan C (2011) A probabilistic framework to detect buildings in aerial and satellite images. IEEE Trans Geosci Remote Sens 49(1):211–221. doi:10.1109/TGRS.2010.2053713
Sjahputera O, Scott GJ, Claywell B, Klaric MN, Hudson NJ, Keller JM, Davis CH (2011) Clustering of detected changes in high-resolution satellite imagery using a stabilized competitive agglomeration algorithm. IEEE Trans Geosci Remote Sens 49(12):4687–4703. doi:10.1109/TGRS.2011.2152847
Sotiras A, Davatzikos C, Paragios N (2013) Deformable medical image registration: a survey. IEEE Trans Med Image 32(7):1153–1190
Sportouche H, Tupin F, Denise L (2011) Extraction and three-dimensional reconstruction of isolated buildings in urban scenes from high-resolution optical and SAR spaceborne images. IEEE Trans Geosci Remote Sens 49(10):3932–3946. doi:10.1109/TGRS.2011.2132727
Stankov K, He D-C (2013) Building detection in very high spatial resolution multispectral images using the hit-or-miss transform. IEEE Geosci Remote Sens Lett 10(1):86–90. doi:10.1109/LGRS.2012.2193552
Sun S, Salvaggio C (2013) Aerial 3D building detection and modeling from airborne LiDAR point clouds. IEEE J Sel Top Appl Earth Obs Remote Sens 6(3):1440–1449. doi:10.1109/JSTARS.2013.2251457
Szeliski R, Zabih R, Scharstein D, Veksler O, Kolmogorov V, Agarwala A, Tappen M, Rother C (2008) A comparative study of energy minimization methods for markov random fields with smoothness-based priors. IEEE Trans Pattern Anal Mach Intell 30(6):1068–1080

Tack F, Buyuksalih G, Goossens R (2012) 3D building reconstruction based on given ground plan information and surface models extracted from spaceborne imagery. ISPRS J Photogramm Remote Sens 67(0):52–64. doi:http://dx.doi.org/10.1016/j.isprsjprs.2011.10.003

Taneja A, Ballan L, Pollefeys M (2013) City-scale change detection in cadastral 3D models using images. In: Proceedings of the IEEE international conference on computer vision and pattern recognition

Tang Y, Huang X, Zhang L (2013) Fault-tolerant building change detection from urban high-resolution remote sensing imagery. IEEE Geosci Remote Sens Lett 10(5):1060–1064. doi:10.1109/LGRS.2012.2228626

Taubenböck H, Esch T, Felbier A, Wiesner M, Roth A, Dech S (2012) Monitoring urbanization in mega cities from space. Remote Sens Environ 117:162–176, ISSN 0034-4257. http://dx.doi.org/10.1016/j.rse.2011.09.015

Tian J, Reinartz P, d'Angelo P, Ehlers M (2013) Region-based automatic building and forest change detection on Cartosat-1 stereo imagery. ISPRS J Photogramm Remote Sens 79:226–239, ISSN 0924-2716. http://dx.doi.org/10.1016/j.isprsjprs.2013.02.017

Turlapaty A, Gokaraju B, Du Q, Younan NH, Aanstoos JV (2012) A hybrid approach for building extraction from spaceborne multi-angular optical imagery. IEEE J Sel Top Appl Earth Obs Remote Sens 5(1):89–100. doi:10.1109/JSTARS.2011.2179792

Tzotsos A, Karantzalos K, Argialas D (2011) Object-based image analysis through nonlinear scale-space filtering. ISPRS J Photogramm Remote Sens 66(1):2–16, ISSN 0924-2716. http://dx.doi.org/10.1016/j.isprsjprs.2010.07.001

Tzotsos A, Karantzalos K, Argialas D (2014) Multiscale segmentation and classification of remote sensing imagery with advanced edge and scale-space features. In: Weng Q (ed) Scale issues in remote sensing. Wiley, Hoboken. doi:10.1002/9781118801628.ch09

Unsalan C, Sirmacek B (2012) Road network detection using probabilistic and graph theoretical methods. IEEE Trans Geosci Remote Sens 50(11):4441–4453. doi:10.1109/TGRS.2012.2190078

Vicente-Serrano SM, Pérez-Cabello F, Lasanta T (2008) Assessment of radiometric correction techniques in analyzing vegetation variability and change using time series of Landsat images. Remote Sens Environ 112(10):3916–3934, ISSN 0034-4257. http://dx.doi.org/10.1016/j.rse.2008.06.011

Vieira MA, Formaggio AR, Rennó CD, Atzberger C, Aguiar DA, Mello MP (2012) Object based image analysis and data mining applied to a remotely sensed Landsat time-series to map sugarcane over large areas. Remote Sens Environ 123:553–562, ISSN 0034-4257. http://dx.doi.org/10.1016/j.rse.2012.04.011

Villa P (2012) Mapping urban growth using soil and vegetation index and Landsat data: the Milan (Italy) city area case study. Landsc Urban Plan 107(3):245–254, ISSN 0169-2046. http://dx.doi.org/10.1016/j.landurbplan.2012.06.014

Villa G, Moreno J, Calera A, Amorós-López J, Camps-Valls G, Domenech E, Garrido J, González-Matesanz J, Gómez-Chova L, Martínez JÁ, Molina S, Peces JJ, Plaza N, Porcuna A, Tejeiro JA, Valcárcel N (2012) Spectro-temporal reflectance surfaces: a new conceptual framework for the integration of remote-sensing data from multiple different sensors. Int J Remote Sens 34:3699–3715. doi:10.1080/01431161.2012.716910

Volpi M, Tuia D, Camps-Valls G, Kanevski M (2012) Unsupervised change detection with kernels. IEEE Geosci Remote Sens Lett 9(6):1026–1030

Volpi M, Tuia D, Bovolo F, Kanevski M, Bruzzone L (2013) Supervised change detection in VHR images using contextual information and support vector machines. Int J Appl Earth Obs Geoinf 20:77–85, ISSN 0303-2434. http://dx.doi.org/10.1016/j.jag.2011.10.013

Wachinger C, Navab N (2012) Entropy and Laplacian images: structural representations for multi-modal registration. Med Image Anal 16(1):1–17, ISSN 1361-8415. http://dx.doi.org/10.1016/j.media.2011.03.001

Wang T-L, Jin Y-Q (2012) Postearthquake building damage assessment using multi-mutual information from pre-event optical image and postevent SAR image. IEEE Geosci Remote Sens Lett 9(3):452–456

Wang F, Wu Y, Zhang Q, Zhang P, Li M, Lu Y (2013) Unsupervised change detection on SAR images using triplet markov field model. IEEE Geosci Remote Sens Lett 10(4):697–701. doi:10.1109/LGRS.2012.2219494

Wegner JD, Hansch R, Thiele A, Soergel U (2011) Building detection from one orthophoto and high-resolution inSAR data using conditional random fields. IEEE J Sel Top Appl Earth Obs Remote Sens 4(1):83–91. doi:10.1109/JSTARS.2010.2053521

Weng Q (2011) Advances in environmental remote sensing: sensors, algorithms and applications. CRC Press/Taylor & Francis, Boca Raton

Weng Q (2012) Remote sensing of impervious surfaces in the urban areas: requirements, methods, and trends. Remote Sens Environ 117(15):34–49

Wilkinson G (2005) Results and implications of a study of fifteen years of satellite image classification experiments. IEEE Trans Geosci Remote Sens 43(3):433–440

Woodcock CE, Allen R, Anderson M, Belward A, Bindschadler R, Cohen W, Gao F et al (2008) Free access to Landsat imagery. Science 320:1011

Wulder MA, Butson CR, White JC (2008) Cross-sensor change detection over a forested landscape: options to enable continuity of medium spatial resolution measures. Remote Sens Environ 112(3):796–809, ISSN 0034-4257. http://dx.doi.org/10.1016/j.rse.2007.06.013

Wulder MA, Masek JG, Cohen WB, Loveland TR, Woodcock CE (2012) Opening the archive: how free data has enabled the science and monitoring promise of Landsat. Remote Sens Environ 122:2–10

Xian G, Homer C (2010) Updating the 2001 national land cover database impervious surface products to 2006 using landsat imagery change detection methods. Remote Sens Environ 114:1676–1686

Xian G, Homer C, Fry J (2009) Updating the 2001 national land cover database land cover classification to 2006 by using Landsat imagery change detection methods. Remote Sens Environ 113(6):1133–1147, ISSN 0034-4257. http://dx.doi.org/10.1016/j.rse.2009.02.004

Yang B, Zhang Y, Luan X (2013) A probabilistic relaxation approach for matching road networks. Int J Geogr Inf Sci 27(2):319–338. doi:10.1080/13658816.2012.683486

Yang X, Lo CP (2000) Relative radiometric normalization performance for change detection from multi-date satellite images. Photogramm Eng Remote Sens 66:967–980

Yetgin Z (2012) Unsupervised change detection of satellite images using local gradual descent. IEEE Trans Geosci Remote Sens 50(5):1919–1929. doi:10.1109/TGRS.2011.2168230

Yousif O, Ban Y (2013) Improving urban change detection from multitemporal SAR images using PCA-NLM. IEEE Trans Geosci Remote Sens 51(4):2032–2041. doi:10.1109/TGRS.2013.2245900

Zanotta DC, Haertel V (2012) Gradual land cover change detection based on multitemporal fraction images. Pattern Recognit 45(8):2927–2937, ISSN 0031-3203. http://dx.doi.org/10.1016/j.patcog.2012.02.004

Zebedin L, Bauer J, Karner K, Bischof H (2008) Fusion of feature and area-based information for urban buildings modeling from aerial imagery. In: European conference on computer vision, vol 5305, Lecture notes in computer science. Springer, Berlin, pp 873–886

Zhang J (2012) Multi-source remote sensing fusion: status and trends. Int J Image Data Fusion 1(1):5–24

Zhang Q, Seto KC (2011) Mapping urbanization dynamics at regional and global scales using multi-temporal DMSP/OLS nighttime light data. Remote Sens Environ 115(9):2320–2329, ISSN 0034-4257. http://dx.doi.org/10.1016/j.rse.2011.04.032

Zhang H, Qi Z-f, Ye X-y, Cai Y-b, Ma W-c, Chen M-n (2013) Analysis of land use/land cover change, population shift, and their effects on spatiotemporal patterns of urban heat islands in metropolitan Shanghai, China. Appl Geogr 44:121–133, ISSN 0143-6228. http://dx.doi.org/10.1016/j.apgeog.2013.07.021

Zhou W, Huang G, Troy A, Cadenasso M (2009) Object-based land cover classification of shaded areas in high spatial resolution imagery of urban areas: a comparison study. Remote Sens Environ 113(8):1769–1777

Chapter 11
Fusion of Airborne Hyperspectral and LiDAR Remote Sensing Data to Study the Thermal Characteristics of Urban Environments

Christian Berger, Frank Riedel, Johannes Rosentreter, Enrico Stein, Sören Hese, and Christiane Schmullius

Abstract This study focuses on the derivation of an urban surface material map to parameterize a 3D numerical microclimate model. For this purpose, fusion of airborne hyperspectral and light detection and ranging (LiDAR) remote sensing data is performed. In a first step, surface materials are extracted from the preprocessed input datasets using a hybrid, three-stage classification approach. The resulting map is then utilized in combination with the LiDAR object height information data to parameterize the microclimate model. To demonstrate the potential of data-driven microclimate modeling, two case studies are presented for selected test sites in the City of Houston, Texas. The results of this study highlight that the synergistic combination of hyperspectral and LiDAR data enables reliable mapping of some of the key input parameters required for urban microclimate modeling. Moreover, classification-based microclimate simulations can reveal the thermal properties of urban neighborhoods under varying conditions and, thus, facilitate the identification of hot spot areas and critical land cover configurations.

Keywords Data fusion • Hyperspectral • LiDAR • Surface material • Mapping • Urban microclimate • Modeling

11.1 Introduction

Over the past decades, the world has faced a continuous and increasingly dynamic urbanization. While city dwellers make up one-half of the world's population today, this share is predicted to add up to 70 % by 2050 (United Nations 2008). The

C. Berger (✉) • F. Riedel • J. Rosentreter • S. Hese • C. Schmullius
Department of Earth Observation, University of Jena, Löbdergraben 32, D-07743 Jena, Germany
e-mail: christian.berger@uni-jena.de; f.riedel@uni-jena.de; johannes.rosentreter@uni-jena.de; soeren.hese@uni-jena.de; c.schmullius@uni-jena.de

E. Stein
Earth Observation Center, German Aerospace Center (DLR), Oberpfaffenhofen, D-82234 Wessling, Germany
e-mail: enrico.stein@dlr.de

M. Helbich et al. (eds.), *Computational Approaches for Urban Environments*, Geotechnologies and the Environment 13, DOI 10.1007/978-3-319-11469-9_11

ever-growing percentage of global urban population comes at a high price. Many cities all over the world are sprawling rapidly, and their spread is associated with alarming rates of land consumption (Scalenghe and Marsan 2009; Angel et al. 2011; Taubenböck et al. 2012). As a result of this process, the natural "skin" of our planet is being successively replaced by man-made surfaces with distinct thermal properties (Sobrino et al. 2012).

It is generally agreed that urban areas and urban expansion affect the climate at the local scale (Oke 1973; Landsberg 1981; Oke 1982; Arnfield 2003; Kalnay and Cai 2003). A very prominent example of this local climate change is a phenomenon called the urban heat island (UHI). It refers to the observation that cities often feature higher air and surface temperatures than their surroundings, especially at night (Howard 1833; Oke 1973, 1982; Voogt 2002). The implications of the UHI effect are diverse and range from changes in precipitation patterns (Changnon 1992; Lowry 1998; Yuan and Bauer 2007) to raises in air pollution (Voogt 2002; Yuan and Bauer 2007), water use (Guhathakurta and Gober 2007), energy consumption (Voogt 2002; Yuan and Bauer 2007; Ewing and Rong 2008), and mortality rates (Curriero et al. 2002; Johnson and Wilson 2009). Considering that the UHI intensity is expected to increase in the future (McCarthy et al. 2010) while, at the same time, more and more people will be exposed to the living conditions in the cities of tomorrow (United Nations 2008), there is an urgent need for up-to-date, spatially explicit urban climatological information which city planners can incorporate into decision-making processes to foster effective management and to safeguard sustainable urban development.

Microclimate modeling is a powerful tool to analyze the thermal characteristics of urban environments at the local scale. However, it requires high spatial resolution, area-wide information on urban surface materials, and object heights. Since these information are still lacking in many urban areas, hypothetical scenarios often represent the only way to parameterize microclimate models unless extensive field surveys are up for debate. As an alternative way of data collection, hyperspectral and light detection and ranging (LiDAR) remote sensing technologies are becoming increasingly available (e.g., NASA Jet Propulsion Laboratory 2014; Cook et al. 2013; LiDAR Online 2014; OpenTopography 2014) and offer unique capabilities for urban surface material and object height mapping. Thus, they hold a great potential for microclimate modeling applications. This study aims at the derivation of an urban surface material map to parameterize a 3D numerical microclimate model by fusion of airborne hyperspectral and LiDAR remote sensing data. To demonstrate the potential of classification-based microclimate modeling, two case studies are presented for selected test sites in the City of Houston, Texas. This chapter is structured as follows. Section 11.2 provides a brief review of the scientific literature related to this study. In Sects. 11.3 and 11.4, the data and methods used to achieve the above goal are presented. Section 11.5 is dedicated to the description and discussion of the study results, and the section "Conclusions" summarizes the findings of this investigation.

11.2 Related Work

Remote sensing has become an increasingly important technology to gain a better understanding of the urban climate (Arnfield 2003; Voogt and Oke 2003; Heldens et al. 2011). However, while a lot of studies have considered the urban climate at the macro- or mesoscale, only little research has been directed toward the use of Earth observation data for urban microclimate analyses. This is partly because these kind of analyses usually focus on spatial scales smaller than 2 km (Helbig et al. 1999) and, thus, require high spatial resolution input data to properly resolve all relevant urban land cover elements (Welch 1982; Woodcock and Strahler 1987; Jensen and Cowen 1999). Thanks to recent technological advancements, this requirement is fulfilled by a growing number of airborne and satellite-based sensors (Ehlers 2009).

Quattrochi and Ridd (1994) were among the first to employ remote sensing imagery for the study of urban microclimates. They investigated the thermal day and night responses of 25 urban surface materials in Salt Lake City, Utah, using airborne Thermal Infrared Multispectral Scanner (TIMS) data. Ben-Dor and Saaroni (1997) examined the microscale structures of the UHI in Tel-Aviv, Israel, based on data provided by a thermal video radiometer mounted on a helicopter. Stone and Norman (2006) combined Advanced Thermal and Land Applications Sensor (ATLAS) data with property tax records of Atlanta, Georgia, to assess the influence of the size and material composition of single-family residential land use parcels on surface UHI formation. Jung et al. (2007) performed a joint analysis of Digital Airborne Imaging Spectrometer (DAIS) and additional thermal data acquired over Gyöngyös, Hungary, to define the relationship between the abundance of urban vegetation and land surface temperature (LST). Rigo and Parlow (2007) utilized satellite images from different platforms, a digital elevation model (DEM), a digital surface model (DSM), and in situ measurements to calculate and model the ground (or storage) heat flux density in Basel, Germany, with three different approaches. Xu et al. (2008) exploited hyperspectral imagery collected by the Operative Modular Imaging Spectrometer (OMIS) as well as topographic and meteorological information to map the spatial variations of turbulent sensible heat flux in Shanghai, China. Sobrino et al. (2012) interpreted LST measurements of the Airborne Hyperspectral Scanner (AHS) along with in situ data of air temperature to define the minimum spatial resolution required to properly estimate the surface UHI effect at the district level of Madrid, Spain.

Most relevant to this study is the research conducted by Heldens (2010) and Heldens et al. (2010, 2012). Their work explored the potential of airborne hyperspectral data and object height information for urban microclimate modeling in the City of Munich, Germany. To this end, hyperspectral data collected by the Hyperspectral Mapper (HyMap) sensor and object heights derived from High Resolution Stereo Camera (HRSC) imagery were utilized to infer an urban surface material map. This material map was later employed to drive ENVI-met 4 (beta), a 3D coupled flow-energy balance model to predict the urban microclimate (Bruse and Fleer 1998; Huttner and Bruse 2009). Thanks to its almost contiguous spectral coverage between 450 and 2,500 nm (Cocks et al. 1998) as well as its high spatial resolution, HyMap imagery was well suited to discriminate between different urban surface materials having distinct reflection and absorption features across its entire spectral range (cf. Heiden et al. 2007; Herold et al. 2007; Franke et al. 2009). In conjunction with the object heights extracted from HRSC data, hyperspectral remote sensing and its derivative products enabled providing the spatial information required for urban microclimate modeling with ENVI-met (Heldens 2010; Heldens et al. 2010, 2012).

The present work aims at evaluating the potential of airborne hyperspectral data and LiDAR-derived object heights for urban surface material mapping and microclimate modeling. In contrast to previous studies, the hyperspectral data cover a much smaller spectral range (380–1,050 nm, Itres Research Ltd. 2013), which complicates the detection of surface materials having spectral key features occurring only at longer wavelengths (e.g., in the short-wave infrared domain; Heiden et al. 2007; Herold et al. 2007). A dedicated approach for the fusion of the above datasets is presented, and the achieved mapping and modeling results are described and discussed accordingly.

11.3 Materials

11.3.1 Study Area

The methods described in this study are applied to hyperspectral and LiDAR remote sensing data that have been acquired over the City of Houston, Texas (29°43′16″ N, 95°21′24″ W). Houston is situated in the American South, close to the Gulf of Mexico, and comprises an area of roughly 1,500 km^2 (US Census Bureau 2013). Due to its location on the Gulf Coastal Plain (Yu et al. 2010), the city lies about 13 m above sea level (The City of Houston 2013) and does not feature significant topography (Streutker 2002). With more than two million inhabitants, Houston constitutes the fourth most populous city in the United States (The City of Houston 2013). According to the Köppen-Geiger classification system (Köppen 1936; Peel et al. 2007), the city's climate can be described as humid subtropical (Cfa). In summer, days with air temperatures above 32 °C and a relative humidity of more than 60 % are not uncommon (National Climatic Data Center, 2012a,b). Since, during that season, winds are often light and offer only little relief (University of Utah 2009), Houstonians are frequently subjected to considerable heat stress. Therefore, it is not without reason that Houston has become one of the most air-conditioned places in the world (Wilson 1992). Given its specific demographic and climatic features, the city represents a suitable area of investigation for the present study.

11.3.2 Data Basis

This study makes use of the Compact Airborne Spectrographic Imager (CASI) and LiDAR data provided within the 2013 IEEE GRSS Data Fusion Contest (Image Analysis and Data Fusion Technical Committee 2013). Both datasets cover the campus and the neighboring urban area of the University of Houston and share a common spatial resolution of 2.5 m. The CASI image was acquired between 12:37:10 and 12:39:40 CDT on June 23, 2012. Captured at an average height of 1,676 m aboveground, it features 144 spectral bands in the 380–1,050 nm region that were calibrated to at-sensor spectral radiance ($\mu W/(cm^2\, sr\, nm)$). The LiDAR data were recorded between 09:37:55 and 10:38:10 CDT on June 22, 2012. The airborne sensor's average height aboveground was 610 m. A DSM with elevation in meters above sea level was derived from the point cloud, registered to the CASI data, and delivered in GeoTIFF format. The nominal vertical accuracy of the DSM is 10–15 cm; its horizontal accuracy is about 20–30 cm.

11.4 Methods

The overall workflow presented in this study consists of three consecutive steps: (i) data preparation, (ii) material mapping, and (iii) microclimate modeling. Figure 11.1 illustrates the role of the datasets being used in the context of each stage of the data fusion approach. After data preparation, surface materials are extracted from the preprocessed CASI and LiDAR data by means of feature fusion (Pohl and van Genderen 1998). The surface material map is then utilized in combination with the object height information provided by the LiDAR data to parameterize a 3D microclimate model for simulating the spatial patterns of urban air temperature at day- and nighttime. In the following sections, the three abovementioned steps are described in more detail.

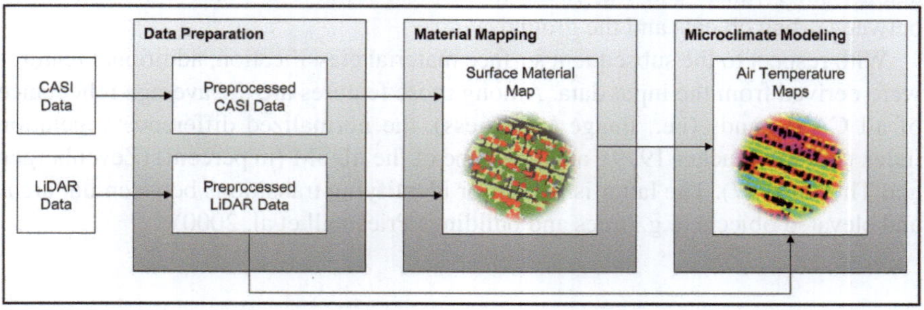

Fig. 11.1 The conceptual workflow of this study

11.4.1 Data Preparation

Preprocessing of the CASI data included the conversion of at-sensor spectral radiance to surface reflectance values using ATCOR-4 (Richter and Schläpfer 2013), spectral smoothing using a Savitzky-Golay filter (Savitzky and Golay 1964), and image clipping to exclude areas affected by clouds and cloud shadows from further analysis. Spectral smoothing is a commonly applied pre-classification step to reduce the noise contained in spectral signatures collected by hyperspectral sensors. It enables an improved identification of spectral key characteristics for urban surface material mapping. In principle, Savitzky-Golay filters approximate the "true" signature of noise-contaminated image spectra by a higher-order polynomial operating within a predefined moving window. As opposed to conventional mean filters, they mostly preserve the position and width of absorption features as well as their absolute minima and maxima. Ideally, the filtering result should reflect a compromise between spectral preservation and smoothing. Using a symmetrical kernel size and a third order polynomial, the selection of filter parameters applied in this study is based on Vaiphasa (2006).

Preprocessing of the LiDAR data involved two further steps. First, a normalized digital surface model (nDSM) was calculated from the LiDAR data. Since Houston is located at the seaside on a flat terrain, a constant elevation value of 13 m was subtracted from the DSM (The City of Houston 2013). The resulting nDSM contains the height of urban objects relative to the ground. Second, the nDSM was smoothed to enable the creation of derivative LiDAR products (e.g., nDSM slope). Even though the calculation of the nDSM is potentially too simplistic at times, one has to consider that the LiDAR DSM was made available in GeoTIFF format only (Image Analysis and Data Fusion Technical Committee 2013). Without having the original LiDAR point cloud, there is hardly any possibility to apply more sophisticated processing techniques to the input data. This is particularly true for existing nDSM generation approaches from the literature since these usually rely on LiDAR raw data. Moreover, the elevation threshold used to generate the nDSM was not only taken from official numbers (The City of Houston 2013) but was also compared against other values and was found to be most suitable to make a discrimination between urban objects and the ground.

With respect to the subsequent surface material classification, additional features were derived from the input data. Among those features are the average reflectance of all CASI bands (i.e., image brightness), the normalized difference vegetation index (NDVI) (Tucker 1979), and the slope of the nDSM (in percent) (Zevenbergen and Thorne 1987). The latter is useful for identifying transitions between flat areas and elevated objects (e.g., trees and buildings Priestnall et al. 2000).

11.4.2 Material Mapping

An overall number of 11 surface material classes were extracted from the data basis. The surface materials were primarily chosen to meet the requirements for urban

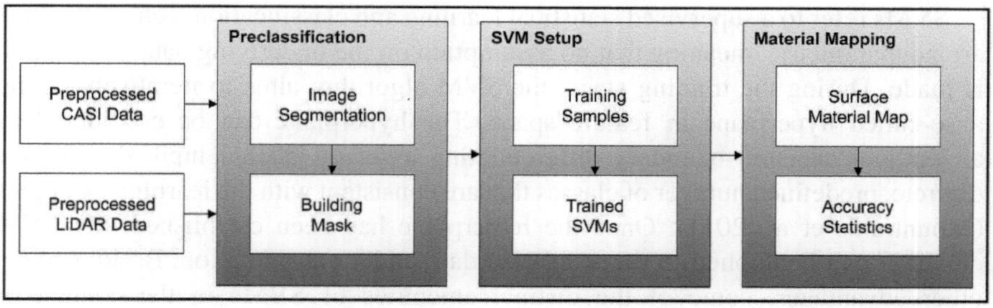

Fig. 11.2 The three-stage approach for surface material mapping

microclimate modeling. However, some target classes were also included because the spectral and spatial resolution of the input data allowed for their classification without introducing too many uncertainties in the final surface material map. For the purpose of material mapping, a hybrid, three-stage classification approach was employed (Fig. 11.2).

In the first stage, a series of segmentation algorithms was applied to the input data (Berger et al. 2013). The goal of the segmentation step was to obtain larger image objects for homogeneous regions, such as patches of grass or parking lots, and smaller image objects for heterogeneous regions, such as densely built-up areas. After the initial segmentation, a rule-based classification of the created image objects was performed (Benz et al. 2004; Trimble Ltd. 2013) to generate a building mask. The mask allowed the exclusion of spectrally similar ground classes (e.g., asphalt roads) from the later classification of roofing materials (Herold et al. 2004; Herold and Roberts 2006; Herold 2007). Image segments were first divided into elevated and non-elevated objects using the LiDAR nDSM. An object height of 2 m served as the separation threshold. The height value was chosen to enable the differentiation between small but elevated objects such as allotment garden cottages and pseudo-elevated objects such as vehicles (Ma 2005; Yu et al. 2010). Subsequently, buildings were separated from all other elevated image objects. Among the features used to derive the building mask were image brightness, the NDVI, and the slope of the nDSM (Berger et al. 2013).

In the second stage, 15 samples of each target class were selected for supervised data classification. The selection was based on spectral profile analyses, literature comparisons, and visual inspection of the CASI and LiDAR data. In addition, the training samples provided by the IEEE GRSS (Image Analysis and Data Fusion Technical Committee 2013) were taken into account. On the basis of these samples, two object-based support vector machines (SVMs) were established (Trimble Ltd. 2013). While the first SVM was trained to attribute one of four roofing materials to each object in the building mask, the second SVM was set up to extract the remaining target classes. Both SVMs were parameterized with a radial basis function and took full advantage of all available or previously generated features of the CASI and LiDAR data.

SVMs refer to a supervised statistical learning and classification technique. They are nonparametric, meaning that no assumption on the underlying data distribution is made. During the training stage, the SVM algorithm aims to iteratively define a so-called hyperplane in feature space. The hyperplane can be considered as an optimal decision boundary differentiating a set of labeled input data into a discrete, predefined number of classes that are consistent with the learning examples (Mountrakis et al. 2011). Once the hyperplane has been established, the SVM classifier can be applied to the unlabeled data under consideration. Besides many other advantages, some of the distinct amenities of SVMs in the context of multidimensional data classification are their ability to incorporate hyperspectral images without the need of any feature reduction procedure (Melgani and Bruzzone 2004) and their relative insensitivity to training sample size and quality (Mountrakis et al. 2011). Considering that the areal percentages of some surface types (e.g., roofing materials and water bodies) are comparatively low in the study area, SVM-based classifiers are well suited for the urban surface material mapping task presented in this work.

In the third stage, the trained SVMs were applied to the input data. To assess the accuracy of the resulting material map, use was made of several data sources including 0.5 m spatial resolution aerial imagery provided by the NOAA digital coast initiative (National Oceanic and Atmospheric Administration 2014), Google Street View, and Google Earth. A random sampling design was chosen comprising at least 20 sample points per target class to assess overall accuracy, errors of commission and omission, as well as the kappa index of agreement (Cohen 1960; Congalton and Green 2009).

11.4.3 Microclimate Modeling

To simulate the urban microclimate, ENVI-met 4 (beta) was used (Bruse and Fleer 1998; Huttner and Bruse 2009). Based on the fundamental laws of fluid mechanics, thermodynamics, and atmospheric physics (Bruse 2000), ENVI-met is a 3D coupled flow-energy balance model to predict the interactions between urban surfaces, vegetation, and the atmosphere for a given test site and time interval (Bruse 1999). Version 4 of the model allows, for the first time, a complete 3D representation of all land cover elements of the urban area considered, including the physical, thermal, and hydrological properties of every house front and model building block (Huttner and Bruse 2009). Hence, the information provided by the urban surface material map and the LiDAR nDSM (i.e., urban object heights) can be fully exploited for urban microclimate modeling using ENVI-met 4. While the model enables the calculation of several climatic parameters, the focus of this study was put on the simulation of urban air temperature to analyze the thermal characteristics of two test sites in the study area under varying conditions. To this end, a three-stage modeling approach was employed (Fig. 11.3).

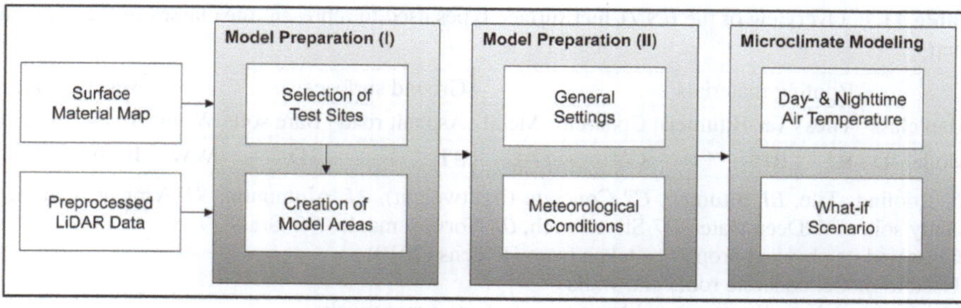

Fig. 11.3 The three-stage approach for microclimate modeling

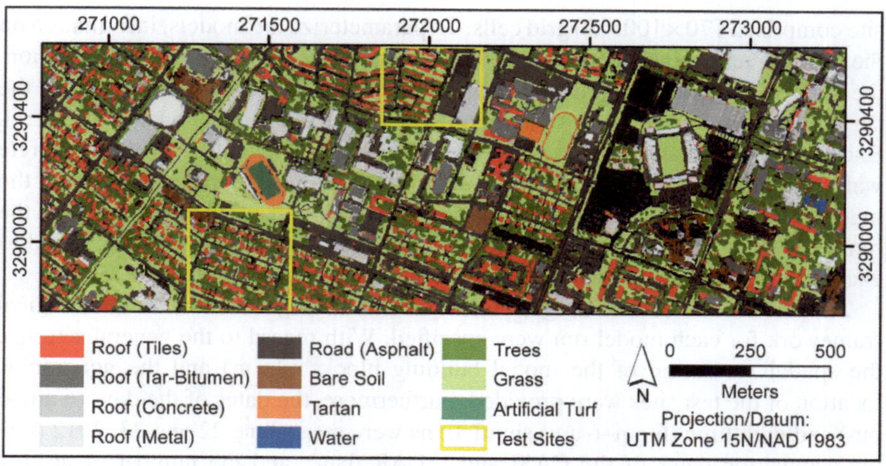

Fig. 11.4 Surface material map of the study area in Houston

In the first stage, two modeling test sites were selected (Fig. 11.4). The first test site is located in the south of the study area (29°43′6.84″ N, 95°21′47.57″ W) and comprises three blocks of single-family homes, with each block being surrounded by streets. This test site was used to model the spatial patterns of day- and nighttime air temperatures. The second test site is situated in the north of the study area (29°43′25.48″ N, 95°21′27.69″ W) and is mainly characterized by residential land use (row development). However, the western part of the test site also features a multilevel parking garage and two large parking lots that lie in-between the residential area and the parking garage. This test site was used to set up a fictional what-if scenario in which the two parking lots were replaced by multilevel parking garages to assess the impact of such an urban planning measure on nocturnal air temperatures in the adjacent residential area.

After their selection, a spatial model of both test sites was created in ENVI-met. Each model area had a spatial resolution of 2.5 m. While the southern test site consisted of 130 × 150 × 30 grid cells in x-, y-, and z-dimension, the northern test

Table 11.1 Overview of the ENVI-met surface types used to represent the classes in the material map

	Roofing materials				Ground surfaces			Vegetation	
Map class	Tiles	Tar-Bitumen	Concrete	Metal	Asphalt road	Bare soil	Water	Trees	Grass
Model ID	R1	BI[a]	C2[b]	AL	ST	SD	WW	B7/07[c]	FG[d]

R1 Roofing: Tile, *BI* Bitumen, *C2* Concrete (lightweight), *AL* Aluminum, *ST* Asphalt road, *SD* Sandy soil, *WW* Deep water, *B7* Silver birch, *07* Norway maple, *FG* Grass
[a]Class and its physical properties taken from Heldens (2010)
[b]Used to model concrete roofs and walls
[c]Used to model small ($\leq$6 m) and large trees (>6 m), respectively
[d]Plant height reduced to 15 cm

site comprised $170 \times 100 \times 30$ grid cells. To parameterize the models, information on the location and height of all land cover elements have to be provided. In addition, the type of vegetation, wall and roofing materials of buildings, as well as the surface material and soil type of non-built-up areas needs to be known (Table 11.1). Except for wall materials and soil types, which were defaulted to thin concrete walls respectively sandy soil, all necessary information were available through the material map and the LiDAR nDSM. After transferring these information from the remote sensing products to ENVI-met, each model area was slightly simplified to exclude smaller misclassifications in the material map from modeling.

As part of the second stage, some general settings and the meteorological framework for each model run were specified. With regard to the general settings, the spatial resolution of the model building blocks (2.5 m) and the geographic location of the test sites were provided. Furthermore, the dates of the day- (6 am–6 pm) and nighttime (6 pm–6 am) simulations were set to June 22 and 23, 2012 (i.e., the acquisition dates of the CASI and LiDAR data), and the number of nesting grids, which are used to minimize modeling uncertainties at the border of each test site, was defined as three. Considering the meteorological framework, ENVI-met requires hourly values of air temperature and humidity as well as the mean wind direction for each modeling date. These information were taken from data of a nearby climate station (John Dunn Helistop, Houston, Texas; Lott et al. 2001). After model preparation, the third and final stage of the approach comprised the actual simulation task. While day- and nighttime simulations of air temperature were carried out for the southern test site, two nighttime simulations were run for the northern test site (what-if scenario).

11.5 Results and Discussion

The results of this study are the urban surface material map and the microclimate modeling outputs. In the following, both results are described and discussed.

11.5.1 Surface Material Map

By using the hybrid, three-stage classification approach presented in this study, hyperspectral and LiDAR remote sensing data of Houston were fused and turned into area-wide information on urban surface materials (Fig. 11.4). Unless extensive field surveys are up for debate, the described mapping framework represents an effective way to properly capture the high degree of spectral and spatial heterogeneity found in the study area. At the same time, the obtained thematic map is able to deliver an indirect overview of the land cover and land use elements of the urban scenery, including industrial areas, single-/double-family homes and other residential districts, roads, sports stadiums, running tracks, urban green spaces, and water bodies. The mean user's and producer's accuracies of the surface material map are 79 and 81 %, respectively. While very high accuracies (≥90 %) are obtained for metal roofs, artificial turf, trees, and water bodies, high to medium accuracies (between 90 and 70 %) are reported for concrete and tile roofs, grass, asphalt roads, and bare soil. The lowest accuracies (≤70 %) are observed only for tar-bitumen roofs and tartan (Table 11.2).

Misclassifications in the map are due to different reasons. To name a few, errors in the building mask (partly originating from the simple calculation of the LiDAR nDSM) compromised the differentiation of roof materials, the spectral similarity between bare soil areas and sealed surfaces caused confusion (Yang et al. 2003; Bauer et al. 2008; Weng 2008; Esch et al. 2009; Elmore and Guinn 2010; Luo and Mountrakis 2010; Leinenkugel et al. 2011), the spatial resolution of the input data hampered feature extraction due to the presence of different target classes within a single pixel (mixed pixels) (Welch 1982; Woodcock and Strahler 1987; Jensen and Cowen 1999; Ben-Dor et al. 2001; Small 2003), and the limited spectral range of the

Table 11.2 Confusion matrix for the surface material map of the study area

		Reference												
		RTi	RTa	RCo	RMe	RAs	BAR	TAR	WAT	TRE	GRA	ART	Total	UA
Classification	Roof (tiles)	16	4	0	0	0	0	0	0	0	0	0	20	0.80
	Roof (tar-bitumen)	4	12	0	2	1	0	0	0	0	1	0	20	0.60
	Roof (concrete)	0	1	13	0	3	0	0	0	0	3	0	20	0.65
	Roof (metal)	0	3	2	30	1	2	0	0	2	0	0	40	0.75
	Road (asphalt)	0	1	0	0	18	0	0	0	0	1	0	20	0.90
	Bare soil	0	0	0	0	2	13	0	0	0	5	0	20	0.65
	Tartan	4	0	0	0	7	1	15	0	2	10	1	40	0.38
	Water	0	0	0	0	0	0	0	20	0	0	0	20	1.00
	Trees	0	0	0	0	0	0	0	0	40	0	0	40	1.00
	Grass	0	0	0	0	0	0	0	0	1	39	0	40	0.98
	Artificial turf	0	0	0	0	0	0	0	0	0	1	19	20	0.95
	Total	24	21	15	32	32	16	15	20	45	60	20	300	–
	Producer's acc.	0.67	0.57	0.87	0.94	0.56	0.81	1.00	1.00	0.89	0.65	0.95	–	–

Overall accuracy = 0.78 | Kappa coefficient = 0.76

CASI data (up to 1,050 nm) impaired the detection of some surface materials with spectral key features occurring only at longer wavelengths (e.g., in the short-wave infrared domain; Heiden et al. 2007; Herold et al. 2007). Apart from these errors, the overall accuracy of the material map is 78.3 %, and the kappa coefficient amounts to 0.76. Therefore, and according to other authors Altman (1991) and Grouven et al. (2007), the mapping result is considered as suitable for the parameterization of urban microclimate models like ENVI-met.

11.5.2 Air Temperature Maps

The simulation results for the southern test site are compiled in Fig. 11.5. The displayed day- and nighttime temperatures were calculated at a height of 1.75 m. The mean wind direction was east-southeast. Both simulations reproduce the typical thermal behavior of urban environments over the course of a warm summer day and night. At daytime, streets, driveways, parking lots, and the air layers above these and similar land use elements are heating up most intensively because they are almost constantly exposed to direct solar radiation and their physical properties facilitate the absorption of high amounts of thermal energy. At nighttime, higher air temperatures prevail in the slipstream of buildings and over broad streets because construction materials release the energy that has been stored during the day with less efficiency than natural surfaces. For both simulations, air temperatures around trees and over larger patches of short vegetation are relatively low due to the cooling effect of plant transpiration. Besides these observations, it is interesting to see that temperatures in the northernmost block of the test site are considerably raised at night. Obviously, the number and density of buildings in this block have exceeded a critical threshold so that the cooling effect of trees and larger patches of short vegetation has almost no impact on air temperature. The block's proximity to the main street in the north further contributes to the observed thermal pattern.

The simulation results for the northern test site are compiled in Fig. 11.6. The displayed nighttime temperatures were calculated at a height of 1.75 m aboveground. In this hypothetical pre-post comparison, two large parking lots are replaced by multilevel parking garages (see the yellow buildings in Fig. 11.6b) to study the effect of such a planning measure on nocturnal air temperatures. The mean wind direction was again east-southeast. The modeling outputs reveal that the residential area in the slipstream of the original parking lots would heavily suffer from increased heat stress due to the construction of further parking garages. According to the simulations, people living in close vicinity to the fictional parking garages would be exposed to raises in nighttime air temperature of up to 2 °C. Moreover, additional model runs indicate that this observation is independent of the height of the new parking garages. This case study highlights that even small-scale changes or modifications of land cover and land use can have severe effects on the urban microclimate and, by implication, also on thermal comfort (Toy et al. 2007),

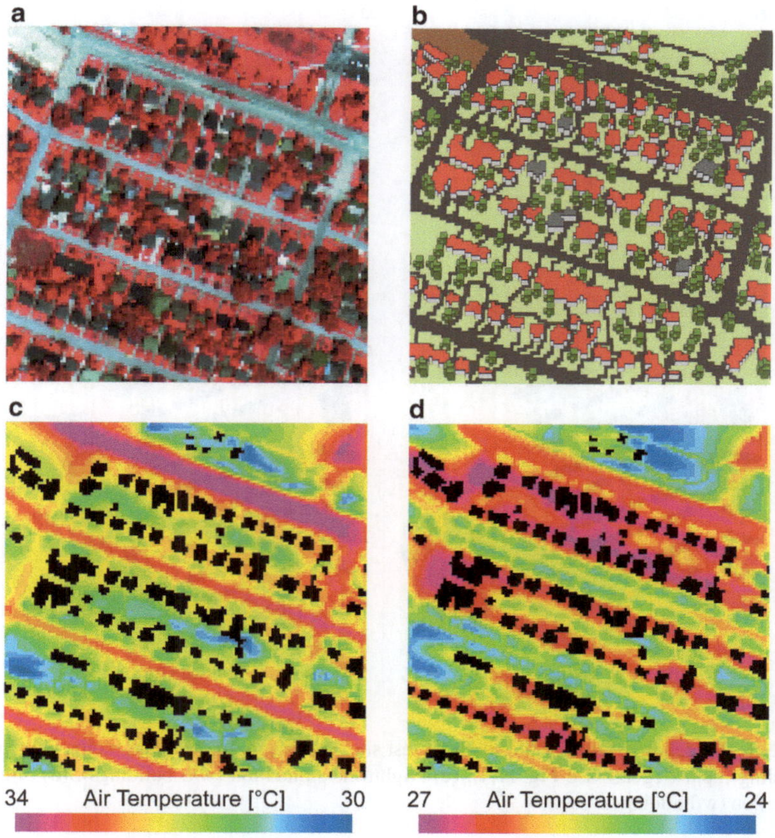

Fig. 11.5 Modeling results for the southern test site. (**a**) Original CASI data (color-infrared). (**b**) Modeling input (legend: see Fig. 11.4). (**c**) Daytime simulation (6 am–6 pm). (**d**) Nighttime simulation (6 pm–6 am)

water and energy consumption (Voogt 2002; Guhathakurta and Gober 2007; Ewing and Rong 2008; Yuan and Bauer 2007), and possibly human health (Curriero et al. 2002; Johnson and Wilson 2009). Although the function of the urban land (i.e., the provision of parking space) was preserved in the above comparison, the difference prior to and after replacing the parking lots by multilevel garages could not be any bigger from an urban climatological perspective.

It should be noted that a validation of the presented model simulations was not carried out in this study. This is mainly because of three reasons. First, the main goal of this contribution was not to test the well-documented performance of ENVI-met (Ali-Toudert et al. 2005; Ali-Toudert and Mayer 2006; Samaali et al. 2007; Skelhorn et al. 2014) but rather to demonstrate that it is technically feasible to parameterize a widely used 3D urban microclimate model by fusion of hyperspectral CASI and LiDAR remote sensing data. The two above case studies were provided to highlight

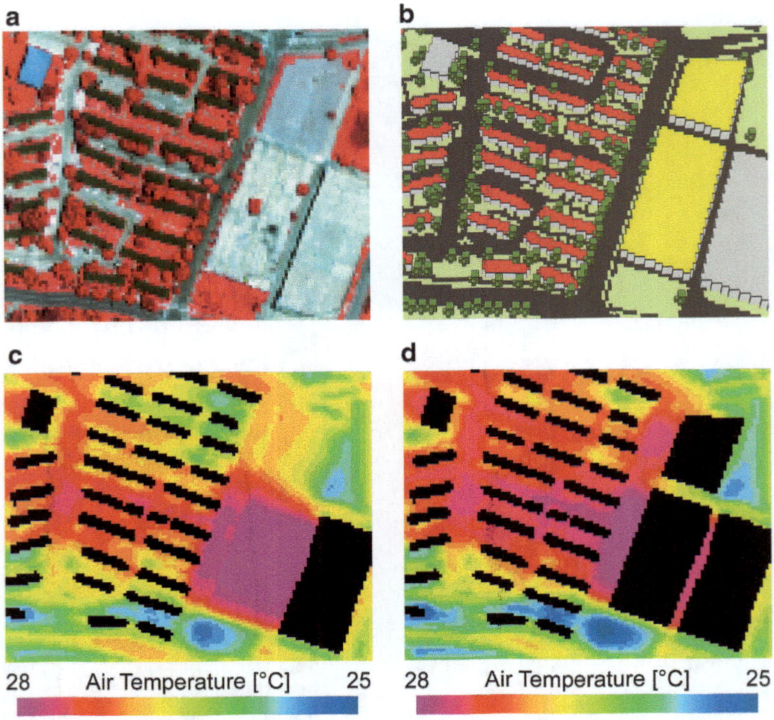

Fig. 11.6 Modeling results for the southern test site. (**a**) Original CASI data (color-infrared). (**b**) Modeling input (legend: see Fig. 11.4). (**c**) Nighttime simulation (w/o buildings). (**d**) Nighttime simulation (w/buildings)

the potential of the described mapping and modeling workflow for urban planning applications. Second, ENVI-met was primarily designed and reportedly works well for making relative, not absolute, comparisons among climate variables and their spatial patterns (Emmanuel and Fernando 2007; Bruse 2009; Heldens 2010). Since the two case studies presented do not compare absolute values of modeled air temperature, the necessity of validating the simulation outputs is debatable. Third, it was practically impossible to objectively assess model performance due to the lack of suitable reference data. Of the information needed to quantify simulation quality, neither multiday air temperature records from a high-density network of climate stations nor thermal imagery acquired by an airborne remote sensing system (e.g., AVIRIS (NASA Jet Propulsion Laboratory 2014) or G-LiHT (Cook et al. 2013)) were available for the modeling test sites.

As an alternative to model validation, one can refer to different aspects of this work suggesting that the obtained simulation results can be considered as reliable. One indication is that the surface material map – one of the most important modeling inputs – has proven to be sufficiently accurate to parameterize ENVI-met (cf. Sect. 11.5.1). Nevertheless, to exclude smaller misclassifications from the simulations, the spatial models of each test site were slightly simplified (cf. Figs. 11.5b and 11.6b). Another point is that version 4 of ENVI-met does not only require the input of certain meteorological framework conditions for a proper model initialization but it also allows for constraining the diurnal profile of specific climate variables during each simulation. This means that it is possible to fit the model to known, temporally variable climatic conditions for a given date and region of modeling. In the case of this study, the absolute values of modeled parameters like air temperature and relative humidity were forced on the basis of hourly measurements taken from a nearby climate station (John Dunn Helistop, Houston, Texas; Lott et al. 2001). Based on these indications, it is reasonable to assume that the obtained modeling results and the conclusions drawn from the relative comparisons made are valid.

Despite some remaining uncertainties, the above examples clearly demonstrate that classification-based modeling holds a large potential to capture, analyze, monitor, and predict the urban microclimate under varying conditions. The simulations enable the identification of hot spot areas that are exposed to increased heat stress and decreased thermal comfort at day and night. In addition, dedicated what-if scenarios facilitate the determination of critical land cover configurations that should be avoided in urban design. The obtained mapping and modeling products therefore represent promising sources of information which decision makers can potentially incorporate into various urban planning activities to foster effective management and to safeguard sustainable urban development.

11.6 Conclusions

Microclimate modeling is a powerful tool to study the thermal characteristics of urban environments at the local scale. However, it requires high spatial resolution, area-wide information on urban surface materials, and object heights that are usually hard to obtain by traditional field surveys. This work aimed at the derivation of an urban surface material map to parameterize a 3D numerical microclimate model by fusion of airborne hyperspectral and LiDAR remote sensing data. To demonstrate the potential of data-driven microclimate modeling, two case studies were presented for selected test sites in the City of Houston, Texas. The results of this study highlight that a synergistic combination of hyperspectral and LiDAR data enables reliable mapping of some of the key input parameters required for urban microclimate modeling. Moreover, classification-based microclimate simulations can reveal the thermal properties of urban neighborhoods under varying conditions and, thus, facilitate the identification of hot spot areas and critical land cover configurations which should be avoided in urban design. Given the ever-increasing availability of hyperspectral and LiDAR data (e.g., NASA Jet Propulsion Laboratory 2014; Cook et al. 2013; LiDAR Online 2014; OpenTopography 2014), it is concluded that spatially explicit microclimate modeling should be an integral part of urban planning to enable making more informed decisions about the future of urban environments.

However, further investigations are still required in order to get a more complete picture of the data and methods under consideration. Among other issues, future studies should comprehensively test the generalization capabilities of the developed mapping procedure (different study areas and/or input data) and rigorously validate the performance of data-driven microclimate modeling. In the long run, this would effectively help promoting hyperspectral and LiDAR remote sensing technologies for urban planning applications.

Acknowledgements The authors would like to thank the Hyperspectral Image Analysis group and the NSF Funded Center for Airborne Laser Mapping (NCALM) at the University of Houston for providing the "grss_dfc_2013" data set used in this study and the IEEE GRSS Image Analysis and Data Fusion Technical Committee for organizing the 2013 Data Fusion Contest. The authors also wish to thank the three anonymous reviewers whose comments helped to substantially improve an earlier version of this chapter.

References

Ali-Toudert F, Djenane M, Bensalem R, Mayer H (2005) Outdoor thermal comfort in the old desert city of Beni-Isguen, Algeria. Clim Res 28(3):243–256

Ali-Toudert F, Mayer H (2006) Numerical study on the effects of aspect ratio and orientation of an urban street canyon on outdoor thermal comfort in hot and dry climate. Build Environ 41(2):94–108

Altman D (1991) Practical statistics for medical research. Chapman & Hall, Boca Raton, p 624

Angel S, Parent J, Civco D, Biel A (2011) Making room for a planet of cities. Policy focus report, ID PF027, Lincoln Institute of Land Policy, Cambridge

Arnfield A (2003) Two decades of urban climate research: a review of turbulence, exchanges of energy and water, and the urban heat Island. Int J Clim 23(1):1–26

Bauer M, Loeffelholz B, Wilson B (2008) Estimating and mapping impervious surface area by regression analysis of Landsat imagery. In: Weng Q (ed) Remote sensing of impervious surfaces. CRC, Boca Raton, pp 3–19

Ben-Dor E, Levin N, Saaroni H (2001) A spectral based recognition of the urban environment using the visible and near-infrared spectral region (0.4–1.1 μm). A case study over Tel-Aviv, Israel. Int J Remote Sens 22(11):2193–2218

Ben-Dor E, Saaroni H (1997) Airborne video thermal radiometry as a tool for monitoring microscale structures of the urban heat Island. Int J Remote Sens 18(14):3039–3053

Benz U, Hofmann P, Willhauck G, Lingenfelder I, Heynen M (2004) Multi-resolution, object-oriented fuzzy analysis of remote sensing data for GIS-ready information. ISPRS J Photogramm 58(3–4):239–258

Berger C, Voltersen M, Hese S, Walde I, Schmullius C (2013) Robust extraction of urban land cover information from HSR multi-spectral and LiDAR data. IEEE J Sel Top Appl Earth Obs Remote Sens 6(5):2196–2211

Bruse M (1999) Simulating microscale climate interactions in complex terrain with a high-resolution numerical model: a case study for the Sydney CBD area (model description). In: Proceedings of the ICUC & ICB, Sydney, pp 1–6

Bruse M (2000) Anwendung von mikroskaligen Simulationsmodellen in der Stadtplanung. In: Bernhard L, Küger T (eds) Simulation raumbezogener Prozesse: Methoden und Anwendung, IfGIprints 9. University Press, Münster, pp 1–21

Bruse M (2009) ENVI-met board. To compare simulated and measured results validly. http://www.envi-met.info/envibbv3/viewtopic.php?f=11&t=142. Cited 6 Jan 2014

Bruse M, Fleer H (1998) Simulating surface–plant–air interactions inside urban environments with a three dimensional numerical model. Environ Model Softw 13(3–4):373–384

Changnon S (1992) Inadvertent weather modification in urban areas: lessons for global climate change. Bull Am Meteor Soc 73(5):619–627

Cocks T, Jenssen R, Stewart A, Wilson I, Shields T (1998) The HyMap airborne hyperspectral sensor: the system, calibration and performance. In: Proceedings of the 1st EARSeL workshop on imaging spectroscopy, Zurich, pp 37–42

Cohen J (1960) A coefficient of agreement for nominal scales. Educational and Psychological Measurement 20(1), 37–46

Congalton R, Green K (2009) Assessing the accuracy of remotely sensed data. Principles and practices, 2nd edn. CRC, Boca Raton

Cook B, Corp L, Nelson R, Middleton E, Morton D, McCorkel J, Masek J, Ranson KJ, Ly V, Montesano P (2013) NASA Goddard's LiDAR, hyperspectral and thermal (G-LiHT) airborne imager. Remote Sens 5(8):4045–4066

Curriero F, Heiner K, Samet J, Zeger S, Strug L, Patz J (2002) Temperature and mortality in 11 cities of the Eastern United States. Epidemiology 155(1):80–87

Ehlers M (2009) Future EO sensors of relevance – integrated perspective for global urban monitoring. In: Gamba P, Herold M (eds) Global mapping of human settlement. CRC, Boca Raton, pp 321–337

Elmore A, Guinn S (2010) Synergistic use of Landsat multispectral scanner with GIRAS land-cover data to retrieve impervious surface area for the Potomac River Basin in 1975. Remote Sens Environ 114(10):2384–2391

Emmanuel R, Fernando H (2007) Urban heat Islands in humid and arid climates: role of urban form and thermal properties in Colombo, Sri Lanka and Phoenix, USA. Clim Res 34(3):241–251

Esch T, Himmler V, Schorcht G, Thiel M, Wehrmann T, Bachofer F, Conrad C, Schmidt M, Dech S (2009) Large-area assessment of impervious surface based on integrated analysis of single-date Landsat-7 images and geospatial vector data. Remote Sens Environ 113(8):1678–1690

Ewing R, Rong F (2008) The impact of urban form on U.S. residential energy use. Hous Policy Debate 19(1):1–30

Franke J, Roberts D, Halligan K, Menz G (2009) Hierarchical multiple endmember spectral mixture analysis (MESMA) of hyperspectral imagery for urban environments. Remote Sens Environ 113(8):1712–1723

Grouven U, Bender R, Ziegler A, Lange S (2007) The kappa coefficient. Dtsch Med Wochenschr 132(23):65–68

Guhathakurta S, Gober P (2007) The impact of the Phoenix urban heat Island on residential water use. J Am Plan Assoc 73(3):317–329

Heiden U, Segl K, Roessner S, Kaufmann H (2007) Determination of robust spectral features for identification of urban surface materials in hyperspectral remote sensing data. Remote Sens Environ 111(4):537–552

Helbig A, Baumüller J, Kerschgens M (1999) Stadtklima und Luftreinhaltung, 2nd edn. Springer, Berlin

Heldens W (2010) Use of airborne hyperspectral data and height information to support urban micro climate characterisation. PhD thesis, Julius-Maximilians-Universität, Würzburg

Heldens W, Esch T, Heiden U (2012) Supporting urban micro climate modelling with airborne hyperspectral data. In: Proceedings of the IGARSS, Munich, pp 1598–1601

Heldens W, Heiden U, Esch T, Dech S (2010) Potential of hyperspectral data for urban micro climate analysis. In: Proceedings of the ESA hyperspectral workshop, Frascati, pp 1–8

Heldens W, Heiden U, Esch T, Stein E, Müller A (2011) Can the future EnMAP mission contribute to urban applications? A literature survey. Remote Sens 3(9):1817–1846

Herold M (2007) Spectral characteristics of asphalt road surfaces. In: Weng Q (ed) Remote sensing of impervious surfaces. CRC, Boca Raton, pp. 237–248

Herold M, Roberts D (2006) Multispectral satellites – imaging spectrometry – LIDAR: spatial – spectral tradeoffs in urban mapping. Int J Geoinf 2(1):1–13

Herold M, Roberts D, Gardner M, Dennison P (2004) Spectrometry for urban area remote sensing – development and analysis of a spectral library from 350 to 2400 nm. Remote Sens Environ 91(3–4):304–319

Herold M, Schiefer S, Hostert P, Roberts D (2007) Applying imaging spectrometry in urban areas. In: Weng Q, Quattrochi D (eds) Urban remote sensing. CRC, Boca Raton, pp 137–161

Howard L (1833) The climate of London deduced from meteorological observations made in the metropolis and at various places around it, vol I–III, 2nd edn. J. Rickerby, London

Huttner S, Bruse M (2009) Numerical modeling of the urban climate – a preview on ENVI-met 4.0. In: Proceedings of the 7th ICUC, Yokohama, pp 1–4

Image Analysis and Data Fusion Technical Committee (2013) 2013 IEEE GRSS data fusion contest. http://www.grss-ieee.org/community/technical-committees/data-fusion/data-fusion-contest/. Cited 6 Jan 2014

Itres Research Ltd. (2013) CASI-1500 hyperspectral imager. http://www.itres.com/products/imagers/casi1500. Cited 6 Jan 2014

Jensen J, Cowen D (1999) Remote sensing of urban/suburban infrastructure and socio-economic attributes. Photogramm Eng Remote Sens 65(5):611–622

Johnson D, Wilson J (2009) The socio-spatial dynamics of extreme urban heat events: the case of heat-related deaths in Philadelphia. Appl Geogr 29(3):419–434

Jung A, Tökei L, Kardevan P (2007) Application of airborne hyperspectral and thermal images to analyse urban microclimate. Appl Ecol Environ Res 5(1):165–175

Kalnay E, Cai M (2003) Impact of urbanization and land-use change on climate. Nature 423:528–531

Köppen W (1936) Das geographische System der Klimate. In: Köppen W, Geiger R (eds) Handbuch der Klimatologie, vol 1, Part C. Gebrüder Borntraeger, Berlin, pp 1–44

Landsberg H (1981) The urban climate. International geophysics series, vol 28. Academic, New York City

Leinenkugel P, Esch T, Gähler M (2011) Settlement detection and impervious surface estimation in the Mekong Delta using optical and SAR remote sensing data. Remote Sens Environ 115(12):3007–3019

LiDAR Online (2014) LiDAR online – worldwide LiDAR data and geoservices. https://www.lidar-online.com/. Cited 6 Jan 2014

Lott N, Baldwin R, Jones P (2001) The FCC integrated surface hourly database, a new resource of global climate data. Technical report 2001-01, National Climatic Data Center, Asheville

Lowry W (1998) Urban effects on precipitation amount. Prog Phys Geogr 22(4):477–520

Luo L, Mountrakis G (2010) Integrating intermediate inputs from partially classified images within a hybrid classification framework: an impervious surface estimation example. Remote Sens Environ 114(6):1220–1229

Ma R (2005) DEM generation and building detection from LiDAR data. Photogramm Eng Remote Sens 71(7):847–854

McCarthy M, Best M, Betts R (2010) Climate change in cities due to global warming and urban effects. Geophys Res Lett 37(9):1–5

Melgani F, Bruzzone L (2004) Classification of hyperspectral remote sensing images with support vector machines. IEEE Trans Geosci Remote Sens 42(8):1778–1790

Mountrakis G, Im J, Ogole C (2011) Support vector machines in remote sensing: a review. ISPRS J Photogramm 66(3):247–259

NASA Jet Propulsion Laboratory (2014) AVIRIS data – ordering free AVIRIS standard data products. http://aviris.jpl.nasa.gov/data/free_data.html. Cited 6 Jan 2014

National Climatic Data Center (2012a) Houston, TX. Average relative humidity (%). http://lwf.ncdc.noaa.gov/oa/climate/online/ccd/avgrh.html. Cited 6 Jan 2014

National Climatic Data Center (2012b) Houston, TX. Mean number of days with maximum temperature 90 degrees F or higher. http://lwf.ncdc.noaa.gov/oa/climate/online/ccd/max90temp.html. Cited 6 Jan 2014

National Oceanic and Atmospheric Administration (2013) NOAA digital coast data access viewer. http://www.csc.noaa.gov/dataviewer/. Cited 6 Jan 2014

Oke T (1973) City size and the urban heat Island. Atmos Environ 7(8):769–779
Oke T (1982) The energetic basis of the urban heat Island. Q J R Meteor Soc 108(455):1–24
OpenTopography (2014) OpenTopography – a portal to high-resolution topography data and tools. http://www.opentopography.org/. Cited 6 Jan 2014
Peel M, Finlayson B, McMahon T (2007) Updated world map of the Köppen-Geiger climate classification. Hydrol Earth Syst Sci 11(5):1633–1644
Pohl C, van Genderen J (1998) Multisensor image fusion in remote sensing: concepts, methods and applications. Int J Remote Sens 19(5):823–854
Priestnall G, Jafaar J, Duncan A (2000) Extracting urban features from LiDAR digital surface models. Comput Environ Urban Syst 24(2):65–78
Quattrochi D, Ridd M (1994) Measurement and analysis of thermal energy responses from discrete urban surfaces using remote sensing data. Int J Remote Sens 15(10):1991–2022
Richter R, Schläpfer D (2013) Atmospheric/topographic correction for airborne imagery. ATCOR-4 user guide, version 6.2.1. German Aerospace Center (DLR), Wessling
Rigo G, Parlow E (2007) Modelling the ground heat flux of an urban area using remote sensing data. Theor Appl Climatol 90(3–4):185–190
Samaali M, Courault D, Bruse M, Olioso A, Occelli R (2007) Analysis of a 3D boundary layer model at local scale: validation on soybean surface radiative measurements. Atmos Res 85(2):183–198
Savitzky A, Golay M (1964) Smoothing and differentiation of data by simplified least squares procedures. Anal Chem 36(8):1627–1639
Scalenghe R, Marsan F (2009) The anthropogenic sealing of soils in urban areas. Landsc Urban Plan 90(1):1–10
Skelhorn C, Lindley S, Levermore G (2014) The impact of vegetation types on air and surface temperatures in a temperate city: a fine scale assessment in Manchester, UK. Landsc Urban Plan 121:129–140
Small C (2003) High spatial resolution spectral mixture analysis of urban reflectance. Remote Sens Environ 88(1–2):170–186
Sobrino J, Oltra-Carrió R, Sòria G, Bianchi R, Paganini M (2012) Impact of spatial resolution and satellite overpass time on evaluation of the surface urban heat island effects. Remote Sens Environ 117:50–56
Stone B, Norman J (2006) Land use planning and surface heat Island formation: a parcel-based radiation flux approach. Atmos Environ 40(19):3561–3573
Streutker D (2002) A remote sensing study of the urban heat Island of Houston, Texas. Int J Remote Sens 23(13):2595–2608
Taubenböck H, Esch T, Felbier A, Wiesner M, Roth A, Dech S (2012) Monitoring urbanization in mega cities from space. Remote Sens Environ 117:162–176
The City of Houston (2013) Houston facts and figures. http://www.houstontx.gov/abouthouston/houstonfacts.html. Cited 6 Jan 2014
Toy S, Yilmaz S, Yilmaz H (2007) Determination of bioclimatic comfort in three different land uses in the city of Erzurum, Turkey. Build Environ 42(3):1315–1318
Trimble Ltd. (2013) eCognition developer 8.7.1. Reference book. Trimble documentation, Munich
Tucker C (1979) Red and photographic infrared linear combinations for monitoring vegetation. Remote Sens Environ 8(2):127–150
United Nations (2008) World urbanization prospects. The 2007 revision. Executive summary, Department of Economic and Social Affairs. Population Division, New York City
University of Utah (2009) Houston, TX. Wind – average speed (mph). Utah and national climate data. http://www.met.utah.edu/jhorel/html/wx/climo.html. Cited 6 Jan 2014
US Census Bureau (2013) Houston (city), Texas. Land area in square miles, 2010. State & County QuickFacts. http://quickfacts.census.gov/qfd/states/48/4835000.html. Cited 6 Jan 2014
Vaiphasa C (2006) Consideration of smoothing techniques for hyperspectral remote sensing. ISPRS J Photogramm 60(2):91–99
Voogt J (2002) Urban heat island. In: Munn T (ed) Encyclopedia of global environmental change, vol 3. Wiley, Chichester, pp 660–666

Voogt J, Oke T (2003) Thermal remote sensing of urban climates. Remote Sens Environ 86(3): 370–384

Welch R (1982) Spatial resolution requirements for urban studies. Int J Remote Sens 3(2):139–146

Weng Q (2008) Remote sensing of impervious surfaces. In: Weng Q (ed) Remote sensing of impervious surfaces. CRC, Boca Raton, pp XV–XXVI

Wilson F (1992) A moment in building. Blueprints X(3) http://web.archive.org/web/20070206075059/http://www.nbm.org/blueprints/90s/summer92/contents/contents.htm. Cited 6 Jan 2014

Woodcock C, Strahler A (1987) The factor of scale in remote sensing. Remote Sens Environ 21(3):311–332

Xu W, Wooster M, Grimmond C (2008) Modelling of urban sensible heat flux at multiple spatial scales: a demonstration using airborne hyperspectral imagery of Shanghai and a temperature-emissivity separation approach. Remote Sens Environ 112(9):3493–3510

Yang L, Huang C, Homer C, Wylie B, Coan M (2003) An approach for mapping large-area impervious surfaces: synergistic use of Landsat-7 ETM+ and high spatial resolution imagery. Can J Remote Sens 29(2):230–240

Yu B, Liu H, Wu J, Hu Y, Zhang L (2010) Automated derivation of urban building density information using airborne LiDAR data and object-based method. Landsc Urban Plan 98(3–4):210–219

Yuan F, Bauer M (2007) Comparison of impervious surface area and normalized difference vegetation index as indicators of surface urban heat Island effects in Landsat imagery. Remote Sens Environ 106(3):375–386

Zevenbergen L, Thorne C (1987) Quantitative analysis of land surface topography. Earth Surf Process Landf 12(1):12–56

Chapter 12
Modeling Urban Land Use Change: Integrating Remote Sensing with Socioeconomic Data

Junmei Tang

Abstract Rapid urban development has stimulated the progress in predicting and evaluating urban landscape evolution. As a result of rapid socioeconomic development, the land use pattern of Houston, TX, has undergone significant changes over the past 30 years. It is essential to simulate urbanization processes in Houston to examine where and to what extent landscape change has occurred and further to understand how and why the change can occur. This research developed two cellular automata (CA) models based on the same remote sensing data source: one was based on the classification from Landsat images and another one incorporated the socioeconomic data with the same classification results. The predicted results from these two models suggested that the incorporation of socioeconomic data improved the accuracy in human-intervened landscapes, such as residential and industrial/commercial area. More socioeconomic data and finer data sources were needed to improve the CA model to predict the heterogeneous pattern within urban areas.

Keywords Urban land use change • CA model • Socioeconomic data • Remote sensing

12.1 Introduction

Rapid urbanization in the past 50 years, triggered by the population growth and migration from rural to urban and suburban areas, presents one of the greatest challenges in environmental, economic, social, political, and cultural research (Antrop 2004; Tang et al. 2012; Tayyebi et al. 2012). The total urban population is 82 % with an estimated 1.2 % annual increasing rate from 2010 to 2015 in the United States (US Census 2011). The motivation to model urban landscape dynamics arises from the process of examining where and to what extent landscape change has occurred, and furthermore, the need to understand how and why the changes can

---

J. Tang (✉)
Geography and Environmental Systems Department, University of Maryland, Baltimore County, Baltimore, MD 21250, USA
e-mail: junmei@umbc.edu

M. Helbich et al. (eds.), *Computational Approaches for Urban Environments*, Geotechnologies and the Environment 13, DOI 10.1007/978-3-319-11469-9_12

occur (Weng 2002; Yang and Lo 2002). One of the greatest challenges in designing effective urban models is that their performances are often limited by the inadequate digital data source over time as well as the consideration of external driver such as socioeconomic development and human disturbance (Pickett et al. 1997; Mcintyre et al. 2000).

Remote sensing data, with the ability to provide large-scale data sources such as historical maps or urban land use maps, has been used as an effective tool in quantitatively measuring urban landscape and modeling urbanization at a relatively large spatial scale (Herold et al. 2003; Tang 2011). Images from satellite sensors provide a large amount of cost-effective multispectral and multi-temporal data to monitor landscape changes and estimate biophysical characteristics of land surfaces (Weng 2002). Many researchers have proposed the routine to combine remote sensing with GIS in urban growth models (Tang 2011; Tayyebi et al. 2013). Significant progress in acquiring remotely sensed data in a higher spatial resolution and developing the spatial geographic process model has widened our research on the process, driving forces, and impacts of the urbanization.

The cellular automata (CA) model, introduced by Tobler in 1979, is one of the most powerful spatial dynamics techniques used to simulate complex urban systems (Batty and Xie 1994). The CA model allows researchers to view the city as a self-organizing system in which the basic land parcels are developed into various land use types. Cecchini and Viola (1990) applied simple decision rules in the CA model to predict the complex, large-scale structure in the urban growth process. Wu (1998) combined the multicriteria evaluation (MCE) and GIS into the CA model to define the transition rules in a visualized environment. Shafizadeh-Moghadam and Helbich (2013) used AHP (analytical hierarchy process) to determine the weight in a Markov chains-cellular automata urban growth model.

The advantages of the CA model in simulating urban spatial process and dynamics (Hillier and Hanson 1984; White and Engelen 1993) have been widely documented because the theoretical abstraction of the CA model and the practical constraints in the real world can be easily related (Batty and Xie 1994; Clarke and Hoppen 1997; Wu and Martin 2002). The model begins from a homogeneous cell-based grid and adjusts itself through the transition rule derived from its local spatiotemporal neighborhood. This makes the CA model suitable to simulate complex and hierarchical structures since more unknown, immeasurable spatiotemporal variables can be incorporated and manipulated in this model. Another advantage in CA simulation is the ability of the model to incorporate proper parameters or weights to model the alternative socioeconomic states in the model development (Clarke and Gaydos 1998; Li and Yeh 2000). With better computer techniques, the CA model is also able to explore more complex human behavior through defining different transition rules (Li and Yeh 2000; Wu and Martin 2002). However, the tension between the simple local transition rule in CA models and the complex, unpredicted social changes in urban landscapes still remains.

In this context, this chapter attempted to develop a spatial-explicitly CA model to simulate urban growth patterns using the classification result from Landsat images and another one incorporated the socioeconomic data with the same classification results. Two CA models were compared to test how the socioeconomic data could improve the urban model simulation in Houston during the last 30 years. Specially, the following research questions were addressed: How the socioeconomic data could be incorporated with remote sensing in the urban growth model? Does the socioeconomic data improve the model? In which classes does this model improve?

12.2 Urban Model Review and Socioeconomic Data in the Model

With the availability of spatial data on a large scale, various sophisticated models, especially after the late 1990s, were developed such as UrbanSim model (Waddell 2002), Markov chain model (Stewart 1994), LUCAS model (Berry et al. 1996), CLUE model (De Kong et al. 1999), area-based model (Lichtenberg 1985; Tayyebi et al. 2011, 2013); CA model (Batty and Xie 1994), Land Transformation Model (Pijanowski et al. 1997, 2014), and agent-based model (Liebrand et al. 1998). The detailed review of these spatial explicit models is listed in Table 12.1.

In terms of the methods to represent the model object, there are vector-based models and grid-based models (Herold 2004), and both of them have been used to incorporate socioeconomic data. Vector-based models use the thematic map as the input data for the model, and the spatial objects are usually defined as homogenous land units. *UrbanSim* is one of land use simulation models for the growth government, regional land use, and transportation planning in the states of Hawaii, Oregon, and Utah (Waddell 2002). Within the context of urban infrastructure and governmental policy, *UrbanSim* represents zonal structure in the urban area to monitor the socioeconomic-related behaviors of households, business, and land developers. Theoretically, *UrbanSim* is an object-oriented model. *What if* model (Klosterman 1999) begins with uniform analysis zones or homogeneous land units generated from the GIS software. Through applying the governmental policies and land use demands, this model derives the aggregating value of the regional condition on the land units. *What if* model projects future land use patterns by balancing the supply, demand, and land sustainable at different locations. *Area-base*d model is a vector-based model used in resource assessments to predict the availability of farm and forest land. Transformed from the regional model (Palmquist 1989), *area-based* model allocates the proportions of a given land use to predefine land use categories using Lichtenberg's (1985) acreage allocation method (Tayyebi et al. 2011, 2013).

Another vector-based model is *Markov* model which predicts future landscape patterns based on the spatial transition probability. Although Markov model is a typical spatial transition model, early Markovian analysis is a descriptive tool to predict land use change on a local or regional scale (Bell 1974; Bourne 1976; Arsanjani et al. 2013). Actually, the Markov model is not a strict vector-based model; it is based on the statistical results from the thematic map. Lopez et al. (2001) used Markov chain to simulate the relationships among a set of urban and

Table 12.1 Detailed comparison of urban models

	Model name	Developer	Purpose	Variables	Strengths	Weakness
Vector based	UrbanSim	Waddell (2002)	Predict land use, physical development, the movement, and location of businesses and households	Socioeconomic, environmental parameters	Simulate the interaction between urban activities and the natural environment	Deterministic model
	What if	Klosterman (1999)	Determine what will happen in land use patterns in the future if policy choices are made	Natural conditions, urban infrastructure	Fully operational model in adapting particular data sets and policy concerns	Lack of a firm theoretical basis
	Area based	Lichtenberg (1985) and Tayyebi et al. (2011, 2013)	Project the proportions of land use using hedonic rent theory and acreage allocation model	Based on Palmquist's hedonic rent theory and acreage allocation model (Lichtenberg 1985)	Easy to incorporate available socioeconomic data, such as age, income, population, and rent	Long-term prediction is not good
	Markov	Bell (1974) and Tang et al. (2007)	Characterize the land use/land cover change	Multi-temporal change, transition probabilities	Mathematically compact, easy to implement	Does not account for spatial context

Grid based	CA	Batty and Xie (1994) and Tang (2011)	Model the spatial structure of urban land use over time	Landscape maps, environmental factors in neighborhood	The spatial factor and ecological aspects are easily to be incorporated	Face challenge in incorporating human decision making
	LUCAS	Berry et al. (1996)	Simulate the landscape change with socioeconomic information and its environmental impacts	Socioeconomic variables, such as transportation network, ownership, population density, etc.	Flexible and interactive computing environment	The patch size is no sense due to the pixel-based method
	LTM	Pijanowski et al. (1997, 2014)	Analyze the land use change and predict land use pattern	Landscape maps, social, political, and environmental factors	Can be applied to multiple scales using a moving scalable window metric	Suppose all the variables constant
	CLUE	De Kong et al. (1999)	Predict future land use	Biophysical drivers and human drivers	Covers a wide variety of biophysical and human factors and ranges multiple spatiotemporal scales	No social or political factors
Grid/vector	Agent based	Liebrand et al. (1998)	Represent a wide variety of entities and its activities	A simulated environment, entities under human decision making	Flexibility and successful in replicating human decision	Difficult to develop and control

social variables in predicting land use/cover change in the urban fringe of Morelia city, Mexico. Weng (2002) demonstrated that the integration of satellite remote sensing and GIS techniques into the stochastic urban modeling was an effective approach for analyzing the direction, rate, and spatial pattern of landscape change in Zhujiang Delta of China. Tang et al. (2007) improved the Markov chain model by incorporating a modified genetic algorithm in the urban boundary expansion for urban simulation. Mathematically, most vector-based models rely on some static equations, and this characteristic provides the potential in integrating the statistical information into the model entities. The major drawbacks of such models are the poor handling in dynamic entities and poor representation of external variables, e.g., the spatial information and socioeconomic factors.

The models developed on grid have more advantage in solving these problems than the vector ones. Land-Use Change Analysis System (*LUCAS*) is a grid-based model which integrates socioeconomic and ecological variables in the multilayered, gridded maps (Berry et al. 1996). This model consists of three subject modules: socioeconomics, which derives the transition probability from the function of socioeconomic driving variables; landscape change, which predicts the landscape maps from the socioeconomic module; and environmental impacts, which estimates the impacts of selected environmental variables from the landscape maps from second modules. Land Transformation Model (*LTM*) (Pijanowski et al. 1997, 2014) applied the spatial rules to land use transitions for each location in the processed spatial layer or grid. It is easy to quantify the contribution of different spatial variables because of its grid format. In order to aggregate the land use change and change drivers, this model adopted the similar method with the Conversion of Land Use and its Effects (*CLUE*) model (De Kong et al. 1999). Both of them apply the variable values in grid format to create a series of future land use patterns over the time. *Cellular automata* model has been proposed and developed to simulate the urban land use model by incorporating various socioeconomic variables, such as dynamic transportation model (Aljoufie et al. 2013) and dynamic population density (Van Vliet et al. 2012).

Agent-based model (Liebrand et al. 1998) is a complex behavior model which used both vector data and raster data. Usually, the raster data is the agents' environment, and the agents, in turn, act on the simulated environment. This model can be applied to a wide variety of simulations, including moving cars, animals, people, or even organizations. The socioeconomic variable, as both agents' status and driving forces, was incorporated into the model to simulate individual activities (An et al. 2005). This model is difficult to develop and control since we need to incorporate the "individual agent" information and predict its potential behaviors.

Generally, a reliable urban growth model should have the following capabilities: (1) providing an appropriate theoretical and technical framework for urban growth; (2) understanding and describing the historical dynamics of urban structures; and (3) exploring and incorporating different economic and social parameters to monitor the urban growth.

12.3 Study Area and Data Preparation

The eastern metropolitan area of Houston, Texas, covering an area of 1,200 km^2, was chosen as the study site (Fig. 12.1). Houston is situated in the northern portion of the Gulf coastal plain, a 60 by 80 km-wide swath along the Texas Gulf Coast, 80 km from the Gulf of Mexico (Moser 1998). This area has experienced rapid urban development since the 1930s after the discovery of oil (Tang et al. 2008) in nearby oil fields. These discoveries made it the largest city in Texas as of 1930 and the fourth largest city in United States since 1990 (Texas State Historical Association 2002). Although the government tried to diversify its economy (Key to the city 2001), the city's unchallenged role as an international center of oil technology, headquarters for a number of the world's largest energy companies, and a strong refining and petrochemical manufacturing base should shore up the local economy

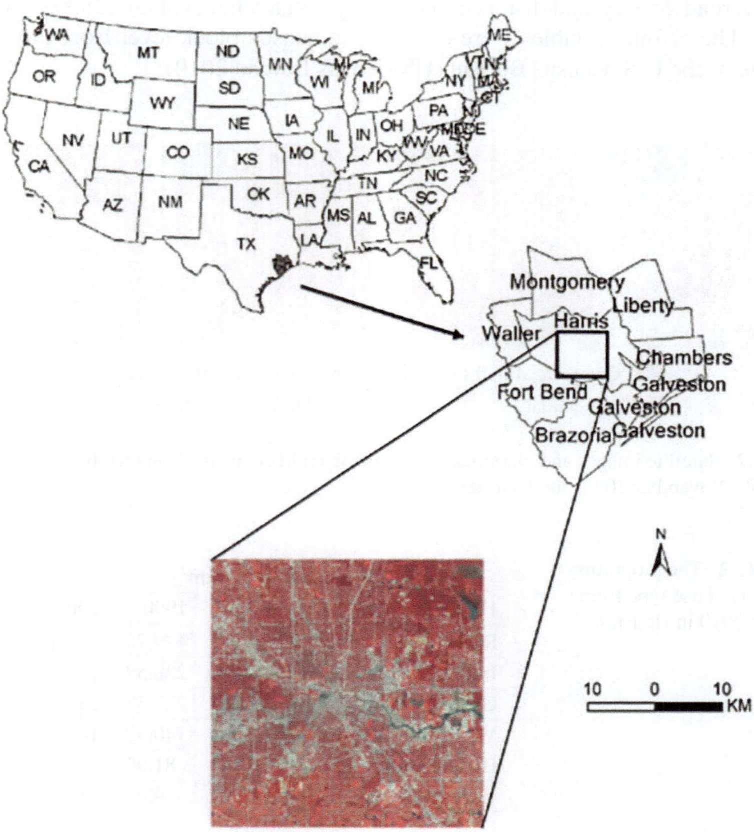

Fig. 12.1 Study site, Houston, Texas, in the United States

of Houston in the near future. The representative land use/land cover classes in this selected region include residential area, commercial/industrial area, transportation, woodland, grassland, and barren/soil.

The satellite Landsat MSS/TM images were collected from 1970s to 2010 in this study. All these images were georeferenced to the Universal Transverse Mercator projection using ENVI. The convention Maximum Likelihood Classification was adopted to obtain four classified landscape maps with six landscape classes for each map. We chose two set of samples around 600 pixels for training samples and test samples, respectively. The selection of separate of training and test samples was guided by the characteristic of each class at different years. The overall accuracy assessment of classified maps was 92 % (1979), 94 % (1990), 96 % (2000), and 95 % (2010). Figure 12.2 shows the detailed proportion of each land use type as shown in Table 12.2.

In order to represent the rapid socioeconomic development in the Houston area, four major socioeconomic variables were collected: population density, house density, road density, and distance to highways (Van Vliet et al. 2012; Aljoufie et al. 2013). These four variables were collected at census block level from the official website of the U.S. Census Bureau (US Census Bureau 2010).

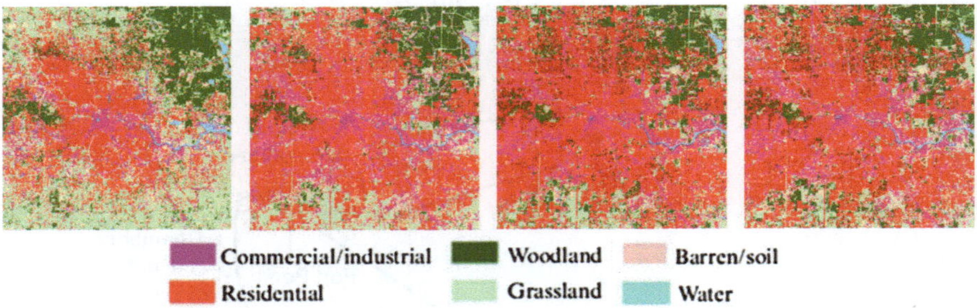

Fig. 12.2 Satellite images and classification results from MLC method on October 1979, December 1990, November 2000, and October 2010

Table 12.2 The proportion of each land use type from 1979 to 2010 in Houston

Houston	Area (km^2) 1979	1990	2000	2010
Residential	312.47	479.78	564.43	572.43
Industrial/commercial	93.50	228.58	198.21	197.07
Grassland	511.34	287.53	235.89	178.63
Woodland	209.68	140.32	184.70	179.43
Barren/soil	70.13	81.98	32.72	87.13
Water	31.73	9.67	11.85	13.11

12.4 Methodology

A cellular automata model was developed to investigate the scenarios of future urban land transformations in Houston. This model started on a 30-m grid and the transition rules were applied to all cells at the same time, and the entire grid was updated at the annual iteration. The transition rules were defined as the difference between the center cell and eight neighbors within 3 × 3 Moore's neighborhood. To determine the state of a cell in a certain time period, the simulation function was written as:

$$\mathrm{S}_{i,j}^{t+1} = a_{\mathrm{N}} \times \mathrm{N}_{i,j}^{t} + a_{\mathrm{M}} \times \mathrm{M}_{i,j} + a_{\mathrm{SE}} \times \mathrm{SE}_{i,j} \tag{12.1}$$

where $\mathrm{N}_{i,j}^{t}$ denotes the diffusion factor regarding its neighborhoods, $\mathrm{M}_{i,j}$ denotes the Markov transition probabilities, $\mathrm{SE}_{i,j}$ denotes the socioeconomic status of each single cell and its neighborhoods; *a* represents the coefficients for these variables.

For a self-organizing CA model, the diffusion factor, Markov transition rules, and socioeconomic status were defined as:

$$\mathrm{N}_{i,j} = \frac{n_{i,j}}{\sum n_{i,j}} \tag{12.2}$$

$$\mathrm{M}_{i,j} = \sum_{k=1}^{k} \frac{\mathrm{N}(i,j) \Big/ \sum_{i=1}^{m} m}{k * m_k} \tag{12.3}$$

$$\mathrm{SE}_{i,j} = \frac{\sum_{n=1}^{n} \left(\frac{d_{i,j}^{n} - \min\left(d_{i,j}^{n}\right)}{\max\left(d_{i,j}^{n}\right) - \min\left(d_{i,j}^{n}\right)} \right)}{n} \tag{12.4}$$

where $n_{i,j}$ is the total number of class *i* surrounding the observed class *j*, N(*i*, *j*) is the observed landscape amount changing from class *i* to class *j* during total *m* years at *k* internal steps, and $d_{i,j}^{n}$ is the different value in the selected four socioeconomic variables between the observed center cell and its *n* neighbors (Fig. 12.3).

Although the socioeconomic data were collected at the last year of simulation, the difference of socioeconomic values between the observed cell and the neighbors was used to determine the socioeconomic factors. Obviously, different socioeconomic variables have different impact weights to the urban land use/land cover change. In order to find the weightiness of each socioeconomic variable, 20 experts in the field of socioeconomic and land use change were invited to assign weights to each variable using the index ranging from 0 to 10 to represent the weight from the highest impact to the lowest impact. The average value of these ratings was shown in Table 12.3.

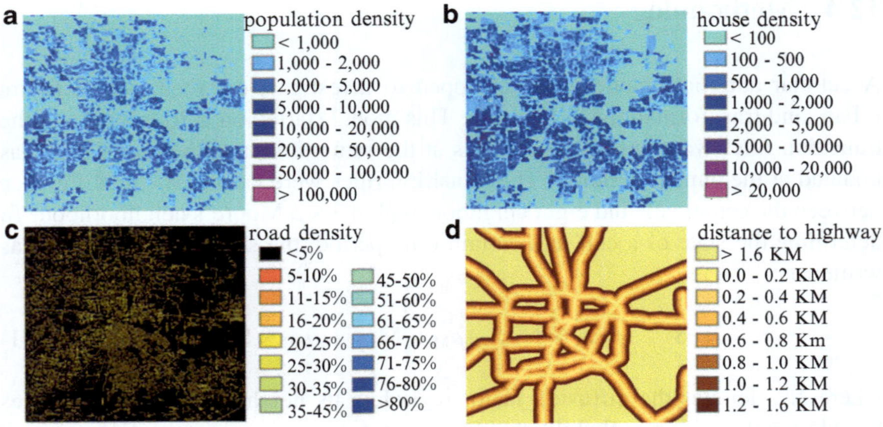

Fig. 12.3 The visualization of the socioeconomic value in Houston (**a**) Population density; (**b**) House density; (**c**) Road density; and (**d**) Distance to highway

Table 12.3 The weight of socioeconomic indices

Houston	Population density	Road density	Distance to highway	House density
Barren/soil	3.67	3.50	3.40	4.00
Industrial/commercial	8.80	8.17	8.42	7.33
Grassland	3.18	2.75	3.36	4.90
Residential	9.46	8.08	6.25	8.92
Transportation	7.66	9.00	8.95	6.82
Woodland	2.82	2.58	3.18	4.70

A critical issue in the CA model is the provision of proper methods to calibrate the CA model to find appropriate coefficients for the diffusion factor, Markov transition rules, and socioeconomic status (Hagen-Zanker and Lajoie 2008; Van Vliet et al. 2011). To calibrate the model, we used the classified Landsat TM image as empirical maps on the following dates: November 5, 1984; July 20, 1990; October 6, 1999; and November 9, 2000. We randomly selected an encoded weight number (ranging from 1 to 10) for each factors, run the CA model using these weight number, and compared the cells simulated in the CA model with the cells located in the empirical maps to choose the weight number with the highest fitness. The CA model was run at yearly intervals to represent one combination until the next calibration year. These steps were repeated until the year of the last calibration map.

For the validation, the model's simulation output was compared to the empirical map, occurring in the same simulated year (Pontius et al. 2004; Pontius and Cheuk 2006) through visual inspection and quantitative evaluation. In this research, we adopted the classified map in October 31, 2011, as an empirical map and overlaid it with the predicted map to generate a black-and-white error image. Meanwhile, an error matrix was built up with the user's and producer's accuracy for each class as well as the overall accuracy and Kappa for the entire landscape.

12.5 Results and Discussion

Since our model was based on actual observation from the last 30 years in Houston, the temporal transition probability matrix is calculated by accumulating the periods from 1979 to 2010. We first calculated the yearly transition matrix between each two subsequent maps between 1979–1990, 1990–2000, and 2000–2010 and then calculated the yearly transition matrix between 1979 and 2010 using Eq. 12.3. The yearly transition probability matrix from 1979 to 2010 is shown in Table 12.4.

Using the yearly transition probability matrices in Table 12.4, we parameterized the Markov transition probability and socioeconomic variable on the census block level into the CA model. Two CA models were built up, one with the socioeconomic variables and another one without. Figures 12.4 and 12.5 show the initial state and simulated pattern of Houston with the socioeconomic variables and without the socioeconomic variables, respectively.

The simulated results from two models have similar pattern in general urban sprawl pattern: fast shrinkage in grassland and woodland and clear outward expansion in residential or industrial/commercial area. This growth pattern could

Table 12.4 Yearly transition probability (%) matrix from 1979 to 2010

Houston (1979–2010)	Residential	Industrial/ commercial	Grassland	Woodland	Barren/soil
Residential	98.07	0.82	0.56	0.42	0.12
Industrial/commercial	1.24	98.01	0.23	0.03	0.4
Grassland	1.85	0.35	97.13	0.53	0.12
Woodland	0.09	0.18	1.19	98.25	0.26
Barren/soil	2.26	2.27	1.26	0.19	94.02

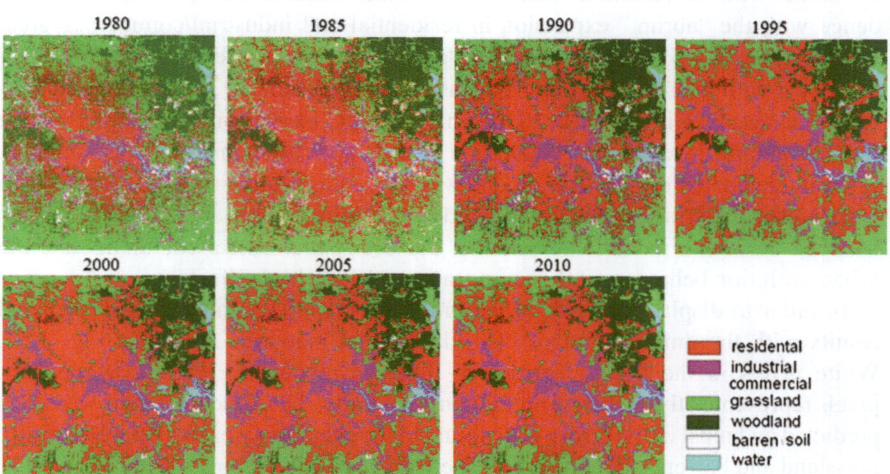

Fig. 12.4 The simulated landscape pattern of Houston with the socioeconomic factors

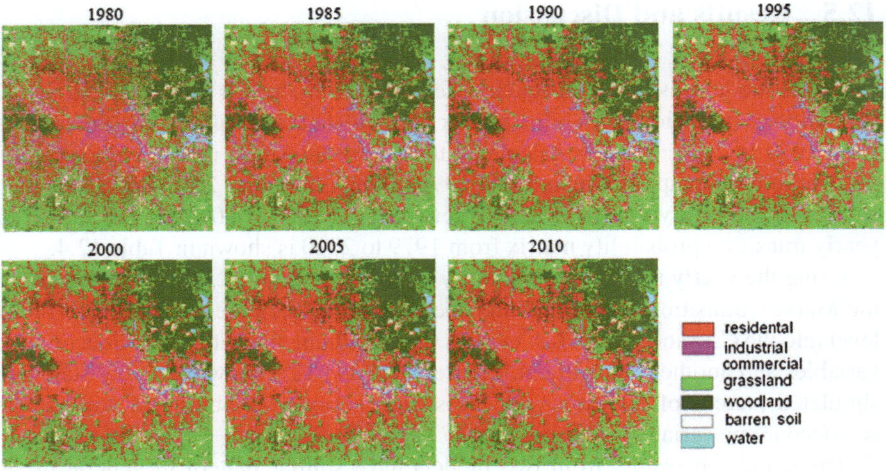

Fig. 12.5 The simulated landscape pattern of Houston without the socioeconomic factors

be observed in the southeastern and northeastern city with a large amount of new residential and industrial/commercial area being built in the last 30 years. Different from other large cities in the United States, Houston did not adopt city zoning laws in its urban planning. Lacking city zoning has led to an abundance of urban sprawl in Houston, resulting in a relatively large metropolitan area and low population density. Land developers inspired the spread of Houston when they built suburbs such as Pasadena (1892), Houston Heights (1892), Deer Park (1892), Bellaire (1911), West University Place (1919), and River Oaks (1922–24).

Although the simulated results from two models have the similar sprawl pattern, the model with the parameterized socioeconomic variables had a better correspondence with the "abrupt" expansion in residential and industrial/commercial area. From Fig. 12.6, we could find that the "abrupt" expansion were simulated well in the model with the socioeconomic data as the larger predicted area in these human-related landscapes in the year 2010. This "abrupt" expansion was caused by the rapid economic development, population growth, and road construction in Houston. The simulated pattern by the model without socioeconomic factors was much tardier, especially in simulating the rapid growth in suburban area. The differences between these two models indicate that the CA spatial model could simulate the urban evolution behaviors with incorporating enough driving factors.

In order to display the error in the predicted map, we compared our predicted results with the empirical maps. The differential map was shown in Fig. 12.7. White pixels in the figure represented the area predicted correctly, while dark pixels represented the incorrect prediction. Generally, the residential areas were best predicted and most of the errors were found in the suburban area, which were mostly grassland and barren/soil landscapes. The woodland was predicted better than other natural landscapes, which might be caused by the large forest reserved area in northeastern Houston in the Sheldon Lake State Park and Dwight D. Eisenhower Park.

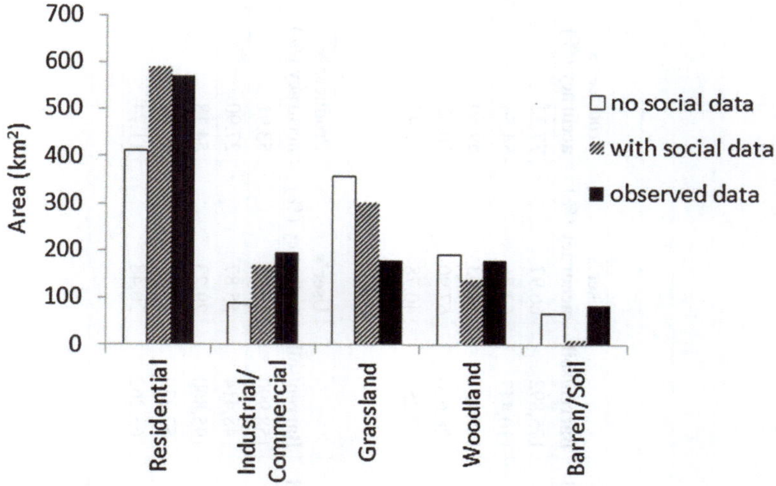

Fig. 12.6 The estimated results from two models in 2010

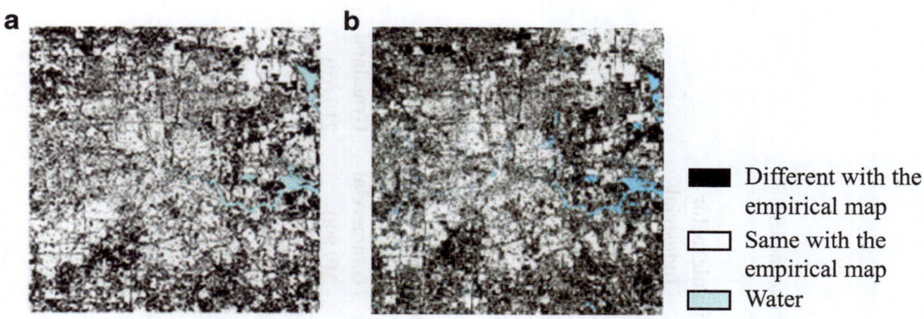

Fig. 12.7 The differential map between the predicted map and empirical map (**a**) with socioeconomic factors and (**b**) without socioeconomic factors

In Fig. 12.7, the predicted result with the socioeconomic data (Fig. 12.7a) was better than the one without the socioeconomic data (Fig. 12.7b) with more white pixels. This could be confirmed in the southwestern Houston, such as Gulfton, Sharpstown, and Bellaire, and southeastern Houston between Deer Park and Pasadena. The incorrect predictions were always also found in the industrial/commercial area in the Southern and Northeastern Houston, such as Missouri City and Jersey village. It was easy to understand since the chosen socioeconomic data, especially the population density and house density, were better to represent the residential area instead of industrial/commercial area.

Further validation of models between the simulated one and predicted one was analyzed through the confusion matrix (Table 12.5). This table showed the comparison results between the simulated result and empirical maps as the value of

Table 12.5 Confusion matrix and the model validation for two models

Empirical map

Predicted result

A: the predicted model with socioeconomic factor							
Houston	**Residential**	**Industrial/ commercial**	**Grassland**	**Woodland**	**Barren/soil**	**User's accuracy (%)**	**Producer's accuracy (%)**
Residential	1,756,004	257,388	230,903	201,696	175,892	66.97	77.32
Industrial/commercial	123,890	467,825	34,540	7,576	110,447	62.86	54.56
Grassland	320,672	109,304	518,946	209,275	185,157	38.63	59.60
Woodland	68,802	15,960	84,838	419,194	28,620	67.90	50.02
Barren/soil	1,688	6,949	1,544	357	7,258	40.78	1.43
Overall accuracy (%), 59.30; **Kappa**, 0.43							
B: the predicted model without socioeconomic factor							
Houston	**Residential**	**Industrial/ commercial**	**Grassland**	**Woodland**	**Barren/soil**	**User's accuracy (%)**	**Producer's accuracy (%)**
Residential	1,095,900	219,890	210,650	150,980	158,990	59.03	53.91
Industrial/commercial	71,801	229,830	48,414	861	48,414	55.87	27.90
Grassland	585,370	162,880	467,240	186,970	195,860	29.23	54.18
Woodland	189,290	44,897	140,520	431,390	52,279	46.97	51.63
Barren/soil	90,344	116,409	33,504	5,363	63,262	20.48	11.74
Overall accuracy (%): 45.56; **Kappa**: 0.27							

user's accuracy and producer's accuracy represented the accuracy for each class and the overall accuracy and Kappa represented the accuracy for the entire landscape. In both models, the best predicted class was the residential area (with 66.97 %/59.03 % user's accuracy and 77.32 %/53.91 % producer's accuracy) and the worst prediction class was barren/soil (with 40.78 %/20.48 % and 1.43 %/11.74 %). The barren/soil class, although had the least area in the study area, were easy to be confused with other classes such as industrial/commercial or residential area. The incorporation socioeconomic data into the model improved the simulation on the residential or industrial/commercial classes which made the barren/soil having the least accuracy.

One disadvantage of incorporating socioeconomic data into the model was the overestimation of residential area in which led to a relative underestimate in the industrial landscape as well as other natural landscapes such as woodland and grassland. This might be improved as more and more socioeconomic data were incorporated as driving forces in the model. The analysis of the model validation showed that the appropriate ancillary parameters were necessary for the CA model to derive a solid result. In fact, the value of the simulation approach lied in its exploratory nature which enabled the improvement of models with additional variables later. Meanwhile, the CA model had an "aggregate" function to smooth the heterogeneous pattern within the urban and suburban area. One solution to solve this problem was to incorporate better data source into the model, such as higher spatial resolution images or sub-pixel classifications, to improve the accuracy of CA models.

12.6 Conclusion

The spatiotemporal CA model of urban landscape patterns using multi-temporal TM and MSS imagery enabled us to characterize the internal structure of landscapes and monitor the landscape dynamics for Houston. Moreover, we also explored the potential of socioeconomic variables to detect how human forces affect the urban spatial pattern.

The CA model, coupled with the Markov transition probability, has indicated the capability of trend projection for the landscape change. This spatiotemporal model provided not only the quantitative description of change in the past but also the direction and magnitude of change in the future. However, based on the experimental results and exploratory analysis, several limitations still exist within the current study:

- Since the modeling process involves the usages of data from multiple sources, the accuracy of prediction result will be closely related to the individual accuracy with each type of data, especially different remote sensing data sources. The development of a robust method to incorporate data in different spatial resolution was still an interesting issue.
- Although the Markov transition probability was calculated on the census block level, it was stationary and unable to accommodate the unpredictable influence

variables, such as the climate, policy, and human disturbance. In addition, the pace of landscape change was usually kept on changing over the entire period.
- In this research, we supposed the relationship between socioeconomic factors, neighborhood effect, and Markov transition probability was linear and deterministic during the calibration. Finding an exact dynamic coefficient between them was still an intricate study in the urban modeling.

Currently, it was not fully conclusive that the CA model based on socioeconomic data was inferior to the one without socioeconomic data, especially for the natural landscapes. It was still necessary to find more sophisticated methods applying to a series of varied landscape to verify this new model.

Most urban landscapes have been influenced by human disturbance, resulting in a heterogeneous mosaic of natural and human-managed patches that vary in size, shape, and arrangements (Turner 1989). The landscape responses to human disturbances are important, however, difficult to be estimated because the landscape-level simulation involved numerous challenging experiments and hypotheses in the development of models (Vaz et al. 2012). These hypotheses are always assumed to make the process model easier to be manipulated, leading to a more homogenous pattern in the predicted result. Thus, it is necessary to relate the homogenous analysis in the model prediction with the heterogeneous analysis in the quantitative landscape method for a comprehensive understanding of the urbanization process. In conclusion, this urban studies show that by incorporating more spatial algorithms into the prediction of landscape change, more accurate long-term landscape changes can be reproduced in the future.

References

Aljoufie M, Zuidgeest M, Brussel M, Van Vliet J, Van Maarseveen M (2013) A cellular-automata based land use and transport interaction model applied to Jeddah, Saudi Arabia. Landsc Urban Plan 112:89–99

An L, Linderman M, Qi J, Shortridge A, Liu J (2005) Exploring complexity in a human-environment system: an agent-based spatial model for multidisciplinary and multiscale integration. Ann Assoc Am Geogr 95(1):54–79

Antrop M (2004) Landscape change and the urbanization process in Europe. Landsc Urban Plan 67:9–26

Arsanjani JJ, Helbich M, Kainz W, Darvishi A (2013) Integration of logistic regression, Markov chain and cellular automata models to simulate urban expansion. Int J Appl Earth Obs 21:265–275

Batty M, Xie Y (1994) From cells to cities. Environ Plan B 21:31–48

Bell EJ (1974) Markov analysis of land use change: an application of stochastic processes to remotely sensed data. Socio Econ Plan Sci 8:311–316

Berry MW, Flamm RO, Hazen BC et al (1996) Lucas: a system for modeling land-use change. IEEE Comput Sci Eng 3(1):24–35

Bourne LS (1976) Monitoring change and evaluation the impact of planning policy on urban structure: a Marlov chain experiment. Plan Can 16:5–14

Cecchini A, Viola F (1990) Eine Stadtbausimulation. Wissenschaftliche Zeltschrift der Hochschule fur Architektur und Bauwesen 36(1):159–162

Clarke KC, Hoppen S (1997) A self-modifying cellular automaton model of historical urbanization in the San Francisco Bay area. Environ Plan B 24:247–261

Clarke KC, Gaydos LJ (1998) Loose-coupling a cellular automata model and GIS: long-term urban growth prediction for San Francisco and Washington/Baltimore. Int J Geogr Inf Sci 12(7):699–714

De Kong GHJ, Verburg PH, Veldkamp A, Fresco LO (1999) Multi-scale modelling of land use change dynamics in Ecuador. Agr Syst 61:77–93

Hagen-Zanker A, Lajoie G (2008) Neutral models of landscape change as benchmarks in the assessment of model performance. Landsc Urban Plan 86:284–296

Herold M (2004) Remote sensing and spatial metrics for mapping and modeling of urban structures and growth dynamics. Ph.D. dissertation, University of California-Santa Barbara

Herold M, Goldstein NC, Clarke KC (2003) The spatiotemporal form of urban growth: measurement, analysis and modeling. Remote Sens Environ 86(3):286–302

Hillier B, Hanson J (1984) The social logic of space. Cambridge University Press, Cambridge

Key to the city (2001) Houston, Harris County, Texas. Available at: http://www.usacitiesonline.com/txcountyhouston.htm. Cited 15 Jan 2014

Klosterman RE (1999) The what if? Collaborative planning support system. Environ Plan B 26(3):393–407

Li X, Yeh AGO (2000) Modelling sustainable urban development by the integration of constrained cellular automata and GIS. Int J Geogr Inf Sci 14(2):131–152

Lichtenberg ER (1985) The role of land quality in agricultural diversification. Ph.D. dissertation, University of California-Berkeley

Liebrand WBG, Nowak A, Hegselmann R (1998) Computer modeling of social process. Sage Publications, London

Lopez E, Bocco G, Mendoza M, Duhau E (2001) Predicting land cover and land use change in the urban fringe: a case in Morelia city Mexico. Landsc Urban Plan 55(4):271–285

Mcintyre NE, Knowles-Yanez K, Hope D (2000) Urban ecology as an interdisciplinary field: differences in the use of "urban" between the social and natural sciences. Urban Ecosyst 4:5–24

Moser DC (1998) Diane Moser Properties. Available online at: http://www.texasbest.com/houston/geograph.html. Cited 15 Jan 2014

Palmquist RB (1989) Land as a differentiated factor of production: a Hedonic model and its implications for welfare measurement. Land Econ 65(1):23–28

Pickett STA, Burch WR, Dalton SE, Foresman TW, Grove JM, Rowntree R (1997) A conceptual framework for the study of human ecosystems in urban areas. Urban Ecosyst 1:185–199

Pijanowski BC, Long DT, Sage SH, Cooper WE (1997) A land transformation model: conceptual elements, spatial object class hierarchies, GIS command syntax and an application to Michigan's Saginaw Bay Watershed. Land use modeling workshop, Sioux Fall, South Dakota, June 3–5, 1997.

Pijanowski BC, Tayyebi A, Doucette J et al (2014) A big data urban growth simulation at a national scale: configuring the GIS and neural network based Land Transformation Model to run in a high performance computing environment. Environ Model Softw 51:250–268

Pontius RG Jr, Cheuk ML (2006) A generalized cross-tabulation matrix to compare soft-classified maps at multiple resolutions. Int J Geogr Inf Sci 20(1):1–30

Pontius RG Jr, Shusas E, McEachren M (2004) Detecting important categorical land changes while accounting for persistence. Agric Ecosyst Environ 101:251–268

Shafizadeh-Moghadam H, Helbich M (2013) Spatiotemporal urbanization process in the megacity of Mumbai, India: a Markov chains-cellular automata urban growth model. Appl Geogr 40:140–149

Stewart W (1994) Introduction to the numerical solution of Markov chains. Princeton University Press, Princeton

Tang J (2011) Modeling urban landscape dynamics using sub-pixel fractions and fuzzy cellular automata. Environ Plann B 38:903–920

Tang J, Wang L, Yao Z (2007) Spatio-temporal urban landscape change analysis using Markov chain and modified genetic algorithm. Int J Remote Sens 28(15):3255–3271

Tang J, Wang L, Yao Z (2008) Analyses of urban landscape dynamics using multi-temporal satellite images: a comparison of two petroleum-oriented cities. Landsc Urban Plan 87(4):269–278

Tang J, Chen F, Schwartz SS (2012) Assessing spatiotemporal variations of greenness in the Baltimore-Washington corridor area. Landsc Urban Plan 105:296–306

Tayyebi A, Pijanowski BC, Tayyebi AH (2011) An urban growth boundary model using neural networks, GIS and radial parameterization: an application to Tehran, Iran. Landsc Urban Plan 100(1):35–44

Tayyebi A, Pekin BK, Pijanowski BC et al (2012) Hierarchical modeling of urban growth across the conterminous USA: developing meso-scale quantity drivers for the land transformation model. J Land Use Sci 8(4):422–442

Tayyebi A, Perry PC, Tayyebi AH (2013) Predicting the expansion of an urban boundary using spatial logistic regression and hybrid raster-vector routines with remote sensing and GIS. Int J Geogr Inf Sci 28:639–659. doi:10.1080/13658816.2013.845892

Texas State Historical Association (2002) Handbook of Texas, online. Available at: http://www.tsha.utexas.edu/handbook/omline/articles/view/HH/hdh3.html. Cited 26 Dec 2009

Tobler W (1979) Cellular geography. In: Gale G, Olsson S (eds) Philosophy in geography. Reidel, Dordrecht, pp 379–386

Turner MG (1989) Landscape ecology: the effects of pattern on process. Annu Rev Ecol Syst 20:171–197

US Census (2010) Population and household. Available at: http://www.census.gov. Cited 20 Sep 2013

U.S. Census (2011) Your gateway to census 2010. Available at: http://en.wikipedia.org/wiki/Urbanization. Cited 15 Jan 2014

Van Vliet J, Bregt AK, Hagen-Zanker A (2011) Revisiting Kappa to account for change in the accuracy assessment of land use change models. Ecol Model 222:1367–1375

Van Vliet J, Hurkens J, White R, Van Delden H (2012) An activity based cellular automaton model to simulate land use dynamics. Environ Plan B 39:198–212

Vaz E, Nijkamp P, Painho M, Caetano M (2012) A multi-scenario forecast of urban change: a study on urban growth in the Algarve. Landsc Urban Plan 104(2):201–211

Waddell P (2002) UrbanSim: modeling urban development for land use, transportation, and environmental planning. J Am Plan Assoc 68(3):297–313

Weng Q (2002) Land use change analysis in the Zhujiang Delta of China using satellite remote sensing, GIS and stochastic modeling. J Environ Manag 64(2):273–284

White R, Engelen G (1993) Cellular automata and fractal urban form: a cellular modeling approach to the evolution of urban land-use patterns. Environ Plan A 25(8):1175–1199

Wu F (1998) Simulating urban encroachment on rural land with fuzzy-logic-controlled cellular automata in a geographical information system. J Environ Manag 53(4):293–308

Wu F, Martin D (2002) Urban expansion simulation of Southeast England using population surface modelling and cellular automata. Environ Plan A 34(10):1855–1876

Yang X, Lo CP (2002) Using a time series of satellite imagery to detect land use and land cover changes in the Atlanta, Georgia metropolitan area. Int J Remote Sens 23(9):1775–1798

Part V
Urban Sensing, Social Networks and Social Media

Chapter 13
Linked Activity Spaces: Embedding Social Networks in Urban Space

Yaoli Wang, Chaogui Kang, Luís M.A. Bettencourt, Yu Liu, and Clio Andris

Abstract We examine the likelihood that a pair of sustained telephone contacts (e.g. friends, family, professional contacts, called "friends") uses the city similarly. Using call data records from Jiamusi, China, we estimate a proxy for the daily activity spaces of each individual subscriber by interpolating the points of geo-located cell towers he or she uses most frequently. We then calculate the overlap of the polygonal activity spaces of two established telephone contacts, what we call *linked* activity spaces.

Our results show that friends and second-degree friends (e.g. friends of friends) are more likely to geographically overlap than random pairs of users. Additionally, individuals with more friends and with many network triangles (connected groups of three friends) tend to congregate in the city's downtown at a rate that surpasses randomness. We also find that the downtown is used by many social groups but that each suburb only hosts one or two groups. We discuss our findings in terms of the need for a better understanding of spatialised social capital in urban planning.

Y. Wang
Department of Geography, University of Georgia, 120K, Geog-Geol Building, Athens, GA 30602, USA

Santa Fe Institute, 1399 Hyde Park Road, Santa Fe, NM 87501, USA

C. Kang
Institute of Remote Sensing and Geographical Information Systems, Peking University, Beijing 100871, China

MIT Senseable City Laboratory, 9-209, 77 Massachusetts Avenue, Cambridge, MA 02139, USA

L.M.A. Bettencourt
Santa Fe Institute, 1399 Hyde Park Road, Santa Fe, NM 87501, USA

Y. Liu
Institute of Remote Sensing and Geographical Information Systems, Peking University, Beijing 100871, China

C. Andris (✉)
Santa Fe Institute, 1399 Hyde Park Road, Santa Fe, NM 87501, USA

Department of Geography, The Pennsylvania State University, 302 Walker Building, University Park, PA 16802, USA
e-mail: clio@psu.edu

M. Helbich et al. (eds.), *Computational Approaches for Urban Environments*, Geotechnologies and the Environment 13, DOI 10.1007/978-3-319-11469-9_13

Keywords Activity space • Daily movement • Call data records • Mobile phones • Social networks • Friendship • Relationships • Cities • Built environment

Abbreviations

CDRs Call data records
GIS Geographic information system(s)
KML Keyhole markup language
LAS Linked activity spaces
POIs Points of interest

13.1 Introduction

In this chapter, we present a methodology that can help elucidate how groups of friends, family and professional contacts use the city. We know that cities are comprised of two interacting components, social networks and physical infrastructure, and that the social dynamics of encounters in urban space form the backbone of city life (Bettencourt 2013). Yet our ability to model social networks and social capital in urban spaces is very limited. This presents a problem because often our behavior results from the influence of others (Salganik and Watts 2008). The establishment, discovery and maintenance of our social ties are guided by the city. These ties will also affect how we use the city: where we choose to meet, live and work.

Within a city, it remains an open question as to whether a citizen benefits most from having his or her social contacts nearby or dispersed. At one extreme, dispersed contacts can expose the ego to new neighborhoods and a variety of urban knowledge (such as finding the quickest post office, the best doctor or an exciting new restaurant), due to their variety of experiences in diverse parts of a city. Yet, it may be more difficult and more expensive to meet spatially dispersed contacts. Having friends in disparate parts of the city is also more likely to lead to a social network where one has few friends "in common" with other friends, which can be a key strength of social networks.

At the other extreme, a socially tight neighborhood forms trusted bonds through multiple channels of social validation (Centola and Macy 2007) and through increased exposure to one another in the outdoors and through neighborhood institutions such as local schools. Proximal social contacts can meet conveniently, benefiting elderly and the mobility challenged and, in some cases, poorer or immigrant communities who likely rely on friends and family for help with amenities such as child care. Yet, in these enclosed neighborhoods, information and social capital from other parts of the city may be less accessible (Granovetter 1983) resulting in missing or unsupportive social systems across the city (Granovetter 1973).

How do urbanites organize their lives to balance their need for information and accessibility with its costs? In order to answer this question, we must measure and model the social network of egos within the urban built environment.

13.1.1 The State of Social/Spatial Modeling

Everything happens somewhere: examining social life as extricated from the influence of the built environment results in an unrealistic view. Yet the methods available for understanding the clustering and dispersion of a set of individual social networks in geography are limited, as social network and urban spatial models have matured in separate domains, and are analyzed in separate spheres, through social network analysis and geographic information systems (GIS), respectively (Andris 2011). Network methods are also rarely used by those who study city form (Sevtsuk and Mekonnen 2012). Social networks represent influences and social capital as graph configurations of nodes (agents) and links (e.g., edges) between nodes where primary metrics are connectivity and embeddedness; alternatively, spatial (e.g., GIS) models are represented in a contiguous topological plane, where adjacency and proximity are primary metrics (Andris 2011).

As a result, social/spatial phenomena are often explained separately by those inclined toward computational sociology or geography, respectively. One example is the study of obesity, where social networks (Christakis and Fowler 2007) and city form (Papas et al. 2007) are examined as causal factors, but not in the same study. To obtain a clearer picture of the mechanisms surrounding obesity, one should consider social ties and the built environment as coincidental factors – as these influences can compound. Similarly, research showing how students use a college campus in space and time via WiFi usage describes the flexibility of meeting places due to mobile computing (Sevtsuk et al. 2009), but could be extended to assess social gatherings in time and place, as do Eagle et al. (2009) on the same college campus, during a similar time period. Eagle et al. (2009) show the temporal social patterns of dyadic (pairwise) relationships in terms of calls, SMS messages and colocation, and alludes to the role of the campus in providing the backdrop for social groups and pairs. When combined with Sevtsuk et al. (2009), this study could provide the social ties within a spatial setting to uncover where friends meet, where they travel on the campus and how these factors can be leveraged to create a better campus environment.

This is not to say that datasets on interpersonal communication and movement have not been embedded into geography; analyses of interplace networks of social flows such as commodities, telecommunications, migration, and commuting are common in computational urban research (examples abound). Yet, these represent place-to-place aggregate flows instead of person-to-person flows and thus do not directly express the decisions of individuals. Small-scale examples of spatially embedded social networks describe gang membership (Radil et al. 2010; Papachristos et al. 2013), transportation (Frei and Axhausen 2011; Arentze et al. 2012), and epidemiology (Emch et al. 2012). We take these initiatives a step further by creating a general method that can respond to patterns of human socialization in a built environment. These studies can elucidate where and when (different types of) relationships form and could be used to advise architects, urban, and transportation planners in creating places that support and create social connectivity.

In working toward this goal, those looking to examine social/spatial problems are aided by the recent proliferation of large datasets evidencing human social contact and movement (such as GPS or cell tower usage records) in the city (Reades et al. 2007). The integration of human movement and activity data, such as information from GPS traces (Gao et al. 2013), check-in data (Cho et al. 2011), online social networks (Scellato et al. 2011), and photo-sharing sites (Crandall et al. 2010; Girardin et al. 2008; Sun et al. 2013), into urban models are providing new windows on how humans use the built environment. Specifically, the use of mobile telephone calls to understand city usage patterns are becoming a cornerstone of modern urban informatics, planning, and transportation (Ratti et al. 2006). We take advantage of mobile telephone call data to test our research questions about the locality or dispersion of social ties in the city.

Further, the relative convenience of colocation for friends can be evaluated. Calabrese et al. (2011) find that in 94 % of telephone calling partners, one partner constantly travels further to meet. On average, the partner traveling further travels 3 times further to meet. This method uses travel time and distance, which is important for logistics. However, we extend this concept by incorporating the built environment into these compromises, to show where in the city friends are likely to meet. By spatially-linking the respective activity spaces of two friends in the GIS, we can better understand how the city is able to provide places for friends to meet, and assess the travel needs to do so—i.e. it is relatively easy for friends with spatially-overlapping activity spaces to meet face-to-face.

13.1.2 Linked Activity Spaces

We use cell phone call data records (CDRs) to model "friendships" (i.e., interpersonal relationships) as a social network, inferred by the frequency calls between two agents, and the sets of locations visited by each member of the social network within the city (i.e., activity spaces). A pair of activity spaces of an ego and alter are called *linked activity spaces* (LAS) if the ego and alter are friends (i.e., contacts) in the dataset. The two activity spaces of friends are modeled within the GIS and spatially analyzed for similarity, via the number of "third places" shared among the pair (following Rosenbaum 2006). Moreover, we analyze the social network as a whole to find whether high-degree *egos* (a.k.a. those with many friends), *triangles* (groups of three agents) and *communities* use the city in significantly similar ways.

We have four main hypotheses for the analysis of LAS. (1) We expect that friends' activity spaces will overlap more often than a random pair of activity spaces, indicating that friends use the city more similarly than a random pair of people. (2) We also hypothesize that egos with high degrees or high clustering coefficients (see Jackson 2010) will be more associated with the city center, as this denser environment tends to have more meeting places, diverse services, commercial areas, and nightlife. (3) In terms of city form and groups, we believe that central areas will play an enhanced role in supporting "clique-like" and modular groups instead of being a mixing pot for many groups. We expect the downtown area to host tight-knit social groups who do not venture to the suburbs often. (4) Finally, we expect that suburban POIs will accommodate individuals from diverse social groups, as these agents are likely visit different parts of the city using automobiles.

This chapter proceeds as follows. We first describe the study area and the setting of the CDR dataset. We then describe, in the methods section, how we delineate each user's activity spaces. We analyze how *linked activity spaces* (LAS) are spatially correlated in an urban environment by shared points of interest (POIs). We conclude with a discussion of the usefulness of this method, its drawbacks, potential applications, and future work.

13.2 Study Area and Dataset

Our study area is the city of Jiamusi, located in northeastern China, with a population 2.5 million (est. 2010). This industrial city serves as a producer of wood pulp and newsprint and participates in the global economy via a thriving international trade harbor. The urban core of Jiamusi is nearly 18 by 10 km in spatial extent, and its residents travel on average 1 km a day (Kang et al. 2012).

13.2.1 Dataset and Sampling

We focus on calls made within the city area and exclude long-distance calls. We use a CDR (call data record) dataset of mobile cell phone calls from an undisclosed mobile phone provider in China.

The original CDR dataset contains nearly 424,000 users over 31 days. Users are anonymized in the dataset. Combined, users make an average of 1,600,000 calls daily. In the 31-day time span of our dataset, each user participates in an average of 328 calls for a total duration of 6.15 hours. Each record of a mobile phone call contains the start time, call duration, and locations of the caller and receiver. The locations are geo-referenced to one of 96 cell towers closest to the mobile phone's location (Table 13.1). The dataset does not include text messages (e.g., SMS).

We process the dataset into two parts: a social network of agents (social network in Table 13.1) and the activity spaces of each agent (spatial patterns in Table 13.1). We filter the network by including only those who use at least three cell towers during the study period in order to eliminate users who may be confined to their home and thus interact with the city differently than a typical mobile user. Also, an individual may have multiple mobile phones, and a phone with fewer than three cell towers used may represent a "secondary" or less frequently used device. In the social network, the *number of calls* is determined between a unique pair of users, and *duration* is the sum of call time between the two users. The network

Table 13.1 Call data record (CDR) variables with original data fields (*top*)

Original table	Caller	Receiver	Caller location (x, y)	Receiver location (x, y)	Start time	Duration
Social network	User 1	User 2	–	–	Number of calls	Total duration between users 1 and 2
Spatial patterns	User 1		Location (x, y)	Location (x, y)		Duration

A social network and spatial data summary table are listed in the *middle* and *bottom* rows, respectively

is undirected in order to reflect each member's inclination to participate in the conversation regardless of the initiator (Calabrese et al. 2011). In other words, records showing that A calls B, or B calls A, are summed to represent a connection between unique, undirected pair A, B. Each pair must have either 10 + mutual phone calls or 10+ min of total call duration in the given month to be considered friends. This process eliminates non-friend calls such as sales calls, as these do not represent persistent relationships. Our resultant dataset has an average of 11.55 calls per friendship connection (with a 95 % confidence interval (c.i.) of [11.35, 11.76]) and an average of 12.52 min for each link (95 % c.i. is [12.22, 12.81]).

The spatial patterns table contains the locations of each user, which are combined to geo-locate a pair of callers in the social network. The coordinates of the cell phone tower where a user places or receives a call are summed and weighted by the number of calls the user places or receives at that cell tower location. We use the resulting set of weighted locations to represent the user's geographic activity pattern (such as Carrasco et al. 2006), which are known to capture "anchor points" (Golledge 1999) such as home and workplace (or school), as they are the most visited locations for the average traveler and, thus, frequent calling points (Schönfelder and Axhausen 2003).

13.2.2 Sampling

We sample the large CDR dataset by selecting a random sample of 150 "seed" users and retrieve their contacts (first-degree ties), second-degree and third-degree ties, in a method similar to Kurant et al. (2011). The number of seed users is calibrated based on our ability to visualize and computationally analyze the resultant dataset. We also choose this method over a random sample of all users (e.g., choosing 20,000 random users and the possible network that might form between them) because the seed method ensures that retrieved nodes have connections (since we select friends, then friends of friends). This method also is able to find groups, whereas in a random sample of the network, nodes may not be connected. This configuration yields a network that is focused on the social interactions of a small sample of users. As a result, this "core" social network does not resemble a complete social network's typical degree distribution (such as Albert and Barabási 2002), traversability, or density (Newman and Park 2003).

13.2.3 *Social Network and Geographic Characteristics*

The degree values for the core network range from 1 to 344, with a mean degree of 43.8 (Fig. 13.1). The diameter of the network is 12. Clustering coefficient values range from 0 to 1, with a mean coefficient of 0.11. The network is visualized in Fig. 13.2. Using Spearman's correlation statistic, we find that users with more contacts make shorter calls, while those with fewer friends speak for longer.

Our dataset includes 96 cell phone towers. The caller' or receiver's location is approximated to the site of the cell tower (which is offset slightly by the telephone provider), although generally the caller or receiver could be found anywhere in the signal radius around the cell tower. We note that it is also possible that a caller's call may be routed to a cell tower that is not the closest to him or her, as the closest cell tower may be saturated with calls or out of service, though we cannot account for such situations. Some towers are used by many subscribers, while others are used by few: the 10 most popular cell phone towers are used by at least 20 % of the population whereas the 40th–91st most popular towers are each used by less than 10 % of the population.

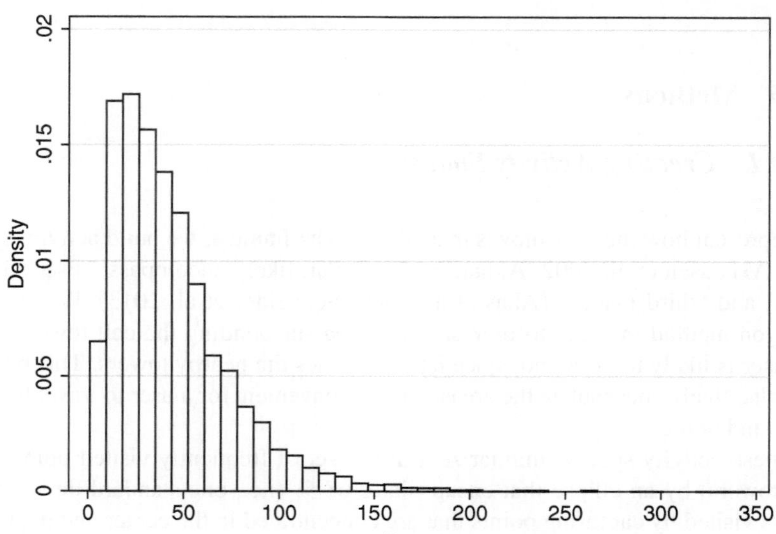

Fig. 13.1 The degree distribution of the calling network, comprised of 8,231 sampled users, is right-skewed with mean 44 and is well described by a log-normal distribution

Fig. 13.2 The aspatial social network is visualized with a force-directed method that places nodes in feature space based on their density of linkages in the Gephi computing environment (Bastian et al. 2009). *Larger* nodes denote higher degree and *smaller* nodes indicate smaller degree

13.3 Methods

13.3.1 Creating Activity Spaces

To represent how the user moves in a city such as Jiamusi, we build activity spaces (e.g. Axhausen et al. 2002; Axhausen 2007) that likely encompass a user's home, work, and "third places" (Ahas et al. 2009; Schneider et al. 2013). We choose a polygon method in order to represent the area surrounding the cell towers where the user is likely to be found, since he or she uses the nearby towers. This polygon will also likely encapsulate the areas that are convenient for a user to travel between work and home.

These activity spaces summarize a user's set of frequently visited points (e.g., cell towers) by an ellipse that encapsulates 68 % (i.e., one standard deviation) of points visited by capturing points that are concentrated in the center and neglecting sparse points in the periphery (as in Carrasco et al. 2006). Ellipses are first centered on the mean geometric center of a user's tower locations (mean of x coordinates and y coordinates; repeating values are allowed if a user visits towers more than once).

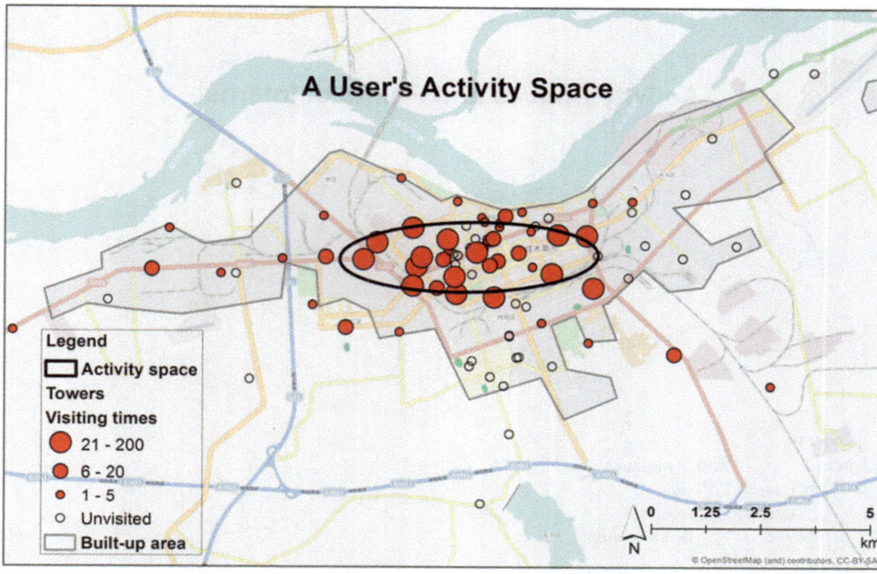

Fig. 13.3 An individual activity space is represented as an ellipse (in *black*). The ellipse captures a user's most frequently used towers, shown as *red circles*, where larger *red circles* indicate more frequent visits

The value of the standard deviation is calculated for all x coordinates to obtain an axis, and y coordinates to obtain a second, perpendicular axis. The ellipse is tilted in a direction that captures the major axis (long edge) of the distribution (see Mitchell 2005).

As mentioned, the ellipse does not typically encompass all visited towers and excludes those not frequently visited, to represent daily activity space (see Fig. 13.3). It captures the essence of the central tendency, dispersion, and direction of the user's travel patterns without including infrequent cell tower usage (such as a traveler's phone call from the airport).

13.3.2 Assessing Overlap of Activity Spaces

After each individual is assigned an activity space, we quantify the similarity between an ego's and an alter's activity spaces. A method for finding whether two activity spaces are similar is not straightforward. The percent overlap between two activity spaces will not account for how much physical area two friends' spaces share. Additionally, using the area that two activity spaces share does not tell us how big their spaces are, i.e., whether this shared area is actually "convenient" relative to their whole activity spaces. Also, with these two methods, we will not

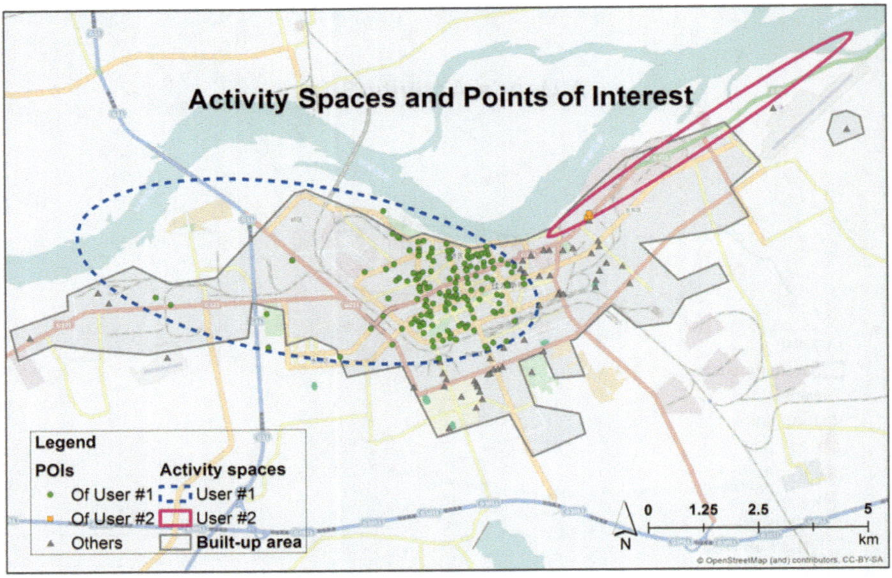

Fig. 13.4 Two single activity space ellipses in *dashed blue* and *solid pink* intersect underlying Points of Interest (POIs). The POIs intersected by the dashed ellipse are *green circles*. Those intersected by the *solid pink* ellipse are *orange squares*. POIs that do not intersect either ellipse are *grey triangles*

be able to understand what amenities and places for meeting each individual has in his or her activity space. We overcome these drawbacks by creating a third layer of relevant points, as suggested by geometric probability theory (Santaló 2004).

Agents' activity spaces are qualified by the points of interest (POIs) they spatially intersect (Fig. 13.4). These POIs are where "optional" activities are likely to occur (Gehl 1987), including services, transport, and recreational areas as a proxy for how agents use the city. POIs are landmarks for social interactions, as third places and the activities performed in third places have been shown to be essential for relationships, social health, and quality of life (Rosenbaum 2006). Friends may visit POIs to dine or to do business.

POIs are selected and digitized by the authors from *Google Maps* (2013) with data provided by *AutoNavi,* which operates under an open-use license. Although it is default to gauge how well this set of geographic information (VGI) reflects actual POIs in the city, it is used as a proxy for all POIs in the city (Coleman 2010; Neis and Zipf 2012). The digitized set of POIs are retrieved as a keyhole markup language (KML) file and analyzed in the Esri *ArcMap 10.1* environment. POIs include the city's recreation spots (including parks, internet cafés, personal wellness centers), commercial centers (restaurants and bars, stores, markets, and shopping centers), public services and institutions (hospitals, post offices, police stations), transportation centers (airports, train stations), and named villages in the suburban area. POIs of similar types (e.g., restaurants) in a 50 m radius are grouped into a single POI to eliminate redundant information.

As mentioned, two activity spaces are considered *linked* if their corresponding nodes are connected via the social network. Thus, friendships are embedded in geographic space through the two activity spaces. Each unique pair of linked activity spaces (LASs) is compared by the number of common POIs shared by both (in absolute number and percent of the user's total POIs). In Fig. 13.5, two pairs of linked activity space ellipses share POIs, where the pair demarcated with a solid blue line shares more POIs. In Fig. 13.6, an ego ellipse is in focus (in solid pink) and shares POIs to various extents with each of his or her alters (in dashed lines). We use a statistical t-test to compare friends' typical number of shared POIs versus that of a random pair of users. Our hypothesis is that friends share more POI points than random pairs.

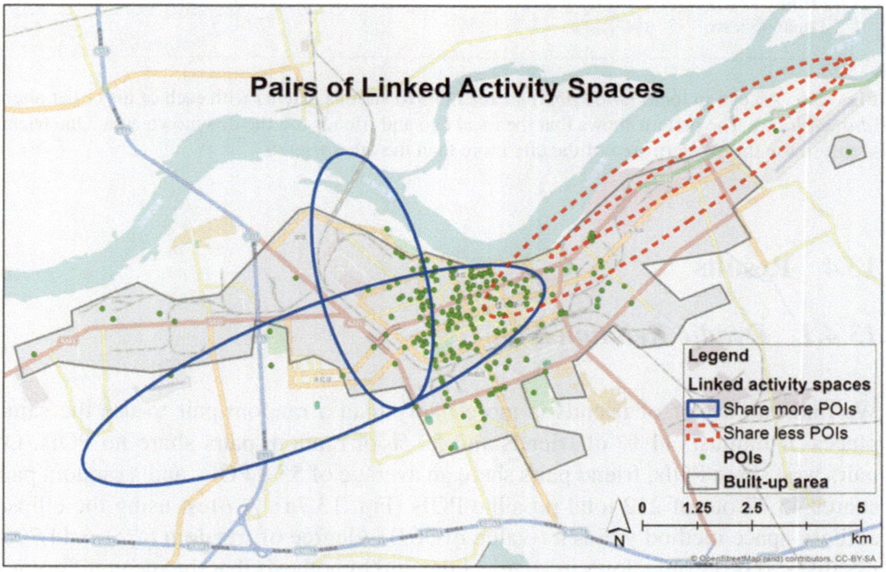

Fig. 13.5 This image shows how linked activity spaces overlap and can be quantified using the number of POIs found in the intersection of activity spaces. These POIs can be a proxy for convenience of shared meeting points. More specifically, this figure shows two pairs of linked activity spaces, in *solid blue* and *dashed red lines*. The pair of *solid blue* ellipses intersects more common POIs than the two *dashed red* ellipses. However, this set of ellipses has a particular place of intersection that is west of the most general POI cluster trend

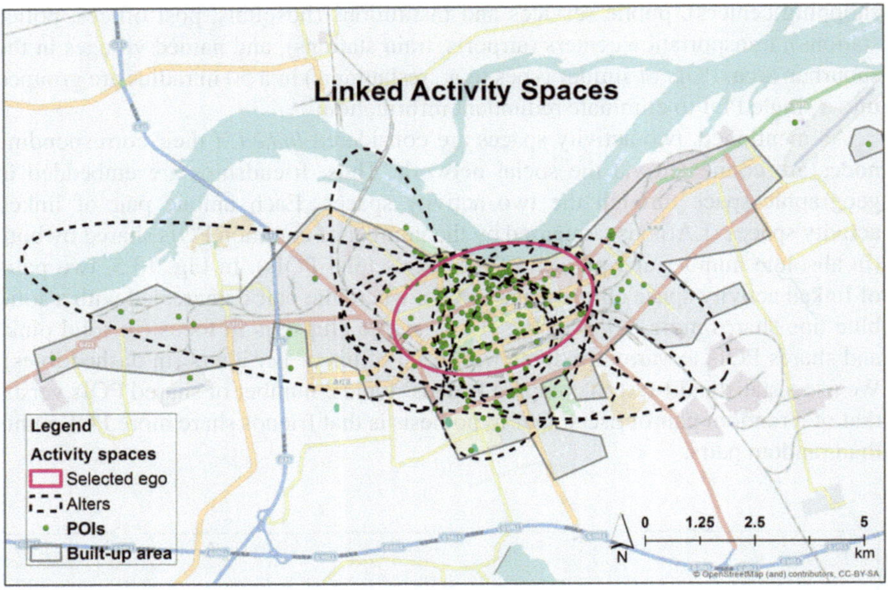

Fig. 13.6 An ego in focus (*solid pink*) shares POIs to various extents with each of his or her alters (*dashed lines*). The pattern shows that the focal ego and friends use the downtown area. One friend seems to use the Western area of the city more than the other friends

13.4 Results

13.4.1 Dyadic Relationships

We find that a pair of friends is more likely than a random pair to use the same places in Jiamusi. 11 % of friends and 50 % of random pairs share no POIs. Of pairs who share POIs, friend pairs share an average of 55.8 POIs, and a random pair shares 45.77 out of 212 total possible POIs (Fig. 13.7a). A *t*-test using the ellipse activity-space method yields a *t*-value of 56.03 (degree of freedom $(df)) = 44{,}706$, $p\text{-value} < 0.001$) and allows us to reject the null hypothesis that the mean difference between these two groups is insignificant, indicating that friends share more POIs than random pairs.

Second-degree friends also utilize urban infrastructure significantly more similarly, in terms of shared POIs, than random users (Fig. 13.7b). The *t*-value is 94.82 ($df = 152{,}070$, $p\text{-value} < 0.001$). The count of POIs shared by second-degree friends on average is 55.23 compared to 45.77 for random pairs.

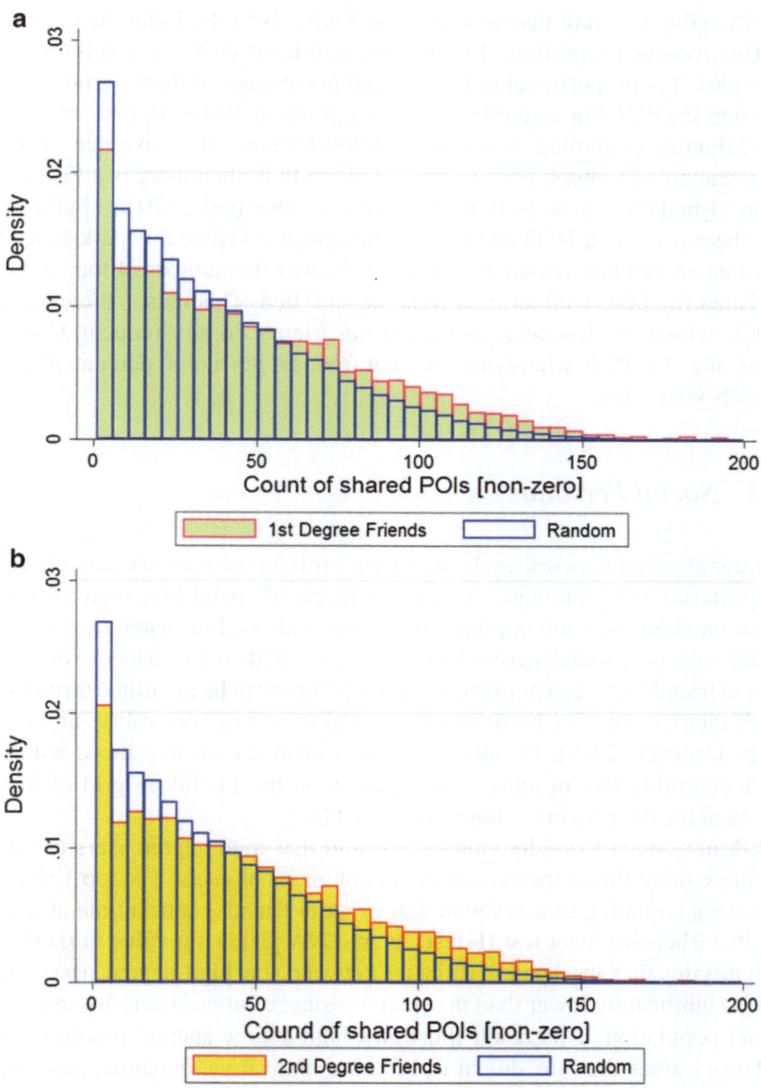

Fig. 13.7 (**a**) The probability density distribution of shared POIs of a random pair (in *blue*) and a linked pair (i.e. friends) (in *red*) show that friends have a greater probability of sharing POIs (as depicted by taller *red bars* towards the tail). We exclude pairs (random or linked) who share no common POIs. (**b**) The distribution of second-degree linked activity spaces shows that second-degree friends share more POIs on average than random pairs

Additionally, it is rare that an ego's alters will visit a POI that the ego does not visit. The *average* proportion of total egos who have visited a specific POI (e.g., "flower park") is proportional to the average percentage of their friends who have also visited the POI. For example, consider a group of 100 egos, each of whom has unique 50 alters (summing to a total of 5,000 friends). If an average of 50 % of one ego's alters (25 alters) have visited a certain POI, then there is a ½ chance the egos has visited there as well. If the average at another park is 20 % of all alters (10 alters), there is about a 1/5th chance that the egos has visited this park as well. The chance that an ego has visited a POI is *1.036 times* the number of total alters who have visited the POI, with an r^2 correlation of 0.968. This means there are few, if any, POIs where one frequents and his or her friends do not frequent. Conversely, there are also few POIs where one does not frequent yet a high percentage of his or her friends visit often.

13.4.2 Social Personas

In traditional social network analysis, a user's role in the network can reflect his or her importance and prominence in various facets of social life, such as providing information about new job opportunities to ones alters. For instance, a figure with a special role in a social network (i.e., a figure with many friends, or who is a "common friend" between poorly connected groups) can be identified through social network metrics such as betweenness centrality, degree centrality, or brokerage statistics (Jackson 2010). In one case, it has been shown that those with higher network centrality live in more central places on the Euclidean grid of longitude and latitude for the network (Onnela et al. 2011).

Confirming our second hypothesis, we find that high-degree users use the city center more often than expected, given a random set of users. The top 1 % of high-degree users (equating to users with 150 or more friends) concentrate at the urban center. A Fisher-Snedecor test (F-test) of ANOVA yields a p-value of 0.006 (95 % c.i.) signifying that the spatial variance between the high-degree users' activity spaces is significantly lower than the spatial variance between activity spaces of the universal population. This result illustrates high-degree agents' proclivity toward high-density areas that are shown to be more innovative, dynamic, and energetic environments (Bettencourt 2013).

We do not find significant spatial patterning with ego characteristics such as clustering coefficient (Jackson 2010), which measures whether one's friends are also friends themselves.

Using Spearman's correlation statistic, we find no significant relationship between user degree or total talking time (call duration) and the size of the user's activity space.

13.4.3 Community and City Form

We hypothesized that central areas play an enhanced role in supporting social communities of friends. The results do not confirm this hypothesis in the sense that the central area of Jiamusi does not favor tight-knit social circles but instead hosts heterogeneous groups of friends.

We define "groups" of friends (i.e., communities) in the Gephi environment (Bastian et al. 2009) with Blondel et al.'s (2008) community detection (modularity) algorithm that assigns each node (friend) to a cluster. This process produces 58 clusters with a modularity value of 0.732. Note that a modularity statistic of 1.0 indicates that communities are partitioned "perfectly," so that a node *i* does not connect with other nodes *j*, if *j* are not in *i*'s modularity group. Smaller modularity values indicate that connections across groups occur more frequently. Roughly, this indicates that, on average, those assigned to a cluster call within the same cluster 73.2 % of the time.

Each network agent (node) is assigned one modularity group. These agents are denoted by their ellipse centroid (geographic centers) in Fig. 13.8. Agents denoted by yellow squares or teal triangles (Fig. 13.8) are examples from two social network clusters with significant spatial clusters that differ from the overall distribution,

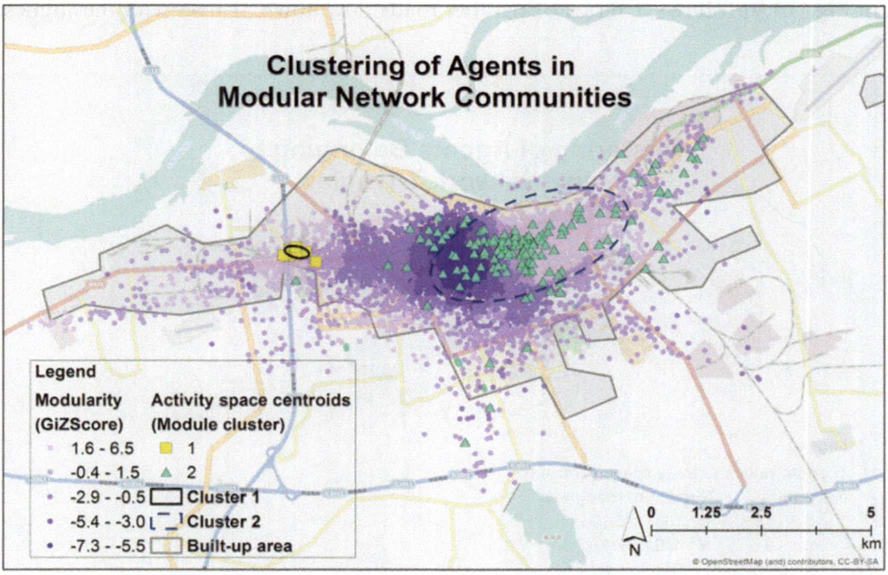

Fig. 13.8 A modularity algorithm is applied to the "aspatial" social network. Then, members of two separate modularity clusters are now mapped in *yellow squares* (denoting one group) or *teal triangles* (denoting a separate group) to show how social network groups use the city. In the downtown core, covered in *dark purple*, clusters are significantly mixed, meaning that many social groups use the downtown but are not confined to the region

deviating westward and north-eastward, respectively. The yellow squares and teal triangle clusters are examples of spatially embedded social groups that are *also* significantly clustered in a way that deviates from the expected spatial distribution of agents the city.

Including these two examples, the spatial distribution of 58 social clusters, in total, does form statistically significant "hot spots" (significantly dense clusters) and "cold spots" (a mixture of modular groups) as shown with the Getis-Ord Gi* statistic (Getis and Ord 1992). Hot spots (light purple areas in Fig. 13.8) contain agents of the same modular group in two major regions. Cold spots (dark purple areas in Fig. 13.8) cover the downtown, signifying that the groups that frequent the downtown are not clustered in the downtown, but have other group members around the city.

Another prominent pattern of community configuration, at a more local scale, is the prevalence of social triangles in the network. A social triangle can be defined as a group three nodes who connect to one another (Latapy 2008) and, pragmatically, will have meeting needs that are different and more complex than those of a dyad but perhaps not as complex as a modular group, which can contain many nodes. In our dataset, agents with the most social triangles cluster in the downtown area. However, this may be an artifact of the high-degree users' downtown, as they are likely to have more social triangles.

The distribution of high-triangle nodes (denoted in green circles and black stars in Fig. 13.9) follows a series of parallel roads downtown. Those with the highest

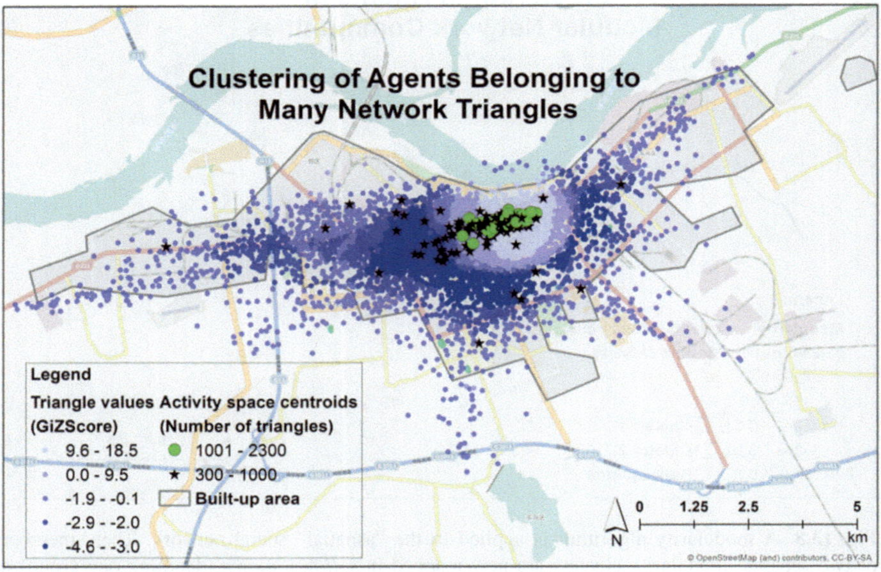

Fig. 13.9 A clustering of agents who are a part of many social triangles (i.e. groups of three friends) congregate toward the urban core (*green circles* and *black stars*). This clustering is statistically significant in one area east of downtown (denoted by *light to dark blue*), via statistical Gi* Z scores

number of social triangles (green circles as activity space centroids) form a tight linear cluster in the core area. This pattern is statistically significant, showing high Getis-Ord GI* Z scores with p-values > .0001 in the area slightly east of the downtown (in light color, Fig. 13.9). P-values are not significant in other parts of the city. Interestingly, centroids covering this neighborhood also saw the most significant modularity clusters (Fig. 13.8).

We also hypothesized that peripheral areas may be more mixed. We reject this hypothesis as the peripheral areas seem to host more cohesive communities than the downtown area. The POIs shared by linked activity spaces are more frequently on the periphery of the city. A POI can have from .01 % to 2.8 % of its pairs shown to be friends. A POI with a high value indicates that it is located in a convenient area for linked activity spaces (i.e. friends) to meet, where POIs with a lower value is more likely to host non-friend individuals. The POI hosting 2.8 % linked activity spaces is located near many popular hotels and the city's largest park – a notable tourist site in the city. This site might be a popular meeting spot for friends, but it may also be the artifact of many local business calls to one another in nearby buildings.

More generally, this ratio increases further from the city center (Figs. 13.10 and 13.11), so that POIs on the outskirts of the city are more likely to host pairs of contacts. This may be because many people have activity spaces that stretch into the downtown for work, but have their social contacts closer to their residences in another part of the city. Though agents with more contacts tend to frequent the dense downtown, given a POI with 100 random people, a user is more likely to find a friend within this 100 people if he or she is at a suburban POI, than an urban

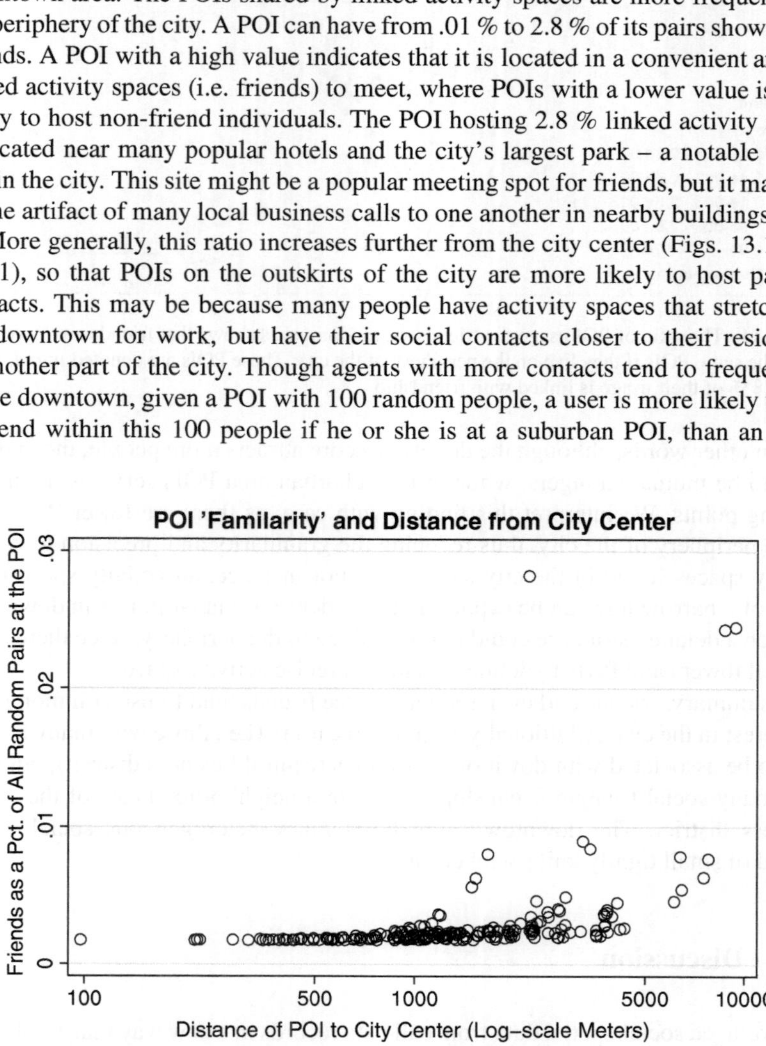

Fig. 13.10 As the POIs are located farther away from the city center there is a higher ratio of friend to non-friend pairs using these points

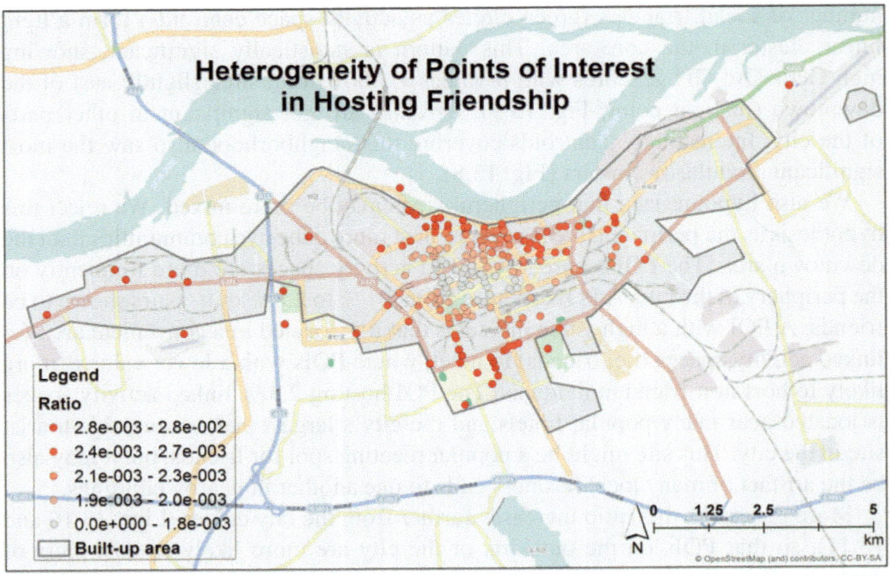

Fig. 13.11 The ratio of POI usage for friends vs. random pairs shows that friends are more likely to use the same POIs if they live on the periphery of the city. These POIs are denoted in *red*, where up to 2.8 % of their usage is linked with friendship

POI. In other words, although the downtown core attracts more people, the majority seem to be mutual strangers, while in the suburban area POIs serve as intentional meeting points. We interpret this finding with care, as there are fewer POI points on the periphery of the city, thus reducing the granularity and precision to capture activity spaces found in the city's outskirts. For instance, an activity space in the shape of a narrow line can be captured via the dense, granular points in downtown, but such a detailed structure could not be defined in the periphery, since there are so few cell towers and POIs to delineate a more precise activity space.

In summary, friends and even second-degree friends tend to use common points of interest in the city. Additionally, high-degree users (i.e., those with many friends) tend to be associated with downtown locations (central business district), but those with many social triangle friendships center in a neighborhood east of the central business district. The downtown core hosts many heterogeneous social groups instead of small tightly knit social clusters.

13.5 Discussion

We leveraged social/spatial data from call data records in a new way that emphasizes social relationships embedded in urban physical space. In this section, we respond to our initial hypotheses regarding how dyads (pairs), social network personas, triads (groups of three), and communities use the city.

First, the LAS method allows us to find the extent to which a pair of friends is more likely to use the same places in the city more than a random pair, so that a user *i* is more likely to have frequented the same POIs as a friend *j* than a random user *r*.

Second, we find that egos with high degrees are inclined toward the city center, while egos with high clustering coefficients show no significant spatial correlation. These results are not necessarily intuitive, as community members in suburban areas might also have high degrees, but do not seem to. Also, users who cling to the urban core (or tight suburban neighborhoods) might also be expected to be part of a number of "cliques" or friend groups; however, this also does not seem to be the case.

Third, we expected that central areas would play an enhanced role in supporting "clique-like" and modular groups, but we found, counter to our expectation, that the downtown was indeed a mixing pot for many groups. We do, however, find two specific neighborhoods that tend to harbor enclosed ("clique-like") social groups. Moreover, triads of friends are likely to use the downtown area.

Also, we had expected that peripheral areas were more mixed as suburbanites often have more access to automobiles and, thus, may not choose to live next to their contacts if they can drive to other parts of the city to visit. Counter to our initial hypothesis, peripheral areas show less frequent mixing of social groups and friends than any other part of the city.

13.5.1 Utility

We find these results useful for theoretical and practical issues in planning. First, in Jiamusi, we find that the ratio of friends' shared POIs to random pairs' shared POIs is 59:46. This ratio can be considered an indicator of clustered socialization. When high, this ratio shows that friends tend to group in certain parts of the city, or use the same amenities in a city. This ratio represents an urban feature that can be compared across cities, over time, and as it correlates with urban features such as population, crime rates or traffic.

Next, a planner can use these findings, for example, to discuss the merits of different models of urbanism. For example, while New Urbanism focuses on neighborhood design, architectural style, and transit-oriented development for high-density walkable cities (Al-hindi and Till 2001; Vanderbeek and Irazabal 2007), and Landscape Urbanism argues that urban design should be flexible and open-ended, (Waldheim 2002), by leveraging existing resources (Cranz and Boland 2004) and preserving wilderness (Yu et al. 2011). The LAS method results can be used to probe the adverse consequences of urban sprawl, such as its challenges for social life (Gehl 1987).

This analysis can be used to plan the location of third places (Rosenbaum 2006), such as restaurants, parks, coffee shops, theaters, and other facilities. Locations could be found by determining places that are convenient for pairs or groups of friends to meet, and combining this with other criteria, such as low traffic or areas known to be safe for pedestrians.

This method can also be used to understand the size, temporal persistence, and location (thus, level of accessibility to other places, environmental quality of the land) of certain places, such as ethnic or working neighborhoods (such as homes near a factory), where members form a dense group of ties – e.g., neighbors are likely to know and depend on one another. The method can show where these neighborhoods are and how they expand and contract over time. This can be useful for investigations of urban social capital (Granovetter 1983), cultural assimilation (De Blij and Murphy 1986), or models of epidemiology or idea spreading.

It is clear that we are only at the beginning of understanding how interpersonal relationships manifest themselves in the built environment. Yet it is a phenomenon; we experience daily as we meet colleagues at work, family at home, and perhaps friends in third places. The tension between the costs of movement in cities and the need for access to the possibilities of the city also guide our decisions about raising families through the choice of neighborhoods and school districts, as well as migration, through the choice of leaving established social circles for new circles (or vice versa).

13.6 Conclusion

A healthy city is built on strong social networks (Gilchrist 2009), but we still do not know what kinds of ties exist in cities and neighborhoods nor the detailed social dynamics that creates and changes them. Because of these limitations, we cannot currently use social network structures (clustered, decentralized, hub-spoke, etc.) as cause or effect variables in assessing planning choices for new or existing neighborhoods and cities. However, as this type of data becomes richer, such studies will become increasingly possible,

Our ability to socialize with others is affected by urban planning and government decisions regarding low-income housing, immigration reform, and health codes, such as the number of people to an urban residence or the choice of building subdivisions versus condominiums (Farber and Li 2013) or narrow versus wide roads (Montgomery 2013). The spatial and social clustering of ties changes with the creation and dissolution of institutions, such as firms, universities, military bases, sports franchises, and religious institutions. Less socialization may also lead to stagnated mobility, not due to a lack of accessibility, but to a reduced need for third places to socialize (Rosenbaum 2006), and visit others' homes.

Urban planners, geographers, government officials, civil engineers, and transportation planners that focus on improving social life in their city may be able to more directly improve residents' quality of life (Cacioppo and Patrick 2008), more so than traditional national level economic stimuli (Montgomery 2013). Instead, planners and geographers have worked toward better urban environments investigating residents' accessibility to amenities (such as hospitals), travel time to work, and social justice issues such as susceptibility to industrial and environmental hazards (Cutter et al. 2003). In addition to these variables, we should emphasize the importance of social capital within the city.

We cannot infer these social patterns from city form, or social networks alone as the connection between these variables is statistical and likely scale dependent. Thus, our task with this chapter was to illustrate how the linked activity space method allows for the integration of information from a social network into geographic space, a combination that is rarely investigated in detail (Andris 2011).

Although we use a call dataset record (CDR) for our analysis, this method can be employed to any dataset that has both evidence of social ties between agents and the geo-location of the agents. Other mechanisms for telecommunications (such as Skype, Google Video/Chat, Viber, and Whats-App) can be substituted for mobile phone calls. If these data are available, it may be worth considering the combination of the CDR dataset for interesting results on which modes of communication are popular in general, or in certain parts of the city, or during certain time frames. We may be able to capture the growth of one mode over another, over a longer time period. One exciting prospect is to see which neighborhoods make more international calls, or calls to other cities.

We do find a number of methodological and pragmatic challenges to this type of research. Many of these challenges stem from the nascent state of big data analysis that will perhaps become more reliable and complete in the future. Nevertheless, there are issues with these data that can be addressed: CDR datasets do not capture an ego with alters who do not appear in the social network. Multiple cell phones per person and multiple people per cell phones do not ensure that the telephone number is a proxy for an individual's communication patterns. Without figures on the provider's market penetration rate is difficult to understand friendships via calls to users who use a different provider. We also note a number of subjective decisions in creating a meaningful sample, such as the minimum number of towers frequented in order to be included in the dataset, the number of seed users, and number of friend "levels" to draw from the networks. None of these issues is a fundamental limitation, so we look forward to future datasets that can overcome some of these difficulties. We hope to see more research on the integration of social networks and urban spaces in the future as a unique window into how urban form and social function shape each other.

Acknowledgments This research was partially supported by the Army Research Office Minerva Program (grant no. W911NF-121A −0097), the John Templeton Foundation (grant no. 15705), the Bill and Melinda Gates Foundation (grant no. OPP1076282), the Rockefeller Foundation, the James S. McDonnell Foundation (grant no. 220020195), the National Science Foundation (grant no. 103522), the Bryan J. and June B. Zwan Foundation, and the University of Georgia.

References

Ahas R, Silm S, Saluveer E, Järv O (2009) Modelling home and work locations of populations using passive mobile positioning data. In: Gartner G, Rehrl K (eds) Location-based services and telecartography II, Lecture notes in geoinformation and cartography. Springer, Heidelberg/Berlin, pp 301–315

Albert R, Barabási A-L (2002) Statistical mechanics of complex networks. Rev Mod Phys 74:7

Al-hindi KF, Till KE (2001) (Re)placing the new urbanism debates: toward an interdisciplinary research agenda. Urban Geogr 22:189–201

Andris C (2011) Methods and metrics for social distance. Dissertation, Massachusetts Institute of Technology

Arentze T, van den Berg P, Timmermans H (2012) Modeling social networks in geographic space: approach and empirical application. Environ Plan A 44:1101–1120

Axhausen KW (2007) Activity spaces, biographies, social networks and their welfare gains and externalities: some hypotheses and empirical results. Mobilities 2:15–36

Axhausen KW, Zimmermann A, Schönfelder S, Rindsfüser G, Haupt T (2002) Observing the rhythms of daily life: a six-week travel diary. Transportation 29:95–124

Bastian M, Heymann S, Jacomy M (2009) Gephi: an open source software for exploring and manipulating networks. In: Proceedings of the international conference on weblogs and social media (ICWSM), San Jose, CA, 17–20 May 2009. AAAI Press, Menlo Park

Bettencourt LMA (2013) The origins of scaling in cities. Science 340:1438–1441

Blondel V, Guillaume J-L, Lambiotte R, Lefebvre E (2008) Fast unfolding of communities in large networks. J Stat Mech Theor E 10:P10008

Cacioppo JT, Patrick W (2008) Loneliness: human nature and the need for social connection. WW Norton & Company, New York

Calabrese F, Smoreda Z, Blondel V, Ratti C (2011) Interplay between telecommunications and face-to-face interactions: a study using mobile phone data. PLOS ONE 7:e20814

Carrasco J A, Miller E, Wellman B (2006) Spatial and social networks: the case of travel for social activities. Paper presented at the 11th international conference on travel behavior research, Kyoto, Japan, 16–20 August 2006

Centola D, Macy M (2007) Complex contagions and the weakness of long ties. Am J Sociol 113:702–734

Cho E, Myers SA, Leskovec J (2011) Friendship and mobility: user movement in location-based social networks. In: Proceedings of the 17th ACM SIGKDD international conference on knowledge discovery and data mining (KDD), San Diego, CA, 21–24 August 2011. ACM, New York

Christakis N, Fowler J (2007) The spread of obesity in a large social network over 32 years. New Engl J Med 357:370–379

Coleman DJ (2010) Volunteered geographic information in spatial data infrastructure: an early look at opportunities and constraints. Paper presented at the GSDI 12 world conference, Singapore, 19–22 October 2010

Crandall DJ, Backstrom L, Cosley D, Suri S, Huttenlocher D, Kleinberg J (2010) Inferring social ties from geographic coincidences. Proc Natl Acad Sci U S A 107:22436–22441

Cranz G, Boland M (2004) Defining the sustainable park: a fifth model for urban parks. Landsc J 23:102–120

Cutter SL, Boruff BJ, Shirley WL (2003) Social vulnerability to environmental hazards. Soc Sci Q 84:242–261

De Blij HJ, Murphy AB (1986) Human geography: culture, society and space. Wiley, New York

Eagle N, Pentland AS, Lazer D (2009) Inferring friendship network structure by using mobile phone data. Proc Natl Acad Sci U S A 106:15274–15278

Emch M, Root ED, Giebultowicz S, Ali M, Perez-Heydrich C, Yunus M (2012) Integration of spatial and social network analysis in disease transmission studies. Ann Assoc Am Geogr 102:1004–1015

Farber S, Li X (2013) Urban sprawl and social interaction potential: an empirical analysis of large metropolitan regions in the United States. J Trans Geogr 31:267–277
Frei A, Axhausen KW (2011) Modeling spatial embedded social networks. Working paper 685: transport and spatial planning, IVT, ETH Zurich, Zurich, Switzerland
Gao S, Wang Y, Gao Y, Liu Y (2013) Understanding urban traffic-flow characteristics: a rethinking of betweenness centrality. Environ Plan B 40:135–153
Gehl J (1987) Life between buildings: using public space (trans: Koch J). Van Nostrand Reinhold, New York
Getis A, Ord JK (1992) The analysis of spatial association by use of distance statistics. Geogr Anal 27:286–306
Gilchrist A (2009) The well-connected community: a networking approach to community development, 2nd edn. Policy Press, London
Girardin F, Calabrese F, Dal Fiore F, Ratti C, Blat J (2008) Digital footprinting: uncovering tourists with user-generated content. IEEE Pervas Comput 7:36–43
Golledge RG (1999) Human wayfinding and cognitive maps. In: Golledge R (ed) Wayfinding behavior: cognitive mapping and other spatial processes. Johns Hopkins University Press, Baltimore, pp 5–45
Granovetter M (1973) The strength of weak ties. Am J Sociol 78:1360–1380
Granovetter M (1983) The strength of weak ties: a network theory revisited. Soc Theor 1:201–233
Jackson MO (2010) Social and economic networks. Princeton University Press, Princeton
Kang C, Ma X, Tong D, Liu Y (2012) Intra-urban human mobility patterns: an urban morphology perspective. Physica A 391:1702–1717
Kurant M, Markopoulou A, Thiran P (2011) Towards unbiased BFS sampling. IEEE J Sel Area Commun 29:1799–1809
Latapy M (2008) Main-memory triangle computations for very large (sparse (power-law)) graphs. Theor Comput Sci 407:458–473
Mitchell A (2005) The ESRI guide to GIS analysis, vol 2. ESRI Press, Redlands
Montgomery C (2013) Happy city: transforming our lives through urban design. Farrar, Straus and Giroux, New York
Neis P, Zipf A (2012) Analyzing the contributor activity of a volunteered geographic information project – the case of OpenStreetMap. Int J Geo-Inf 1:146–165
Newman ME, Park J (2003) Why social networks are different from other types of networks. Phys Rev E 68:036122
Onnela JP, Arbesman S, González M, Barabási AL, Christakis N (2011) Geographic constraints on social network groups. PLoS ONE 6:e16939
Papachristos A, Hureau D, Braga A (2013) The corner and the crew: the influence of geography and social networks on gang violence. Am Sociol Rev 78:417–447
Papas MA, Alberg AJ, Ewing R, Helzlsouer KJ, Gary TL, Klassen AC (2007) The built environment and obesity. Epidemiol Rev 29:129–143
Radil SM, Flint C, Tita G (2010) Spatializing social networks: using social network analysis to investigate geographies of gang rivalry, territoriality, and violence in Los Angeles. Ann Assoc Am Geogr 100:307–326
Ratti C, Williams S, Frenchman D, Pulselli RM (2006) Mobile landscapes: using location data from cell phones for urban analysis. Environ Plan B 33:727–748
Reades J, Calabrese F, Sevtsuk A, Ratti C (2007) Cellular census: explorations in urban data collection. IEEE Pervas Comput 6:30–38
Rosenbaum MS (2006) Exploring the social supportive role of third places in consumers' lives. J Serv Res 9:59–72
Salganik MJ, Watts DJ (2008) Leading the herd astray: an experimental study of self-fulfilling prophecies in an artificial cultural market. Soc Psychol Q 71:338–355
Santaló LA (2004) Integral geometry and geometric probability. Cambridge University Press, Cambridge

Scellato S, Noulas A, Lambiotte R, Mascolo C (2011) Socio-spatial properties of online location-based social networks. In: Proceedings of the international conference on weblogs and social media, Barcelona, 17–21 July 2011, pp 329–336
Schneider CM, Belik V, Couronné T, Smoreda Z, González MC (2013) Unravelling daily human mobility motifs. J R Soc Interface 10:20130246
Schönfelder S, Axhausen KW (2003) Activity spaces: measures of social exclusion? Arbeitsbericht Verkehrs- und Raumplanung, 140, Institut für Verkehrsplanung und Transportsysteme (IVT), ETH Zürich, Zürich
Sevtsuk A, Mekonnen M (2012) Urban network analysis: a new toolbox for ArcGIS. Rev Int Géomatique 22:287–305
Sevtsuk A, Huang S, Calabrese F, Ratti C (2009) Mapping the MIT campus in real time using WiFi. In: Foth M (ed) Handbook of research on urban informatics: the practice and promise of the real-time city. IGI Global, Hershey, pp 326–337
Sun Y, Fan H, Helbich M, Zipf A (2013) Analyzing human activities through volunteered geographic information: using Flickr to analyze spatial and temporal pattern of tourist accommodation. In: Krisp JM (ed) Progress in location-based services. Springer, Heidelberg/Berlin, pp 57–69
Vanderbeek M, Irazabal C (2007) New Urbanism as a New Modernist movement: a comparative look at Modernism and New Urbanism. Tradit Dwell Settl Rev 19:41–57
Waldheim C (2002) Landscape urbanism: a genealogy. Praxis 4:10–17
Yu K, Wang S, Li D (2011) The negative approach to urban growth planning of Beijing, China. J Environ Plan Man 54:1209–1236

Chapter 14
Using Non-authoritative Sources During Emergencies in Urban Areas

Emily Schnebele, Christopher Oxendine, Guido Cervone, Celso M. Ferreira, and Nigel Waters

Abstract During emergencies in urban areas, it is paramount to assess damage to people, property, and environment in order to coordinate relief operations and evacuations. Remote sensing has become the de facto standard for observing the Earth and its environment through the use of air-, space-, and ground-based sensors. These sensors collect massive amounts of dynamic and geographically distributed spatiotemporal data daily and are often used for disaster assessment, relief, and mitigation. However, despite the quantity of big data available, gaps are often present due to the specific limitations of the instruments or their carrier platforms. This chapter presents a novel approach to filling these gaps by using

E. Schnebele (✉)
GeoInformatics and Earth Observation Laboratory (GEO), Department of Geography and Institute for CyberScience, The Pennsylvania State University, University Park, PA, USA
e-mail: eschnebe@gmu.edu

C. Oxendine
Department of Geography & Environmental Engineering, United States Military Academy, West Point, NY, USA
e-mail: christopher.oxendine@usma.edu

G. Cervone
GeoInformatics and Earth Observation Laboratory (GEO), Department of Geography and Institute for CyberScience, The Pennsylvania State University, University Park, PA, USA

Research Application Laboratory, National Center for Atmospheric Research, Boulder, CO, USA
e-mail: cervone@ucar.edu

C.M. Ferreira
Department of Civil & Environmental and Infrastructure Engineering, George Mason University, Fairfax, VA, USA
e-mail: cferrei3@gmu.edu

N. Waters
GeoInformatics and Earth Observation Laboratory (GEO), Department of Geography and Institute for CyberScience, The Pennsylvania State University, University Park, PA, USA

The University of Calgary, Calgary, Canada
e-mail: nwaters@gmu.edu

M. Helbich et al. (eds.), *Computational Approaches for Urban Environments*, Geotechnologies and the Environment 13, DOI 10.1007/978-3-319-11469-9_14

non-authoritative data including social media, news, tweets, and mobile phone data. Specifically, two applications are presented for transportation infrastructure assessment and emergency evacuation.

Keywords Infrastructure assessment • Evacuation • Remote sensing • Inundation modeling • Social media • Geospatial analysis • Big data

14.1 Introduction

Never in the history of humankind have we known so much about our planet. Never in the history of humankind have we had such easy access to data. Never in the history of humankind has our civilization been so much at risk.

Hazards pose a constant threat to the development and sustainment of our infrastructure and our society. Hazards can be natural, anthropogenic, or technological. They are, respectively, events that naturally occur, events resulting from human activities or accidents, or the catastrophic collapse of infrastructure, such as roads, communication networks, or power grids, which are needed for our society to function.

A single catastrophic event can claim thousands of lives; cause billions of dollars of damage; trigger an economic depression that might directly or indirectly affect the entire world; destroy natural landmarks; cause tsunamis, floods, and landslides; render a large territory uninhabitable; and destabilize the military and political balance in a region (Cutter 1993; Alexander 2002; Wisner et al. 2004). Such potential catastrophic consequences are due to the emergence of megacities and the proliferation of nuclear power plants and nuclear waste storage facilities, high dams, and other facilities whose destruction poses an unacceptable risk of global reach (Freudenburg et al. 2008; Casti 2012). Thus, the study of natural hazards and of the processes that govern their occurrence has become a fundamental challenge for the survival of our civilization.

Advances in our ability to observe the Earth and its environment through the use of air-, space-, and ground-based sensors has led to the collection of massive amounts of dynamic and geographically distributed spatiotemporal data. Numerical models are initialized with these high-resolution observations to forecast the future or to simulate the past, generating simulations that can be several orders of magnitude larger than the initial observations. Remote sensing data from air- and space-borne platforms have also become the de facto standard for providing high-resolution information for the assessment, relief, and mitigation of damaged areas during and after emergencies caused by natural, human-made, and technological disasters (Jensen and Cowen 1999; Voigt et al. 2007). However, due to limitations in orbital revisit time, sensor characteristics, and the presence of clouds, there may be gaps in these remote sensing data.

This chapter presents applications for data collected from non-authoritative sources to fill the gaps in remote sensing data during disasters and emergencies. Non-authoritative sources include data volunteered by citizens (also known as volunteered geographic information or VGI (Goodchild 2007; Sui and Goodchild

2011; Sui et al. 2013)) or collected for purposes other than disaster assessment, such as traffic cameras or mobile phone locations. This general class of data, often voluntarily contributed and made available, can consist of pictures, videos, sounds, text messages, etc. Due to the spread of the Internet to mobile devices, an unprecedented and massive amount of data have become available, often geolocated and often in real time. These sources provide a large, rapidly changing, dynamic dataset that not only complements remote sensing observations but also adds an additional, subjective view of how people perceive and react to hazards.

Although non-authoritative data are often published without scientific intent, and usually carry little scientific merit, it is still possible to mine mission critical information. For example, volunteered photos and videos about natural hazards have emerged as a data source during crises and hazardous events to derive local meteorological information, capture and record the physical features of an event, and identify and document flood height (De Longueville et al. 2009; Hyvärinen and Saltikoff 2010; Poser and Dransch 2010). During Hurricane Sandy, geolocated pictures and videos searchable through Google provided early emergency response with ground view information.

Mining these massive amounts of "big data," it is possible to reconstruct a spatiotemporal human terrain that provides knowledge when remote sensing data are unavailable or incomplete. Additionally, non-authoritative data may provide unique knowledge that is not possible to acquire solely from remote sensing instruments.

This chapter discusses the fusion of remote sensing and non-authoritative sources to assess road infrastructure and plan evacuations in an urban environment during emergencies. Two specific applications are discussed:

1. An assessment of New York City transportation infrastructure during and after Hurricane Sandy using crowdsourced remote sensing imagery, numerical models, social media, and ground observations
2. Identification of evacuation routes during emergencies in New York City using traffic information and mobile phone data

14.2 Transportation Infrastructure Assessment

The first application presented in this chapter is a damage assessment of roads during and after Hurricane Sandy in New York City. Multiple sources of data are combined including aerial images contributed by the Civil Air Patrol (CAP), numerical inundation models, VGI harvested from social media, and ground observations.

The utilization of data from multiple sources can help provide a more complete description of a phenomenon. For example, data fusion is often employed with remote sensing data to combine information of varying spatial, temporal, and spectral resolutions as well as to reduce uncertainties associated from using a single source (Zhang 2010). The fused data then provides new or better information than would be available from a single source (Pohl and Van Genderen 1998).

The incorporation of multiple data sources or methods for improved performance or increased accuracy is not limited to the field of remote sensing. Boosting, a common machine learning technique, has been shown to be an effective method for generating accurate prediction rules by combining rough, or less than accurate, algorithms together (Freund et al. 1999). While the individual algorithms may be singularly weak, their combination can result in a strong learner. Furthermore, redundancies in observations provide an increase in the confidence of observations or estimates, while data from multiple sources can provide information when they might not do so if used in isolation.

Figure 14.1 shows the stacked layer approach used to fuse heterogeneous data at different spatial, temporal, and radiometric resolutions. The data are first processed in a GIS environment by resampling them at the highest resolution available using

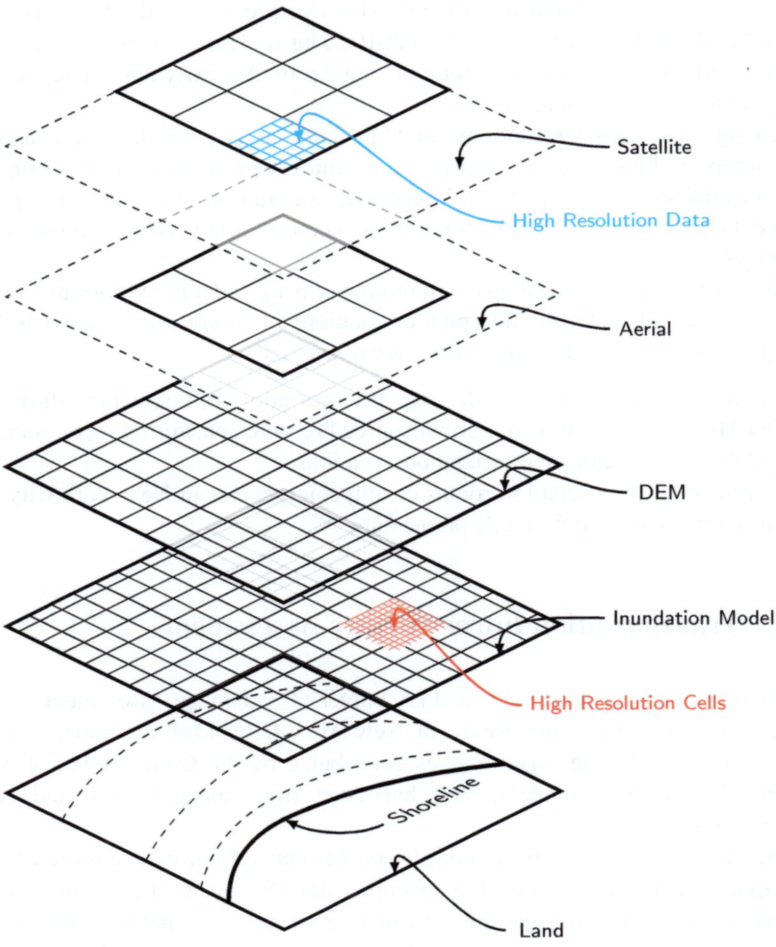

Fig. 14.1 Stacked layer approach used in the methodology. Data comes in different formats and with different resolutions

spatial statistical algorithms (Zhang et al. 2014). For example, points which identify flooding (e.g., photos) are plotted, georeferenced, and then smoothed using a kernel interpolation to create a GIS layer of estimated flood extent. This is a task performed for each data source, resulting in multiple, individual flood extent layers. The analysis is then performed on the fused layers by applying statistical or machine learning algorithms to classify the data and identify anomalies.

Fusing data from multiple sources leads to an improved damage estimation and an increased understanding of the sequence of events that leads to transportation infrastructure failure. In this example, non-authoritative data are used in two different scenarios:

1. Damage assessment during an event
2. Damage assessment after the event

It is assumed that ground truth data might not be available. The novelty of this study is the development of a methodology that takes advantage of "citizens as sensors" (Goodchild 2007) and of various other data, including remote sensing and numerical models, not necessarily designed to be used during emergencies to improve damage assessment. These non-authoritative or nontraditional sources are used to create additional layers which augment traditional sources when they may be lacking or incomplete. The result is shown in the bottom layer, where a flood hazard map is generated. The resulting flood hazard map is then paired with a high-resolution road network to create a road damage map.

14.2.1 Data Sources

14.2.1.1 Remote Sensing Data

High-resolution remote sensing data are routinely used to assess damage during and after hazards, in both urban and rural areas. Two or more images are acquired for an area showing the differences before and after the hazard.

The Civil Air Patrol, the civilian branch of the US Air Force, was tasked with collecting aerial photos of the US East Coast following the impact of Hurricane Sandy in October 2012. Within days of the storm making landfall, hundreds of missions were flown by volunteers from Cape Cod, MA, to Cape May, NJ. From these missions, thousands of aerial photos of the coastline were generated, including those documenting heavily flooded areas.

14.2.1.2 Numerical Surge Model

Recent improvements in understanding the physics of storm surge combined with rapid increases in High Performance Computing (HPC) power have led to the development of physics-based, high-resolution numerical models capable of predicting and simulating hurricane storm surge with reasonable accuracy in coastal areas.

The Sea, Lake, and Overland Surge from Hurricane model (SLOSH) (Jelesnianski et al. 1992) developed by the National Weather Service (NWS) has been extensively used by decision makers to predict storm surge inundation for planning and emergency management, and it is currently the NWS official operational forecast model for storm surge. Several other numerical models have been developed over the years to calculate water levels and currents resulting from hurricane storm surges along the continental shelves and coasts.

Among others are the Advanced Circulation (ADCIRC) model developed by Luettich and Westerink (2004), the fully nonlinear Finite Volume Coastal Ocean Model (FVCOM) developed by Chen et al. (2003), and the Semi-implicit Eulerian-Lagrangian Finite Element (SELFE) model developed by Zhang and Baptista (2008). Recently, the Coastal and Ocean Modeling Testbed (COMT) compared the models' prediction skills (Kerr et al. 2013) and concluded that they all, except SLOSH, generated similar predictions for Hurricane Ike in 2008 and Hurricane Rita in 2005, thus demonstrating the maturity level of storm surge model development.

For this study, a lower-resolution/faster computational time numerical mesh was used to simulate the Hurricane Sandy storm surge in order to maintain similarity to models used in operational forecasts (e.g., Advanced Flooding Guidance System [ASGS]) to represent information that would be available to decision makers before a hurricane landfall. The coupled version of the two-dimensional depth integrated version of the Advanced Circulation (ADCIRC) model and the SWAN wave model (Dietrich et al. 2011) was used to simulate hurricane storm surge along the coast. The ADCIRC model (Luettich and Westerink 2004) is a finite element, shallow water model that solves for water levels and currents at a range of scales and is widely used for storm surge modeling (e.g., Ferreira et al. 2014). This version of the program solves the Generalized Wave Continuity Equation (GWCE) and the vertically integrated momentum equations. SWAN is a third-generation spectral wave model that computes random, short crested wind-generated waves and wave transformation in the near shore and inland waters (Booij et al. 1999). For storm surge simulation, ADCIRC is forced by the wind and pressure fields and the wave radiation stress resulting from the wave model. Tides and river inflow can also be added as a boundary.

The East Coast Mesh (ECM2001) presented by Mukai et al. (2002) was utilized with approximately 250,000 nodes and a resolution of approximately 1.2 km along the study area. ADCIRC allows for the use of an unstructured finite element mesh with variable resolution along the model domain. The hurricane surge model was forced by wind and pressure fields developed by a parametric asymmetric wind model (Mattocks and Forbes 2008) that computes wind stress, average wind speed, and direction inside the Planetary Boundary Layer (PBL) based on the National

Hurricane Center (NHC) Advisory Archive for Hurricane Sandy (NOAA 2013a) track data and meteorological conditions (e.g., central pressure, forward speed, and radius to maximum wind).

To simulate the Hurricane Sandy storm surge, a simulation was run for October 18th until the 28th including tides (tidal potential components M2, S2, N2, K2, K1, O1, and Q1) but neglecting river inflows. The simulations were performed under the HPC environment provided by the Extreme Science and Engineering Discovery Environment (XSEDE) supported by the National Science Foundation (NSF). Results were recorded every 15 min around the study region at every model node and at NOAA Tides and Currents stations (NOAA 2013b). The model results generally overestimate the measured water levels most likely due to the differences between the hypothetical asymmetrical wind and pressure fields and the actual storm conditions.

The spatial flood levels were calculated using a Digital Elevation Model (DEM) with a 1 arc-second resolution from the National Elevation Dataset (NED) for the study region (USGS 2013). The maximum water levels for each model node were extracted for the 29th and the 30th of November and converted to the NAVD88 vertical datum. A spline interpolation with tension was applied to create a maximum water level surface for the study region according to the methodology suggested by Berenbrok et al. (2009). Finally, the water levels were subtracted from the DEM to calculate the spatial flooded extent (Fig. 14.2).

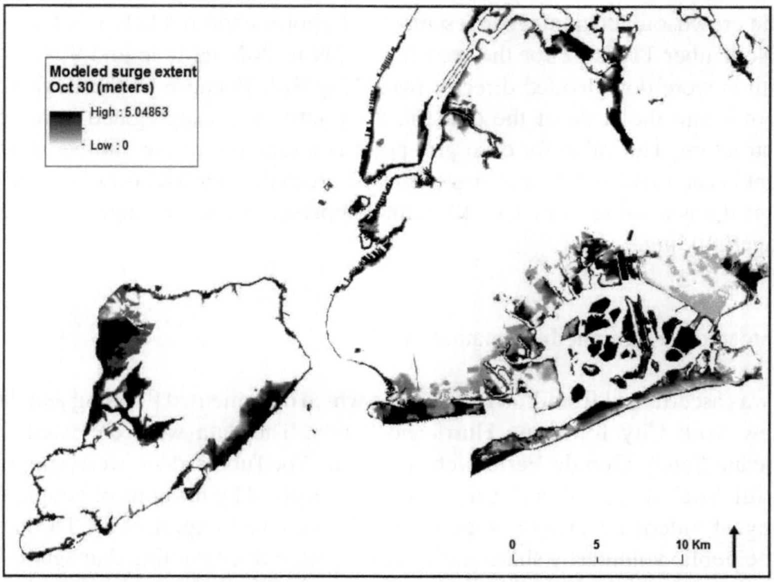

Fig. 14.2 Modeled surge extent for October 30 at 1:00pm

14.2.1.3 Non-authoritative Data

Non-authoritative data are not produced or distributed from necessarily trusted sources and often lack any assertion of verification or accuracy. However, regardless of varying levels of certainty or trustworthiness, non-authoritative sources can provide valuable, real-time, on-the-ground information during disasters when traditional sources are unavailable, lacking, or slow to respond. For example, following the Fukushima nuclear disaster in 2011, the Japanese public supplemented authoritative government sensors with user-generated content. Individuals throughout the country bought personal Geiger counters and contributed to a crowdsourced Geigermap.[1]

Crowdsourced Damage Assessment

Remote sensing data (photos) acquired by the Civil Air Patrol were assessed for damage by thousands of people across the world. The photos were placed on a Hurricane Sandy Google Crisis Map website (Fig. 14.3) for the public to assess visible damage through a crowdsourcing portal supported by MapMill. This yielded a large damage assessment dataset generated from crowdsourced, non-authoritative, nontraditional sources. The photos were also made available online through a Federal Emergency Management Agency (FEMA) website for residents to search by street address to see what, if any, damage their homes may have sustained.

The crowdsourced damage assessments of photos captured between October 31 and November 11, 2012 for the area from 33N to 26N latitude and 90W to 84W longitude were downloaded directly from MapMill. Because of the large volume of photos and the scale of the domain, the photos were aggregated into a 500 m grid structure. The value for each grid point is a function of the number of images present in each grid and their average crowdsourced damage assessment. As a result, each grid has a value from 1 to 10, with 1 representing no damage and 10 severe damage/flooding.

Volunteered Geographic Information (VGI)

VGI was ascertained from YouTube videos which documented flooding and damage in New York City following Hurricane Sandy. The data were collected from a Hurricane Sandy Google Earth website where YouTube videos were supplied by Storyful. YouTube, a video-sharing website, is utilized by millions of people for the sharing of videos covering a wide range of topics and experiences. Through this site the public voluntarily shares information, often documenting damage resulting from natural hazards. The videos were provided with geolocated information and were visually assessed by the authors. The small number of videos ($n = 15$) did

[1] http://japan.failedrobot.com/

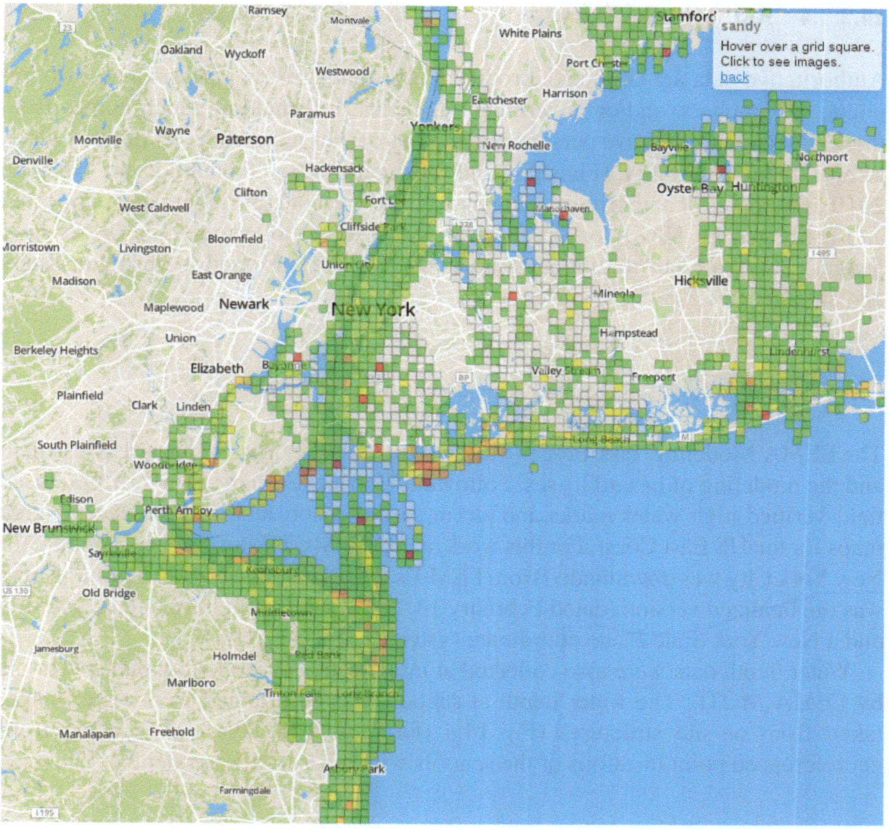

Fig. 14.3 Crowdsourced assessments for the Civil Air Patrol data. Damage assessment: *red* = high, *yellow* = medium, *green* = none

not require any crowdsourcing or automated assessment. Furthermore, it is shown in Schnebele and Cervone (2013) that even a small number of properly located VGI data can help improve flood assessment. Each location corresponding to a video point was assigned a value of 10 (severe damage/flooding).

Photos ($n = 25$) which documented flooding within the study domain were downloaded using the Google search engine and were also visually assessed by the authors. The point locations were georeferenced to create a GIS layer of flooded locations. Each point was assigned a value of 10 (severe damage/flooding).

Twitter, a popular social networking site, is often utilized by the public to share information about their daily lives through micro-blogging. Arizona State University's TweetTracker provided Twitter data for this project (Kumar et al. 2011). Tweets generated in the New York City area extending from 40.92N to 40.54N latitude and 73.75W to 74.13W longitude from October 26 to November 3, 2012 containing the word "flood" were used to provide a temporal framework.

14.2.1.4 Authoritative Data

Authoritative data are collected, produced, and managed by professional cartographers, geographers, and/or government agencies. Information which comes from these official, authoritative persons or agencies carries a certain level of trust which affords them credibility (Flanagin and Metzger 2008; Goodchild and Glennon 2010). Examples of authoritative data may include remote sensing imagery collected and calibrated by NASA or stream flow information collected from USGS river gauges. These are our traditional sources of data and information during disasters and emergencies.

Federal Emergency Management Agency (FEMA)

The FEMA Modeling Task Force (MOTF) consists of experts in hazard assessment and the modeling of hazard losses. Following Hurricane Sandy, FEMA MOTF used field-verified high water marks and storm surge sensor data to create storm surge maps for the US East Coast. For this work, a FEMA MOTF storm surge shapefile for New York City was downloaded from FEMA's GeoPlatform website. The surge map was the finalized version (dated February 14, 2013) with a 1 m horizontal resolution and a New York State Plane coordinate system (Fig. 14.4 (right)).

Water depth data were also collected at inundated New York City public schools by FEMA MOTF. The water depth at schools was ascertained from water marks taken from on-site structures (Fig. 14.4 (left)). A GIS layer was created from georeferenced point locations of the schools with measured water depths.

United States Geological Survey (USGS)

Water height collected by the USGS storm-tide monitoring provides an additional source of authoritative ground data. These official measurements were taken at

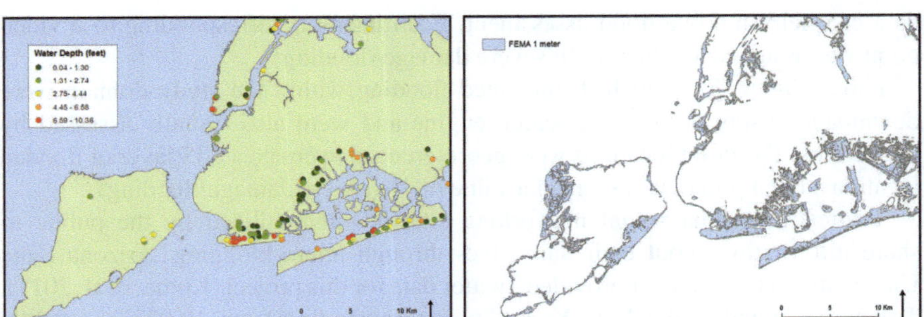

Fig. 14.4 FEMA water depth measured at public schools (*left*) and official flood inundation map (*right*)

different locations throughout the domain. Water height modeled for each point is interpolated using a spline function to create a water height surface. A DEM with a 1 arc-second resolution from the National Elevation Dataset (NED) is subtracted from the water height surface to create a water depth layer (USGS 2013).

United States Census Bureau

A 2012 TIGER/line® shapefile of road networks for the New York City area was downloaded from the US Census Bureau and was georeferenced to New York State Plane coordinates.

14.2.2 Damage Assessment During Emergencies

After individual data layers are generated from available remote sensing and authoritative and non-authoritative data, they are integrated together using an artificial neural network machine learning algorithm. Artificial neural networks are nonlinear data modeling tools for discovering patterns in data from a series of inputs (Atkinson and Tatnall 1997). The network consists of interconnected nodes comprising an input layer, a hidden layer, and an output layer (Fig. 14.5). In this research, the nodes of the input layer consist of the flood identification layers created during preprocessing, and the output layer is a flood assessment surface. The hidden layer nodes, or neurons, are the computational units of the network. The neuron receives the inputs and produces responses. Benediktsson et al. (1990) defines the simplest formal model of the neuron, where the output value is approximated by the function

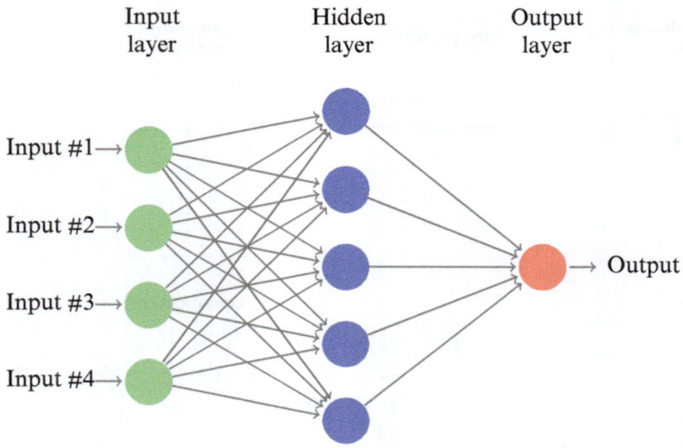

Fig. 14.5 Depiction of an artificial neural network

$$o = K\phi\left(\sum_{j=1}^{n} w_j x_j - \theta\right) \quad (14.1)$$

where K is a constant, ϕ is a nonlinear function, w_j are the weights assigned by the network, and θ is a threshold. The network takes inputs x and produces a response o_i from the output units i. The outputs are either $o_i = 1$ if the neuron i is active for the input x or $o_i = 0$ if it is inactive. The network learns the weights through iterative training and will converge when there is no change from one iteration to the next.

The trained network can then be used for the classification of a new dataset. A feedforward artificial neural network was implemented for this work using the **R** statistical package (Venables and Ripley 2002).

The goal is to classify each pixel as being flooded or not flooded. The neural network classifier is trained using the data layers from October 29 and tested on the October 30 layers. Figure 14.6 illustrates the training of the neural network classifier (blue lines) using available sources of remote sensing and authoritative and non-authoritative data from October 29. The data are first preprocessed (e.g., georeferenced and interpolated) to create individual flood extent estimations which are fed into the neural network to create a classifier. The operational step uses this classifier along with data collected from the subsequent day, October 30. These data (green lines) are preprocessed and then passed through the trained classifier to create a flood extent map.

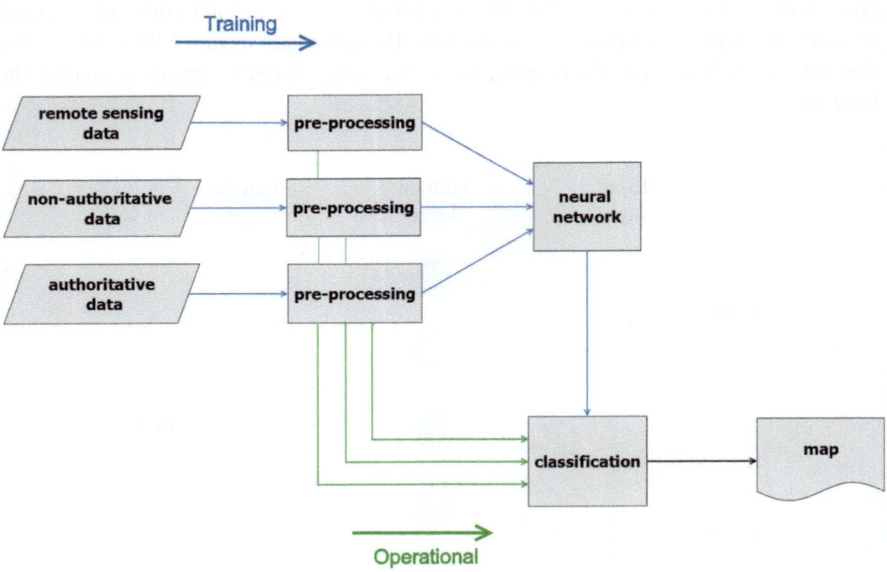

Fig. 14.6 Illustration of the application of a neural network classifier. The classifier is created from training data (*blue lines*) and is then used to create a flood extent map by passing data from a subsequent day (*green lines*) through the classifier

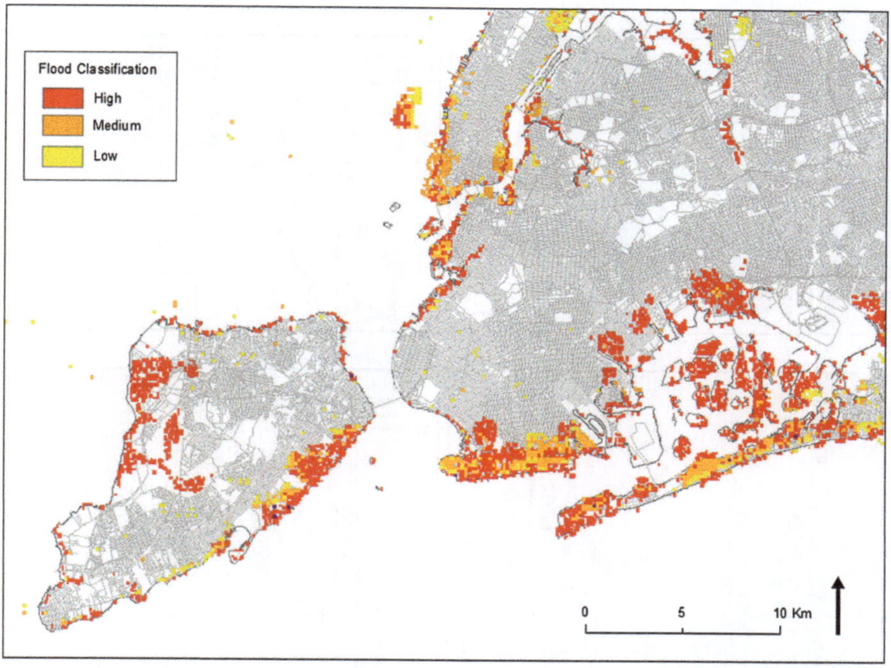

Fig. 14.7 Classification of flooding (high, medium, low flood severity/damage) in New York using an artificial neural network

Because the inundated schools, USGS, and Civil Air Patrol data represented maximum flood extent, it was possible to generate only one layer from each dataset; therefore, these data were used for both days. The initial training and testing datasets produced results indicating flooding along the coastlines of New York City with the greatest damage identified in lower Manhattan and southern edges of Brooklyn and Queens (Fig. 14.7).

14.2.3 Damage Assessment After Emergencies

After an emergency, remote sensing and volunteered data can be employed to provide a damage assessment. In this particular work, the official FEMA flood map is color coded to show not only which areas have been flooded but also which areas have been most affected. In addition, the damage assessment surface is then used to identify roads which may be compromised or may require site inspections (Schnebele et al. 2013).

Crowdsourced data (CAP photos) and VGI (YouTube videos), which are illustrated in Fig. 14.8b, are fused together using a kriging interpolation. Kriging allows for spatial correlation between values (i.e., locations/severity of flooding)

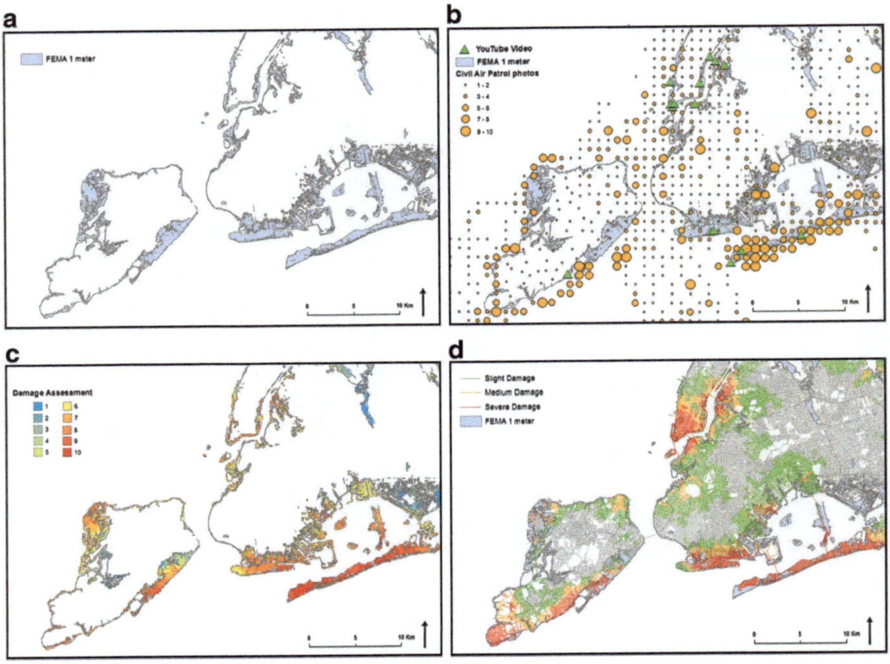

Fig. 14.8 Storm surge extent generated by FEMA and the locations of Civil Air Patrol photos and geolocated videos (**a** and **b**). Flood damage assessment generated from non-authoritative data and the subsequent classification of potential road damages (**c** and **d**)

to be considered and is often used with Earth science data (Oliver and Webster 1990; Olea and Olea 1999; Waters 2009). Ordinary kriging generated a strong interpolation model. Cross-validation statistics yielded a standardized mean prediction error of 0.0008 and a standardized root-mean-squared prediction error of 0.9967. Figure 14.8c illustrates the damage assessment, with values ranging from 1 (no damage) to 10 (severe damage), created from the interpolated surface which is clipped to the boundaries of the FEMA surge extent (Fig. 14.8a) demonstrating how non-authoritative sources can be used to add value to the FEMA map.

Ground information in the form of geolocated videos (Fig. 14.9) enhances the non-authoritative dataset by providing flood information not conveyed in the CAP photos. As illustrated in Fig. 14.8b, the locations of the videos (green triangles) did not coincide with the locations of photos rated as medium/severe damage (larger orange circles, values 7–10). Reasons for this disparity may include flooding captured on video had receded before the Civil Air Patrol flights or were captured at night or flooding may have occurred in areas which were not in a flight path or were unable to be seen from aerial platforms (i.e., flooding in tunnels, under overpasses). By using multiple data sources, flood or damage details not captured by one source can be provided by another.

Fig. 14.9 Example of YouTube video documenting flooding

Overall, there is a very good agreement between the flood extent from FEMA and the assessment generated with the proposed methodology. Figure 14.10 provides examples of agreement between photos identifying flooding/damage and the FEMA-generated flood extent, while Fig. 14.11 includes examples where the locations of flooding or damage did not agree between the Civil Air Patrol and the FEMA data. These areas were located along coastal edges, and therefore a lack of spatial precision in the data is most likely the cause of the discrepancies.

Sources of error in non-authoritative data, such as incorrect information (false positive/negative) or improper geolocation, needed to be considered. Incorrect information can be mitigated by including visually verified photos/videos and the application of multiple sources. Crowdsourcing, in particular, can increase accuracy and enhance information reliability compared to single-source observations (Giles 2005). Geolocation errors can be reduced with automation.

Sparse data or data skewed in favor of densely populated or landmark areas makes the use of non-authoritative data sources especially challenging. Increasing data volume and integrating authoritative data into the methodology can yield increased confidence and include underrepresented areas. Table 14.1 compares and summarizes some features of each type of data. Although non-authoritative data can provide timely, local information often in large volume, they are often viewed with uncertainty. Conversely, the verification and authentication of authoritative data yield trusted results at the cost of time.

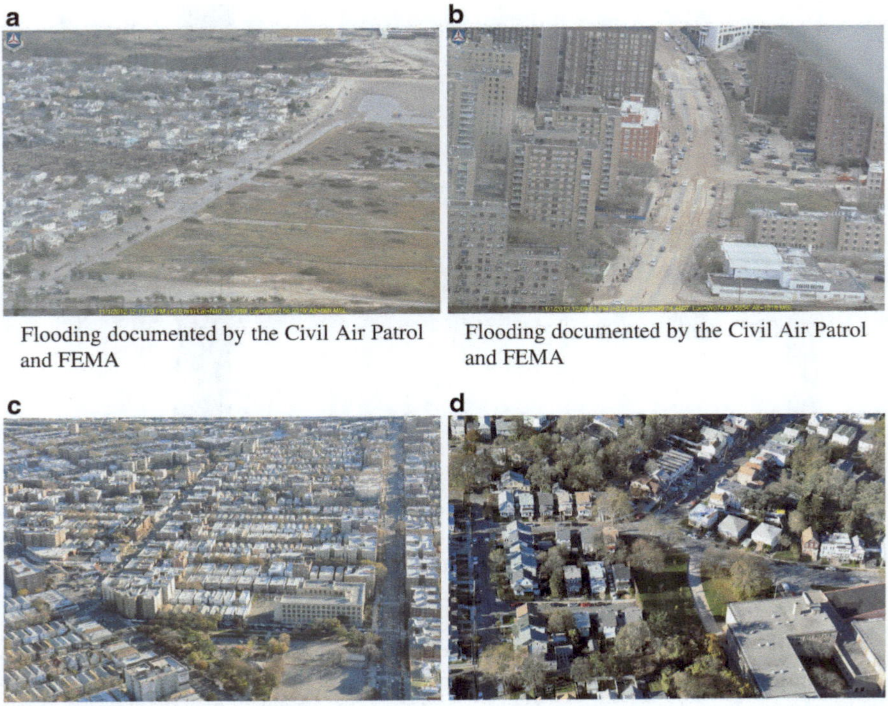

a Flooding documented by the Civil Air Patrol and FEMA

b Flooding documented by the Civil Air Patrol and FEMA

c No flood damage documented by the Civil Air Patrol and FEMA

d No flood damage documented by the Civil Air Patrol and FEMA

Fig. 14.10 Agreement between Civil Air Patrol photos and FEMA evaluation for flooded (**a** and **b**) not flooded (**c** and **d**)

14.2.3.1 Road Damage Map

In Fig. 14.8c, the damage assessment is limited to the FEMA-generated surge extent for the sake of comparison. For the classification of road damage, the non-authoritative assessment is not limited by the FEMA boundary. The fusion of the non-authoritative data predicted flooding and damage outside the FEMA surge boundary, so the full damage assessment was utilized for the road classification. A road network from the TIGER/line® shapefile was layered over the damage assessment surface. Road damage was then classified based on the underlying damage assessment (Fig. 14.8d).

By using the damage assessment surface along with a high-resolution road network layer, roads which may have severe damage can be identified at the street level. This allows authorities to prioritize site inspections, task additional aerial data collection, or identify routes which may be compromised. The identification of potential damage to transportation infrastructure is also crucial to the planning of evacuation routes during and after emergencies.

a

Flooding documented by the Civil Air Patrol but not estimated by FEMA

b

Flooding documented by the Civil Air Patrol but not estimated by FEMA

c

Flooding estimated by FEMA but not confirmed by the Civil Air Patrol

d

Flooding estimated by FEMA but not confimed by the Civil Air Patrol

Fig. 14.11 Disagreement between Civil Air Patrol photos and FEMA evaluation for flooded (**a** and **b**) not flooded (**c** and **d**)

Table 14.1 Comparison between non-authoritative and authoritative data

	Non-authoritative data	Authoritative data
Benefits	Volume	Reliable
	Real time	Verified
	Citizens as sensors	Authenticated
Challenges	Sampling bias	Slow
	Unconfirmed	Unavailable

14.2.3.2 Temporal Assessment

For this study, Twitter data were used to provide a temporal rather than spatial assessment. Although tweets were geolocated using TweetTracker, uncertainty in their location did not allow for a study at a street resolution. However, they provide precise temporal information that can be used to understand the progression of the surge extent over time. To understand the temporal progression is crucial during and after flood events and is very hard to understand using remote sensing instruments, due to their inherent carrier limitations. Twitter data can effectively be used to

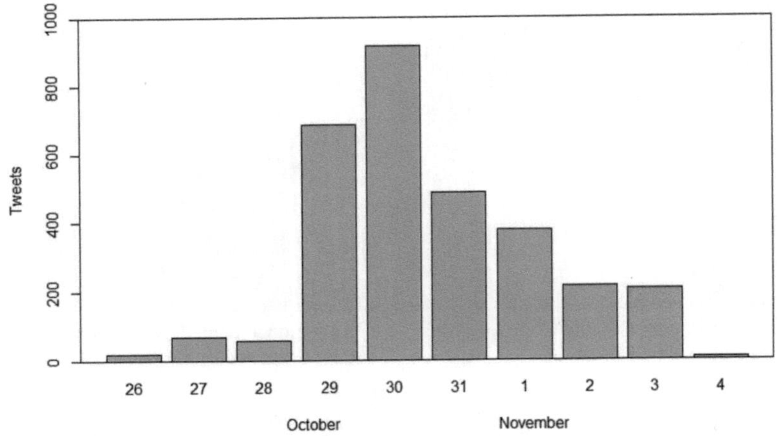

Fig. 14.12 Progression of tweets mentioning the word "flood" in the New York City area

overcome this limitation because of its high temporal resolution. For example, the peak in the number of tweets containing the word "flood" occurs on October 29 and 30, 2012 immediately before and during landfall of Hurricane Sandy in New York City the night of October 29th (Fig. 14.12). Having an indication of event timing can be a very effective aid for emergency managers and response initiatives.

14.3 Evacuations During Emergencies

Emergency evacuations during natural and anthropogenic disasters are time sensitive and require detailed spatiotemporal information for emergency planners to mitigate evacuee risk. One of the primary goals of emergency personnel is to reduce the likelihood of injury and death for citizens within an evacuation zone. First responders are entrusted with providing accurate evacuation information to the public, especially to special populations (e.g., children, elderly, and disabled people are at increased risk during evacuations). Fusing non-authoritative and authoritative information provides increased situational awareness for crisis response personnel. This information enables them to better understand where incidents are occurring, how many people are at risk, and where to allocate resources. The ability to minimize response time is critical to minimizing loss of life and limb.

This section discusses the importance and challenges of evacuations, mobile phone data and its collection, issues of privacy with mobile phone data, and how the use of mobile phone data in evacuations can enhance situational awareness of emergency response planners.

Emergency response personnel can use non-authoritative data, such as near real-time mobile location data, to increase their situational awareness and to redirect resources as needed within their jurisdiction. Figure 14.13 shows a subset of

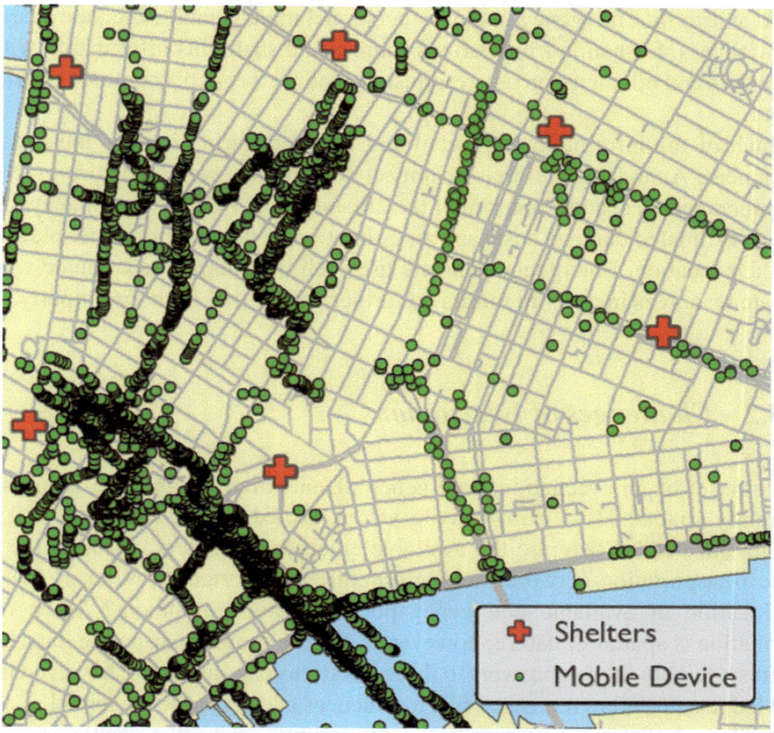

Fig. 14.13 Emergency shelters and example of mobile phones geolocated in New York City. Emergency response personnel can utilize mobile phone data to answer unknown questions. For example, where are citizens located within the evacuation zone, or are there areas where additional police forces should be positioned to reduce congestion?

mobile phone data (over 15,000 phones) collected from OpenCellID in 2011 for NYC as well as the locations of five emergency shelters (OpenCellID 2011). In this example, mobile phone locations provide enhanced situational awareness of spatial and temporal population fluctuations, unlike census data which is a static representation of data collected during the last census. Using mobile phone data, as shown in Fig. 14.13, emergency response personnel can determine evacuation routes to the closest shelters, determine where to position police officers to reroute traffic to reduce congestion, or estimate how many citizens are at risk in an evacuation zone.

14.3.1 Importance of Evacuations

Emergency evacuations often occur with limited or no planning. Notifying the correct individuals based on their location and the type of risk is key to minimizing loss of life and limb. Emergency response personnel manage and execute evacuations. They collect information from various sources and share this information with first responders and other emergency response personnel through a common operational picture.

Much of the information collected and required during a crisis have a spatial component. Analyzing the area of impact for a crisis event provides emergency responders with a spatial footprint of the affected area. Further spatial analysis may provide estimated damage costs and an estimation of the number of people affected and their location. Emergency planners use such information to evaluate where to allocate resources in order to minimize further loss of life, limb, or property.

14.3.2 Challenges of Evacuations

At the onset of an evacuation, emergency planners have many questions that are inherently spatial: How many people were affected by the incident, and where are they located? Where are the available emergency shelters? What is the status of the transportation network including public transportation? What is the status and location of available emergency personnel and equipment? Much of this information is spatial in nature; however, it may be found in disparate databases or systems. Quickly finding answers to these questions can save lives; however, limited knowledge can lead to increased risk to evacuees and response personnel.

One key challenge in evacuations is determining where the people are located. Determining the location of individuals and the risks they may be exposed to is of the utmost importance for planning an emergency response. Populations fluctuate during the day as people travel to work, school, recreational activities, sporting/cultural events, etc. Their location is constantly changing and not always predictable. Currently, the allocation of resources is based on the expertise of first responders, information that has been collected about the situation, and situational awareness of emergency planners.

14.3.3 Mobile Phone Data

During emergencies, traditional mobile phones provide voice and SMS services; however, smartphones also provide enhanced services through mobile applications. Some of these applications include social media networks, notification services, mapping, navigation, e-mail, Internet access, photo and video capture, and crowd-sourcing applications. Individuals use these to improve their situational awareness and share information with others who may be affected by the crisis.

Mobile phone companies collect information from individual mobile phones in the United States and internationally. Policies for how long the data is stored vary by mobile provider. Mobile phone data that are collected may include date, time, latitude, longitude, identification number of the cell phone, and signal strength. After these data are collected, they may be aggregated for additional analysis, cleansed of personally identifiable information, and sold for use in other applications, such as navigation, traffic services, or business development.

For example, mobile phone data are often used to improve navigation by monitoring congestion along interstates and highways. The average speed of mobile phones is compared with the speed limit along a section of highway to determine which routes are flowing normally and which are congested. Older models of mobile phones determine their location by triangulating between mobile phone towers, whereas most modern phones now use GPS to determine their location. This information provides individuals with the ability to find the shortest or quickest route to their destination using mobile phone data. Further analysis of mobile phones that are traveling along highways provides a method to estimate the number of vehicles traveling along a route at different time periods. This information is useful for transportation departments for traffic planning, for businesses for determining locations for new franchises, and for numerous other applications.

14.3.4 Issues of Privacy with Mobile Location Data

Although the collection of near real-time mobile location data to support emergency response greatly increases situational awareness of response personnel and planners; citizens are often concerned with the government or individuals using such information to invade their privacy, track their movement, or in an investigation following an attack.

Personally identifiable information (PII) includes information such as name, social security number, phone number, home address, etc. PII is not needed when collecting real-time mobile location data to support emergency response. Protective measures such as stripping PII from data before mobile phone companies share it with government officials during emergencies provide a measure of privacy. Another measure for protecting individual privacy is through the aggregation of data. In metropolitan areas, mobile phone location data could be aggregated to the nearest road intersections with an estimate of the number of people near that intersection. These data would reduce privacy concerns while also providing invaluable data to emergency response personnel. In suburban or rural areas, data could be assigned by census block or tract, or the spatial and temporal accuracy of the data could be reduced to reduce privacy concerns while still providing increased awareness to response personnel.

With the increasing use of mobile phone data in the private sector to support business development, government policies must be developed to address the application of mobile phone data in large-scale emergencies. Policies on the application of mobile phone location data in emergency response are limited to nonexistent. Although citizens are concerned with privacy, they are also concerned with an accurate and efficient response by emergency response personnel.

14.3.5 Application of Mobile Phone Data in Evacuations

Fusing mobile phone data with existing data sources provides increased situational awareness and fills existing gaps in other data. During large-scale dynamic events, especially in high population density areas, the common operational picture is further supplemented with mobile phone data.

One major challenge for emergency response personnel and planners is to determine where people are located at various times of the day. This challenge becomes even more complicated with limited communications or a complete loss of communications. During emergencies with limited to no communications, it is possible for emergency personnel to use archived data to estimate the density of populations. Previously collected mobile data that accounts for diurnal population change and population change during major events, such as sporting events, parades, festivals, etc., provide response personnel with prediction models to use in their planning. This information can prove very important in international response to disasters where population data are often unreliable.

Providing assistance at the right location is often a challenge for emergency response personnel. Often limited crowdsourced or social media data are available in developing countries; however, mobile phones are much more prevalent and provide a source of data on movements of people within the country. Information on the movement of individuals within a country enables the government and nongovernment organizations to prioritize their response based on spatial and temporal data.

As smartphones become more prominent in developing counties, it is necessary to devote increased efforts to increase awareness of mobile applications and their contributions to citizens. These tools provide citizens and emergency personnel with increased information and damage estimates on their response areas.

Finally, mobile phones provide an opportunity for evacuees to provide data to emergency personnel through text messaging, social media, or phone calls. Although there may be a loss of power, mobile phones usually work on battery power for several hours after an incident. This provides an alternate means of communication and enhances situational awareness when other communication methods are limited or nonexistent.

Mobile phones provide emergency personnel with enhanced situational awareness; however, new policies for protecting the privacy of individuals with respect to mobile phone data are necessary to limit concerns for PII. Standards for collection of data and who can handle data are necessary to protect citizens. Additional research is needed to evaluate how mobile phone data is represented in Geographic Information Systems (GIS). This research should address and refine database storage methods and analysis of mobile phone data.

14.4 Conclusions

During disasters and emergencies in urban areas, timely and accurate information of road networks, infrastructure conditions, and locations of citizens is crucial. But this information can be limited, incomplete, lengthy to acquire, or out of date. Although not necessarily created or posted with the intent of being used for scientific research, non-authoritative data can be harvested during disasters and emergencies to provide timely, on-the-ground information. Although often viewed with uncertainty because of concerns related to, for example, producer anonymity and lack of authoritative verification, these data often provide relevant information that may be difficult for authorities to collect. Non-authoritative sources also provide an additional layer of subjective information which can indicate the severity of potentially dangerous events, as well as how citizens are reacting to the developing danger or coping with damage resulting from a disaster. By presenting varied applications, such as damage assessments and emergency evacuations, the practicality of non-authoritative sources and how they can add value to official data is demonstrated. It is hoped that one day this research will help save lives.

Acknowledgements Work performed under this project has been partially supported by the Office of the Assistant Secretary for Research and Technology, US Department of Transportation award # RITARS-12-H-GMU (GMU #202717) and also partially funded by the Office of Naval Research (ONR) award #N00014-14-1-0208 (PSU #171570) and the Office of Naval Research (ONR) award #N00014-14-1-0208 (PSU #171570). DISCLAIMER: The views, opinions, findings and conclusions reflected in this presentation are the responsibility of the authors only and do not represent the official policy or position of the USDOT/OST-R, or any State or other entity.

References

Alexander DE (2002) Principles of emergency planning and management. Oxford University Press, Oxford/New York

Atkinson PM, Tatnall A (1997) Introduction neural networks in remote sensing. Int J Remote Sens 18(4):699–709

Benediktsson JA, Swain PH, Ersoy OK (1990) Neural network approaches versus statistical methods in classification of multisource remote sensing data. IEEE Trans Geosci Remote Sens 28(4):540–552

Berenbrok C, Mason RR, Blanchard SF (2009) Mapping Hurricane Rita inland storm tide. J Food Risk Manag 2:76–82

Booij N, Ris RC, Holthuijsen LH (1999) A third generation wave model for coastal regions: model description and validation. J Geophys Res 104:7649–7666

Casti JL (2012) X-events: the collapse of everything. HarperCollins, New York

Chen C, Liu H, Beardsley R (2003) An unstructured grid, finite-volume, three-dimensional, primitive equations ocean model: application to coastal ocean and estuaries. J Atmos Ocean Technol 20:159–186

Cutter SL (1993) Living with risk: the geography of technological hazards. Edward Arnold, London

De Longueville B, Smith R, Luraschi G (2009) OMG, from here, I can see the flames!: a use case of mining location based social networks to acquire spatio-temporal data on forest fires. In: Proceedings of the 2009 international workshop on location based social networks, Seattle. ACM, pp 73–80

Dietrich J, Zijlema M, Westerink J, Holthuijsen L, Dawson C, Luettich R, Jensen R, Smith J, Stelling G, Stone G (2011) Modeling hurricane waves and storm surge using integrally-coupled scalable computations. Coast Eng 58:45–65

Ferreira CM, Irish J, Olivera F (2014) Uncertainty in hurricane surge simulation due to land cover specification. J Geophys Res-Oceans 119(3):1812–1827

Flanagin A, Metzger M (2008) The credibility of volunteered geographic information. GeoJournal 72(3):137–148

Freudenburg WR, Gramling R, Laska S, Erikson KT (2008) Organizing hazards, engineering disasters? Improving the recognition of political-economic factors in the creation of disasters. Soc Forces 87(2):1015–1038

Freund Y, Schapire R, Abe N (1999) A short introduction to boosting. Jpn Soc Artif Intell 14(771–780):1612

Giles J (2005) Internet encyclopaedias go head to head. Nature 438(7070):900–901

Goodchild M (2007) Citizens as sensors: the world of volunteered geography. GeoJournal 69(4):211–221

Goodchild MF, Glennon JA (2010) Crowdsourcing geographic information for disaster response: a research frontier. Int J Digit Earth 3(3):231–241

Hyvärinen O, Saltikoff E (2010) Social media as a source of meteorological observations. Mon Weather Rev 138(8):3175–3184

Jelesnianski CP, Chen J, Shaffer WA (1992) SLOSH: sea, lake, and overland surges from hurricanes. National Oceanic and Atmospheric Administration, Technical report NWS 48:1–77

Jensen JR, Cowen DC (1999) Remote sensing of urban/suburban infrastructure and socio-economic attributes. Photogramm Eng Remote Sens 65:611–622

Kerr PC et al (2013) U.S. IOOS coastal and ocean modeling testbed: inter-model evaluation of tides, waves, and hurricane surge in the Gulf of Mexico. J Geophys Res Oceans 118:5129–5172

Kumar S, Barbier G, Abbasi MA, Liu H (2011) TweetTracker: an Analysis tool for humanitarian and disaster relief. In: Fifth international AAAI conference on weblogs and social media (ICWSM), Barcelona

Luettich R, Westerink J (2004) Formulation and numerical implementation of a 2D/3D ADCIRC finite element model version 4. www.adcirc.org

Mattocks C, Forbes C (2008) A real-time event-triggered storm surge forecasting system for the state of North Carolina. Ocean Model 25:95–119

Mukai AY, Luetich JWR, Mark D (2002) East coast 2001, a tidal constituent database for Western North Atlantic, Gulf of Mexico and Caribbean Sea. Coastal inlets research program report, coastal and hydraulics laboratory ERDC/CHL TR-02-24. US Army Corps of Engineers. Engineer Research and Development Center, Vicksburg

NOAA (2013a) National Oceanic and Atmospheric Administration, Atlantic basin hurricane database (HURDAT). http://www.aoml.noaa.gov/hrd/hurdat/. Accessed July 2013

NOAA (2013b) National Oceanic and Atmospheric Administration tides and currents. http://tidesandcurrents.noaa.gov/. Accessed 15 Mar 2013

Olea RA, Olea RA (1999) Geostatistics for engineers and earth scientists. Kluwer Academic, Boston

Oliver MA, Webster R (1990) Kriging: a method of interpolation for geographical information systems. Int J Geogr Inf Syst 4(3):313–332

OpenCellID (2011) OpenCellID. http://opencellid.org/. Accessed Feb 2012

Pohl C, Van Genderen J (1998) Review article multisensor image fusion in remote sensing: concepts, methods and applications. Int J Remote Sens 19(5):823–854

Poser K, Dransch D (2010) Volunteered geographic information for disaster management with application to rapid flood damage estimation. Geomatica 64(1):89–98

Schnebele E, Cervone G (2013) Improving remote sensing flood assessment using volunteered geographical data. Nat Hazards Earth Syst Sci 13:669–677

Schnebele E, Cervone G, Waters N (2013) Road assessment after flood events using non-authoritative data. Nat Hazards Earth Syst Sci 14:1007–1015

Sui D, Goodchild M (2011) The convergence of GIS and social media: challenges for GIScience. Int J Geograph Inf Sci 25:1737–1748

Sui D, Elwood S, Goodchild M (2013) Crowdsourcing geographic knowledge: volunteered geographic information (VGI) in theory and practice. Springer, Dordrecht/New York

USGS (2013) U.S. Geological survey national elevation dataset. http://ned.usgs.gov/. Accessed 04 Aug 2013

Venables WN, Ripley BD (2002) Modern applied statistics with S, 4th edn. Springer, New York. http://www.stats.ox.ac.uk/pub/MASS4, ISBN 0-387-95457-0

Voigt S, Kemper T, Riedlinger T, Kiefl R, Scholte K, Mehl H (2007) Satellite image analysis for disaster and crisis-management support. IEEE Trans Geosci Remote Sens 45(6):1520–1528

Waters N (2009) Representing surfaces in the natural environment: implications for research and geographical education, Ch 3. In: Mount NJ, Harvey GL, Aplin P, Priestnall G (eds) Representing, modeling & visualizing the natural environment: innovations in GIS 13. CRC Press, Boca Raton/London/New York, pp 21–39

Wisner B, Blaikie, P, Cannon T, Davis I (2004) At Risk: natural hazards, people's vulnerability and disasters, 2nd edn. Routledge, New York

Zhang J (2010) Multi-source remote sensing data fusion: status and trends. Int J Image Data Fusion 1(1):5–24

Zhang Y, Baptista AM (2008) A semi-implicit Eulerian-Lagrangian finite-element model for cross-scale ocean circulation. Ocean Model 21:71–96

Zhang J, Atkinson P, Goodchild M (2014) Scale in spatial information and analysis. CRC, Boca Raton

Chapter 15
Towards a Comparative Science of Cities: Using Mobile Traffic Records in New York, London, and Hong Kong

Sebastian Grauwin, Stanislav Sobolevsky, Simon Moritz, István Gódor, and Carlo Ratti

Abstract This chapter examines the possibility to analyze and compare human activities in an urban environment based on the detection of mobile phone usage patterns. Thanks to an unprecedented collection of counter data recording the number of calls, SMS, and data transfers resolved both in time and space, we confirm the connection between temporal activity profile and land usage in three global cities: New York, London, and Hong Kong. By comparing whole cities' typical patterns, we provide insights on how cultural, technological, and economical factors shape human dynamics. At a more local scale, we use clustering analysis to identify locations with similar patterns within a city. Our research reveals a universal structure of cities, with core financial centers all sharing similar activity patterns and commercial or residential areas with more city-specific patterns. These findings hint that as the economy becomes more global, common patterns emerge in business areas of different cities across the globe, while the impact of local conditions still remains recognizable on the level of routine people activity.

Keywords Big data • City Science • Cellphone networks • Urban analysis • Urban planning

---

S. Grauwin (✉) • S. Sobolevsky • C. Ratti
Senseable City Lab, Massachusetts Institute of Technology, 77 Massachusetts Avenue, Cambridge, MA 02139, USA
e-mail: sgrauwin@mit.edu; stanly@mit.edu; ratti@mit.edu

S. Moritz
Ericsson Research, Stockholm, Sweden
e-mail: simon.moritz@ericsson.com

I. Gódor
Ericsson Research, Budapest, Hungary
e-mail: istvan.godor@ericsson.com

M. Helbich et al. (eds.), *Computational Approaches for Urban Environments*, Geotechnologies and the Environment 13, DOI 10.1007/978-3-319-11469-9_15

15.1 Introduction

As digital technologies are becoming more and more widespread, big data created by recording the digital traces left behind human activities become a powerful mean to study various aspects of human behavior. Many of these aspects can be described with telecommunications data which nowadays become global. The exploration of these data provides new perspectives, revealing characteristic usages and regular dynamic patterns at both the individual and collective scale. At the same time, the increasing urbanization of the world's population is deeply affecting urban environments, and it is crucial to develop theoretical frameworks as well as real-time monitoring systems to understand how the individual dynamics shape the structure of our cities in order to make better planning decisions.

In the past years, several studies have shown that it was possible to use telecommunication data to get a fresh view at the spatiotemporal dynamics within a city. In a now-famous paper, Eagle and Pentland (2006) showed that it was possible to decompose mobile phone activity patterns of university students into regular daily routines and that these routines were linked to each student's major and also to employment levels. Building upon this work, González et al. (2008) studied the trajectory of 100,000 anonymized mobile phone users to reveal statistical regularities in human trajectories. This paper, along with other seminal work (Candia et al. 2008; Song et al. 2010), has since generated a research field dealing with human mobility as understood from digital traces (Kang et al. 2013).

In parallel, focusing on records aggregated on spatial locations rather than on individuals, new approaches have been initiated to describe urban landscape based on mobile phone usage patterns (Jacobs-Crisioni and Koomen 2012; Loibl and Peters-Anders 2012; Ratti et al. 2006; Reades et al. 2007, 2009; Calabrese et al. 2011; Sun et al. 2011), to explore the issue of regional delineation (Amini et al. 2014; Kung et al. 2013; Ratti et al. 2010; Sobolevsky et al. 2013), to estimate population density (Girardin et al. 2009; Kang et al. 2012; Rubio et al. 2013; Vieira et al. 2010), or to identify social group and social events (Traag et al. 2011). In particular, by measuring mobile phone data on a 500m by 500m "pixel" grid in Rome (Italy), Reades et al. (2009) especially used a variant of principal component analysis to cluster these pixels into regions with similar patterns of usage and made a qualitative link between these patterns and the number of businesses on the corresponding areas.

This last paper is an example of a line of research dealing with the identification of a specific land use type (Caceres et al. 2012; Calabrese et al. 2010). Other papers have focused on methods to build classification of several land use types based either on (voice calls or SMS) mobile phone patterns (Andrienko et al. 2013; Becker et al. 2011; Pei et al. 2013; Soto and Frías-Martínez 2011; Toole et al. 2012), taxi trip data (Liu et al. 2012), or Twitter data (Frias-Martinez et al. 2012). These studies used different types of methods, from simple clustering to advanced neural network models. A common feature of these papers is that they are limited to the study of a single spatial entity (in general a city) that they study through one type of digital data.

This statement raises some questions: is the behavior detected on one type of mobile phone activity independent of the other type, i.e., is it the same to look at calls, SMS, or even data transfers? How does the results compare between multiple cities? What are the signatures of the mobile network usage in major US, European, or Asian cities, and how do they compare?

This chapter takes advantage of an unprecedented multimodal collection of counter data recording the number of calls, SMS, request, and data transfer resolved both in time and space in three cities – New York, London, and Hong Kong – to investigate such questions. After presenting the data in Sect. 15.2, Sect. 15.3 will investigate the spatial repartition of activity in these three cities. Then in Sect. 15.4, we will focus on people's behavior by investigating the dynamics of the activities on both a local and city scale. In Sect. 15.5, we will show how we can use a clustering algorithm to automatically detect and classify such patterns either within one city or across all of them. Finally, we will discuss in Sect. 15.6 how our work can help us understand the changing nature of modern cities and especially what common features can be captured in the patterns of human behavior: in these global cities, what is the respective influence of city-specific and global factors on human life?

15.2 Materials and Methods

15.2.1 Geographical Background

Figure 15.1 shows a map of the three cities studied in this chapter: New York, London, and Hong Kong. Greater London is divided into 33 "district boroughs," the central ones being referred to as *Inner London*, while the peripheral ones are referred to as *Outer London*. The historic heart of London, the *City of London*, is a major business and financial center, where many banking and insurance institution headquarters are located. While the city has a low resident population (around 7,000), over 300,000 people commute and work there every day, mainly in the financial service sector. New York is divided in 59 "community districts," gathered into five boroughs. The boroughs of Queens, Brooklyn, Staten Island, and Bronx are mainly residential. Manhattan is a major financial and decision-making center (the UN headquarters, Times Square, and the Empire State Building are located there). It has also one of the highest population densities in the world, with around 27,000 residents per square kilometers. Finally, Hong Kong territory consists of 4 regions split into 18 districts. Due to the mountainous nature of its land, less than 25 % of Hong Kong's territory is urbanized: the urban development concentrates on Kowloon peninsula, the northern edge of Hong Kong Island, and a few settlements throughout the New Territories.

Overall, London, New York and Hong Kong are comparable in terms of size and population density (see Table 15.1). They are also all Alpha cities according to the GaWC nomenclature of world cities (Beaverstock et al. 1999) which ranks cities based on their connections with others in domains such as business, finance, law, media, art, fashion, research, technology, education, tourism, and entertainment. As Alpha cities, they are more integrated within the global economy than any other city.

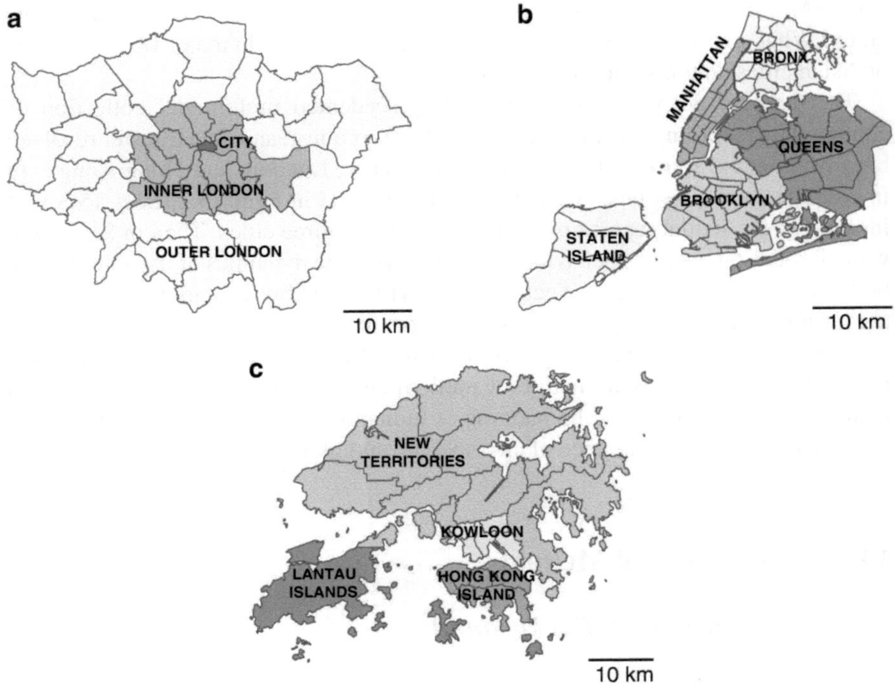

Fig. 15.1 Administrative maps of (**a**) Greater London, (**b**) New York, and (**c**) Hong Kong

Table 15.1 Background information on the scale of the studied cities

City	Area (km^2)	Population	Density (pop/km^2)
Greater London	1,572	~8,200,000	5,206
New York	1,214	~8,350,000	6,865
Hong Kong	1,104	~7,000,000	6,405

Source: wikipedia, 2013

Despite their apparent differences (in terms of history, culture, or weather conditions to name only the obvious ones) and geographical distances, one can expect emerging similarities between these cities due to globalization. They are therefore perfect candidates for tracking universal communication patterns.

15.2.2 Data Gathering and Preprocessing

Our analysis is based on aggregate 3G mobile traffic data (including all kinds of devices like phones, tablets, etc.) supplied by several operators, and corresponding to a statistically significant part of the total 3G mobile traffic of the covered areas, corresponding to several million subscribers in all studied cities (precise penetration rates cannot be given for confidentiality reasons). Data were collected between April 1 and July 7, 2013 at fifteen-minute intervals across the three cities at cell level. While the whole city of New York is covered, the dataset of Greater London is mostly concentrated on the Inner London districts. The Hong Kong measurements cover urban zones, while no data is available for the unpopulated mountainous parts.

The three months of data consist in counter data recording the numbers of calls, SMS, and requests (for data communication initiated either by the users or some applications running in the background in their mobile devices) as well as the amount of data uploaded and downloaded by subscribers (measured in bytes and thereafter denoted by "UL Data" and "DL Data"). The provided data was aggregated at the cell level by the data providers and therefore did not reveal any individual user information. Before receiving the data, the actual numbers were obfuscated by using a secret scaling factor, such that we have only access to normalized amounts of each type of counter data.

In addition to mobile phone data, we gathered various shapefiles, census data, and land use data from open access sources.[1] We thus obtained land use data of different nature and with different number of categories across the different cities (9 categories for London, 6 for New York, 24 for Hong Kong). To better compare the results obtained in our three cities, we converted these original categories into seven land use types that best match them: High-Density Residential, Low-Density Residential, Business and Commercial, Mixed (Residential and Commercial), Infrastructures, Parks, and Other. Details of the procedure are available on request.

The location of the cells' recording mobile phone activity is given as longitude/latitude pairs – the service area of a cell having a typical radius varying from around 100 m (in dense area) to several kilometers (in rural zones), while the census and land use data are provided in polygonal zones corresponding to administrative divisions in the cities. In order to study the mobile phone usage patterns in the different cities and their relationships to census and land use data, we chose to transform the spatial representation of the different datasets by projecting them on uniform lattice grids of 500 m by 500 m "pixels."

To reduce the bias induced by the attribution of the activity within a cell's service area to a single pixel location, we used a smoothing procedure: we defined the activity on one pixel as the mean of activities of all cells within a 1,500 m by 1,500 m square centered on the pixel center. Census data were similarly projected on the grid by interpolating demographic data and most significant land use on each grid cell. The length of 500 m was chosen after testing different grid sizes. It proved to be coarse enough to reduce noise level and detailed enough to explore spatial patterns of activities within the cities. At the end of this procedure, the mobile phone traffic data was projected on around 2,700 pixels in Greater London and around 3,000 pixels in New York and Hong Kong.

---

[1] such as http://data.london.gov.uk/ for London, https://nycopendata.socrata.com/ for New York or http://www.census2011.gov.hk/ for Hong Kong.

15.3 Spatial Repartition of Activity

A simple question that can be addressed with mobile communication data is the tracking of where people are or where they go.

The first way to investigate this question is displayed on Fig. 15.2a–c, which shows the spatial repartition of total request activity as recorded in our datasets. We chose to focus on the request activity since it passively tracks people: even if you don't make any call or send any SMS, your mobile device will still produce network background traffic such as social network synchronization, weather updates, news feeds, etc.

The colors used on the maps of Fig. 15.2a–c emphasize the inhomogeneities of activity repartition, by showing the share ρ of activity on each 500m by 500m pixel normalized by the total activity recorded. A value of $\rho = 1$ hence corresponds to a pixel with average total activity, $\rho > 1$ to a pixel with higher than average activity, and $\rho < 1$ to a pixel with lower than average activity.

The spatial repartition of ρ in the three cities shows a center/periphery dichotomy. In Greater London, we observe a concentric organization, with a very strong activity level in the City of London, and decreasing levels as one moves away from the

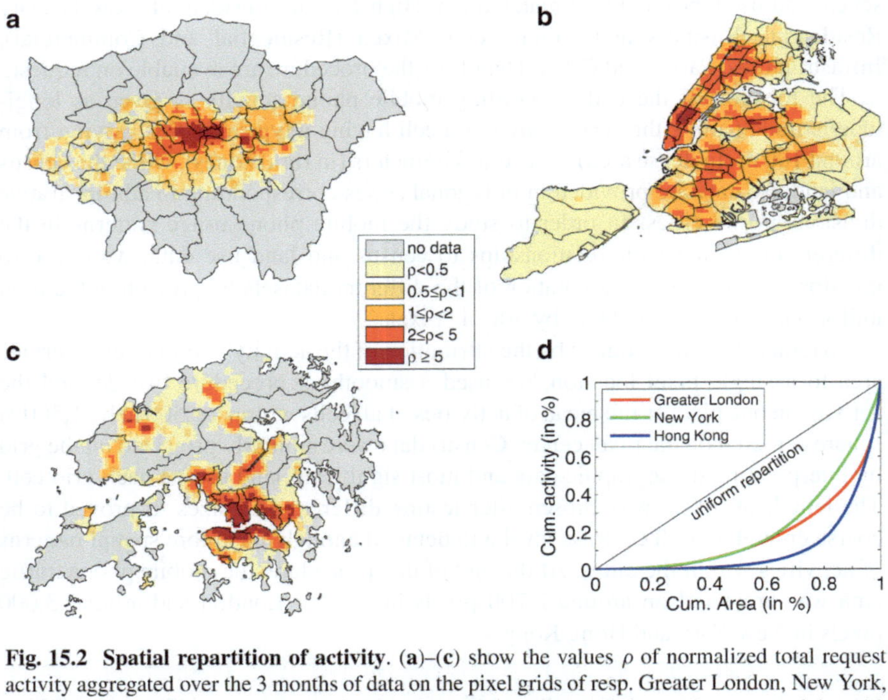

Fig. 15.2 Spatial repartition of activity. (**a**)–(**c**) show the values ρ of normalized total request activity aggregated over the 3 months of data on the pixel grids of resp. Greater London, New York, and Hong Kong. (**d**) Lorentz curves showing the variation of cumulative total request activity with the cumulative area of coverage, the pixels being ranked by increasing total request activity

city center. In New York, the organization is polycentric, with one center in the middle part of Manhattan and another one in Queens. Finally, in Hong Kong we observe one big center of activity divided up between Kowloon and the northern part of Hong Kong Island and secondary centers in the newly developed zones of the New Territories. Low activity zones correspond to the limits with the mountainous areas.

Lorentz curves depicting the variation of cumulative total request activities with the cumulative areas of coverage are presented on Fig. 15.2d. These curves, typically used in economy and ecology to describe inequality in wealth or size Lorenz (1905), describe here unequal repartition of request activity in space. Hong Kong is obviously the most inhomogeneous city (with less than 2 % of request activity in the 50 % less active pixels and 64 % of activity concentrated in the top 10 % most active pixels), followed by London (around 9 % of activity in the 50 % less active pixels and 49 % of activity in the top 10 % most active pixels) and New York (around 9 % of activity in the 50 % less active pixels and 38 % of activity in the top 10 % most active pixels).

A commonly used quantitative measure of inequality, the Gini coefficient, can be defined from a Lorentz graph as the area between the bisector line (corresponding to a uniform repartition) and the Lorentz curve, normalized by the area between the bisector line and the x axis (corresponding to the most inhomogeneous case where all activity is concentrated on one pixel) (Gini 1912). The Gini coefficients of the request Lorentz curve of Fig. 15.2d as well as those corresponding to other types of activity are reported in Table 15.2. Interestingly, the Gini coefficient depends only slightly on the activity type and strongly on the city. The measure of spatial inhomogeneity could thus be done on any type of activity. Again, we find here that the most inhomogeneous city is Hong Kong, followed by London and then New York.

Although this first analysis suffers from some limitations – such as the mismatch between the area covered by our dataset in London and the area within the official boundaries of Greater London – it already provides good insights on the way people interact with their cities. Maps of mobile phone activities could steadily become a complementary tool to more classical maps of population or employment densities obtained through extensive surveys and help urban planners make decisions based on accurate population repartition.

Table 15.2 Gini coefficients For each activity type λ, the Gini coefficient $G^{\lambda} \in [0, 1]$ measures the inhomogeneity of the spatial repartition of the three-month aggregated activity. The higher the coefficient, the more unequal the spatial repartition is

City	$G^{\text{DL Data}}$	$G^{\text{UL Data}}$	G^{Request}	G^{Calls}	G^{SMS}
Greater London	0.608	0.640	0.649	0.618	0.606
New York	0.546	0.555	0.576	0.549	0.523
Hong Kong	0.768	0.765	0.784	0.802	0.781

15.4 Exploring Temporal Patterns

15.4.1 Typical Week Signature

To minimize the impact of special events on the datasets, we followed the procedure presented in Reades et al. (2007) to extract average "typical week" timelines for each pixel at each 15-min interval, using the three-month period. The values of the typical weeks for a given 15-min time interval were calculated as the average of the same intervals from the whole measurement. For example, the typical number of calls for 12:00 to 12:15 on the typical Monday was taken as the average number of calls from 12:00 to 12:15 on every Mondays available in the dataset. Civic holidays, considered as special events, were excluded from the computation to avoid the introduction of unnecessary noise. We use these mathematical notations:

- $A_i^\lambda(t)$, to measure activity of type λ (number of calls, SMS or request, volume of data upload or download) within a given pixel i at time $t \in [1, 672]$ (since the measurements are taken every 15-min, one week comprises $7 \times 96 = 672$ time intervals)
- $A_{\text{city}}^\lambda(t)$, to measure activity of type λ (number of calls, SMS or request, volume of Data upload or download) within a given city at time $t \in [1, 672]$

In order to better compare the relative dynamic patterns across the cities' locations (e.g., recognizing locations with similar patterns up to multiplicative constant due to higher active population density), it is useful to normalize these values by the typical amplitude of activities on each pixel. We thus define the signature S_i^λ of activity type λ on location i thanks to a mean normalization:

$$S_i^\lambda(t) = A_i^\lambda(t)/\langle A_i^\lambda \rangle_t, \tag{15.1}$$

where $\langle \ldots \rangle_t$ denotes an average over the 672 individual 15-min time intervals of a typical week. Similarly, the signature S_{city}^λ of activity type λ at the city scale is given by

$$S_{\text{city}}^\lambda(t) = A_{\text{city}}^\lambda(t)/\langle A_{\text{city}}^\lambda \rangle_t. \tag{15.2}$$

As an illustration of the computation of signatures, Fig. 15.3a displays the mean-normalized call timeline in Greater London over the three-month observation period. This timeline shows daily variations – with peaks of activity during the days and drops of activity at nights – and weekly variations, with daily peaks significantly lower during weekends than during workdays. Figure 15.3b then shows how this timeline can be decomposed into a repeating typical week pattern and a residual part. The residual part accounts for special events, such as the occurrence of civic holidays (notice the lower amount of calls on April 1, May 6, and May 27, respectively, Easter, May Bank, and Spring Bank holidays in London), and general

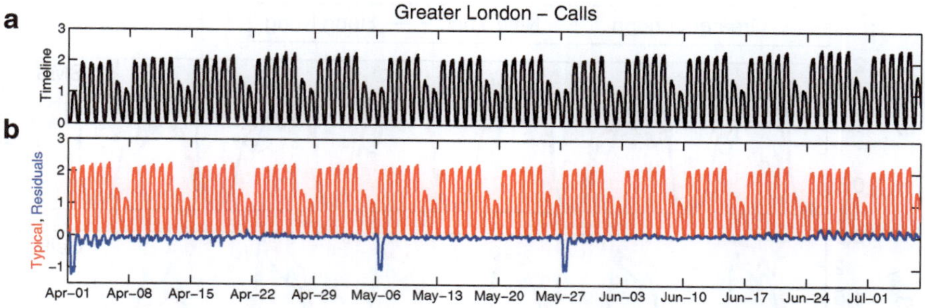

Fig. 15.3 Decomposition of Greater London "calls" timeline. (**a**) The mean normalized timeline can be decomposed into (**b**) a repeating weekly pattern (the city calls signature $S^{\text{Calls}}_{\text{London}}$, in *red*) and a residual part (in *blue*)

trends. For example, we observe a slight overall increase in calls in late spring /early summer which may be due to the arrival of tourists in London at this time of the year.

Overall, the typical week signature captures the main temporal patterns within the cities, by reducing noise and getting rid of long-term trends. Incidentally, it can provide a good predictive baseline of expected mobile device usages. In the example presented in Fig. 15.3, the average absolute ratio between residuals and timelines is approximatively equal to 8 %, but is typically included between 5 and 10 % for the different activity type either in whole cities or at each pixel level. Based on this timeline, an operator could detect irregular operations, as also anticipate upcoming special events.

15.4.2 Comparing Cities' Signatures

Let us first explore the temporal patterns at the macro, citywide scale. The city signatures of the different activity types are displayed on Fig. 15.4 which emphasizes their similarities and differences.

All three studied cities display a broadly comparable rhythm, common to all components of activity. Mobile activity rapidly ramps up in the morning between 6 and 10 AM, followed by rather steady activity levels within the day and a slower decrease of activity at night between 9 PM and 2 AM. The same pattern appears on workdays and with somewhat slower amplitude on weekends. On workdays, we can also observe small peaks of activity at commuting hours and at lunch times. It thus seems reasonable to associate this common rhythm to a simple daily cycle, corresponding to people waking up, going to work, having lunch, and then heading back home in the evening.

Let us now turn to the differences appearing on Fig. 15.4 when comparing the cities.

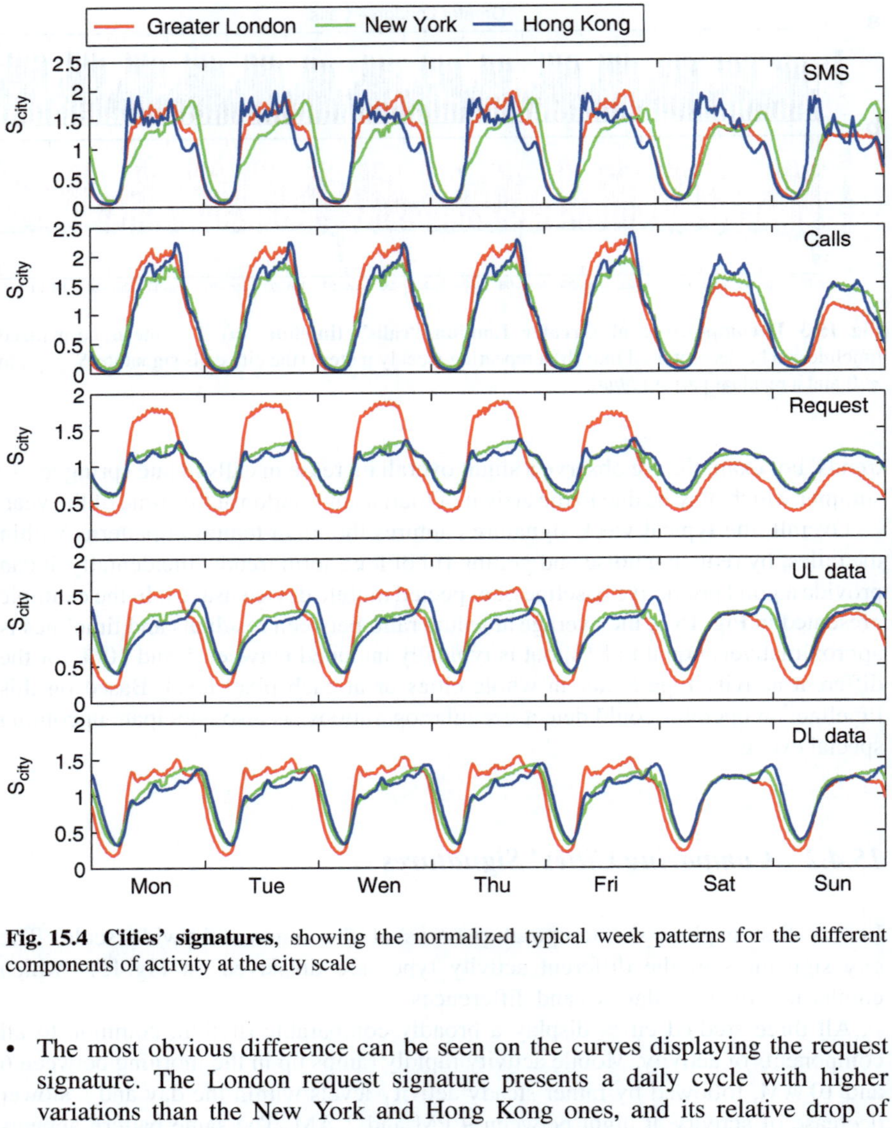

Fig. 15.4 Cities' signatures, showing the normalized typical week patterns for the different components of activity at the city scale

- The most obvious difference can be seen on the curves displaying the request signature. The London request signature presents a daily cycle with higher variations than the New York and Hong Kong ones, and its relative drop of activity in the weekend compared to workdays is also more important. The cost of mobile data plans being higher in London than in other parts of the world, our educated guess is that Londoners use (cheaper) Wi-Fi whenever available to connect their mobile devices. Since these connections are not recorded in our dataset, one can expect a specific negative bias in the data when people typically switch from the operator mobile network to Wi-Fi network: in the evening and on the weekend, when they are at home.
- The early evening drops of UL and DL data signatures in London can be explained in the same way. Londoners do not necessarily go to bed or stop to use

their mobile devices earlier, but more probably use their home Wi-Fi connections more often.
- In both New York and Honk Kong, the daily shape of all activity types is slightly asymmetric, the maximum activity being reached in the evenings. Surprisingly, there is an exception with the SMS activity in Hong Kong that appears to drop earlier than the other activities in the evenings. Rather than a disinterest from Hong Kongers towards texting in the evening, our guess is that the Hong Kong SMS signature reveals a great specific interest towards texting during the day. Text messaging is indeed known to be particularly popular in Asia, where companies use text messages to confirm deliveries and provide alerts, updates, or infotainment.
- We also observe peaks of SMS activity on evenings in New York. These peaks could be explained by the important use of SMS for media voting (e.g., on TV show polls) in the USA like America's Got Talent or X Factor.

At first glance, New York and Hong Kong, while located almost at opposite places of the globe and having different cultural background, may surprisingly appear to have more similar signatures than New York and London which share a common cultural and linguistic background. However we have seen that these signatures are shaped by many different (technological, economical, or cultural) factors and that their interpretation must reflect the multiple influences on people's behaviors. Space is also an important factor to take into account since people do not behave in the same way depending of where they are.

15.4.3 Comparing Local Signatures

In this section, we go further in our exploration of what we can learn about people's behaviors based on their communication patterns. After comparing cities' signatures, we will now compare local signatures from different locations within a same city.

As shown on Fig. 15.5a, we selected five different 500m by 500m pixels corresponding to specific locations in London: one pixel in the City of London (the financial center of London), one centered on Piccadilly Circus (a public space in Westminster close to major shopping, entertainment, and touristic area), one on Camden Market (a popular market place where crafts, clothings, and fast food are sold, especially on weekends), one on Newham (a residential area), and one on Ealing Broadway (a travel hub, part of the National Rail and London Underground networks).

Figure 15.5b shows the signatures of all activity types on each of these five locations, while Figure 15.5c highlights the specificities of each location's signature by displaying their deviation from the whole Greater London's signature.

Similarly to the city level, these five sample signatures display a comparable day/night cycle, but also present specific characteristics revealing the nature of the corresponding locations.

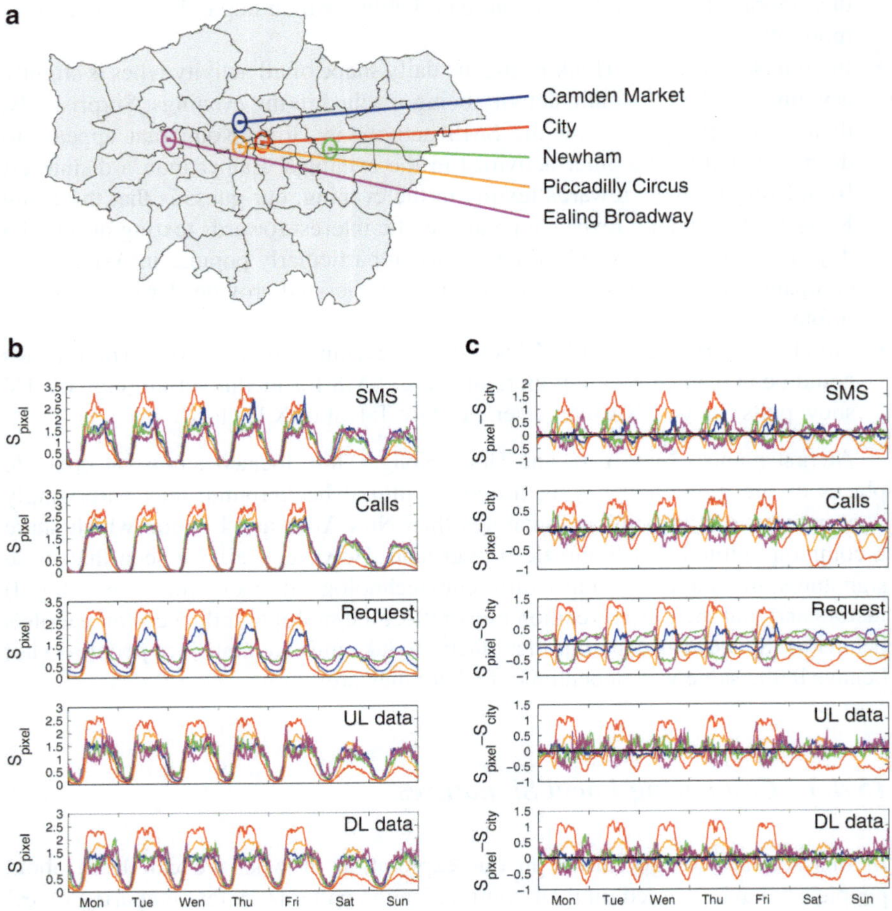

Fig. 15.5 **Local signatures** (**a**) We selected five 500m by 500m grid pixels in specific locations of Greater London. (**b**) The first series of plots shows these locations' signatures in the different components of activity. (**c**) The second series of plots shows the differences between local and whole city signatures. Colors in the plots match those on the map indicating the locations of the selected pixels

- The signatures of the pixel within the City of London show patterns typical of a business area: high amount of activity during working hours and very low activity in the evening from Monday to Friday and huge weekday-to-weekend activity ratios. These signatures also display sharp transitions from low to high activity level at the beginning of working hours and from high to low level at the end of working hours.
- The Piccadilly Circus signatures also show a large amount of activity during working days, but also show significant activity during the weekend. The morning transitions from no activity to some activity are rather smooth (notice in

particular the dips on Fig. 15.5c revealing that this area is less active than other parts of the city in early morning). All these characteristics go in line with the commercial and touristic nature of the place.

- The Camden Market signatures have average patterns during working hours and are typically characterized by a peak of SMS activity during the workday evenings (consistent with the recreational nature of this location where people may gather to share a drink) and high level of request activity around lunch time and in the early afternoon in the weekend (in line with the popularity of the markets).
- The signatures of the Newham pixel are specifically characterized by low weekday-to-weekend activity ratios (suggesting a constant population rate over the week) and specific nonzero request activity at night (the automatic update of the mobile devices revealing that people are sleeping at those locations). These characteristics straightforwardly reveal the residential nature of this location.
- Finally, the signatures in Ealing Broadway are rather comparable to Newham's one – revealing the residential nature of the area – but also show specific peaks of request activity during morning and evening commuting hours, in line with the commuter hub property of the Underground station.

These observations show a correspondence between mobile traffic signatures and the nature of the places studied, suggesting that we could infer the nature of each area based on its signatures. As a side perspective, the identification of the central business district and the residential areas can offer insights on the nature of commuting flows. Urban planners are already aware of the spatial relationship between business district and residential areas, but the visualization of such properties on such a precise spatiotemporal scale has been possible only for a few years thanks to the advancement of digital data gathering.

15.5 Cluster Analysis

15.5.1 Principles

In the previous section, we focused on single pixels, and we have shown how their signatures could reveal human dynamics' features at the local level. In this section, we investigate the question of whether we can use mobile device traffic data to detect large areas with homogeneous properties. Our goal is to group local pixels according to the similarity of their signatures and use these groups to map the urban spatiotemporal structure of the cities.

Among the many different clustering techniques to extract clusters of pixels with similar signatures, we chose a K-means approach, used in many previous studies (Andrienko et al. 2013; Pei et al. 2013; Reades et al. 2007). This approach ensures that each pixel of a cluster has a signature as much like the one the other members of the clusters and as different as possible from the signature of the pixels in the other clusters. Starting from the signatures $\{S_i^\lambda\}$ of the pixels, the signature of a cluster C is defined as the average signature of the pixels within that cluster:

$$S_C^\lambda = \langle S_i^\lambda \rangle_{i \in C} \tag{15.3}$$

Combining all activity types, the K-means algorithm aims at minimizing the quantity E_K measuring the total distance between the locations' signatures and their cluster's signature:

$$E_K = \sum_{k=1}^{K} \sum_{i \in C_k} \text{dist}(i, C_k), \tag{15.4}$$

where the distance $\text{dist}(i, C)$ between a pixel i and a cluster C is defined as

$$\text{dist}(i, C) = \sum_{\lambda} \sum_{t} \left(S_i^\lambda(t) - S_C^\lambda(t) \right)^2, \tag{15.5}$$

and K is a pre-imposed number of clusters. This simple quantity does not take into account the temporal structure of the signatures (the order of the different time intervals does not matter), but in the following, it will prove to deliver consistent results. Each pixel is characterized here by a 3,360-dimensional feature vector (5 signatures of different type, each being valued on 672 time intervals).

A notable drawback of the K-means algorithm is the difficulty to determine the "best" number K of clusters, whose value can depend on the shape and scale of the distribution of points in a dataset and the desired clustering resolution. Different ad hoc techniques to make that decision exist, most of them based on finding the value of K best balancing the search for minimizing the intra-cluster distance E_K and maximizing the intercluster distances. There is however no consensus on the best method to use, and the correct choice of K may also often rely on the researchers' expert opinion and search for interpretable results. We guided our choice by looking at local maxima of the silhouette index (Rousseeuw 1987). All cities presented local maxima for $K = 2$ clusters (corresponding roughly to city centers and city suburbs), $K = 6$, and larger values of K which vary with the studied city. All results presented in the following have been obtained for $K = 6$, which is the most relevant case. We indeed found that allowing a larger number of clusters mostly added clusters concentrated of very few pixels in areas with very low mobile phone activity and without any regular signatures.

15.5.2 Revealing the Spatial Structures of Cities

We conducted an independent K-means clustering analysis for each city. As we previously stated, the best cluster size distribution and interpretability was achieved for $K = 6$ in each case. Figures 15.6a–15.8a show the spatial projections of the

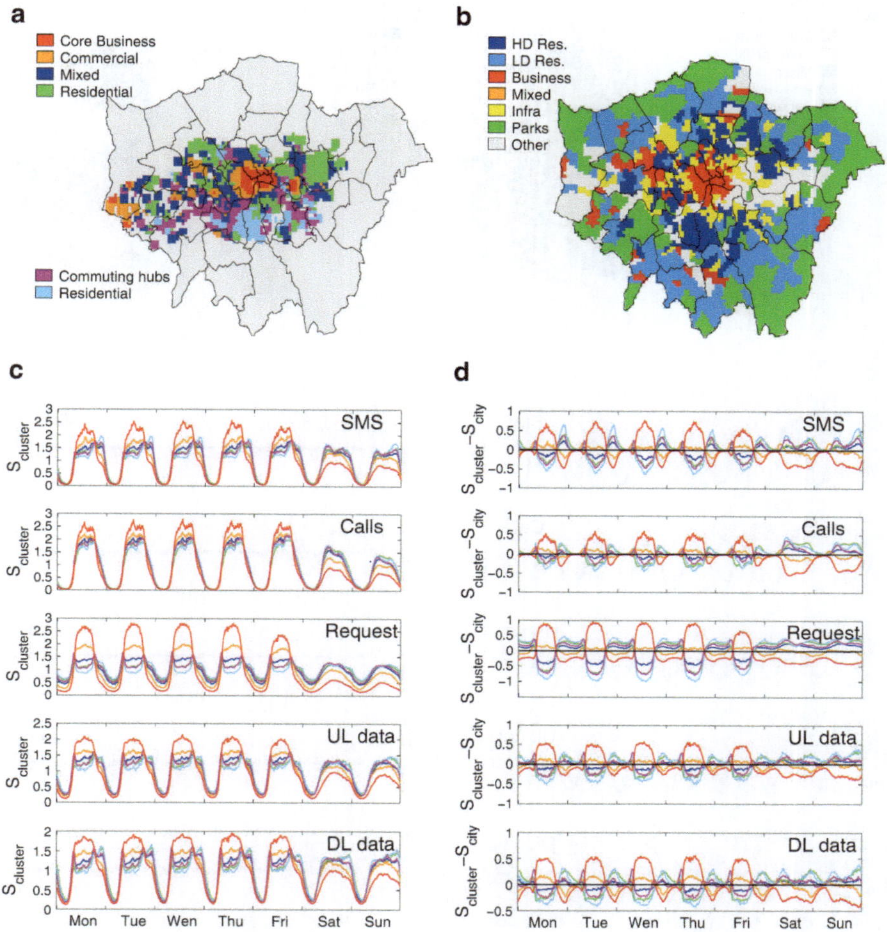

Fig. 15.6 Greater London clusters. (**a**) Spatial projection of $K = 6$ clusters, with their interpretation in the legend (see details in main text). (**b**) Actual land use maps as extracted from census data. (**c**) Signatures of the clusters in the different components of activity. (**d**) Deviations of the signatures compared to the whole city signatures displayed on Fig. 15.4. Colors on the signature plots match those on the cluster map (**a**), and *gray areas* correspond to zones with no recorded data

clusters on a map of the cities. First of all, it is worth noting that the clusters are made of spatially cohesive groups of pixels shaping a concentric-like structure within the cities. The signatures of the clusters as well as their deviation from their city signature are displayed on Figs. 15.6c,d–15.8c,d, and Table 15.3 lists the share of total activities occurring within the surface covered by the clusters. To better understand the nature of the clusters and their relation with standard land use classification, Figs. 15.6b–15.8b display land use maps of the cities built from extracted census data (see Sect. 15.2.2). Finally, Table 15.4 lists the average

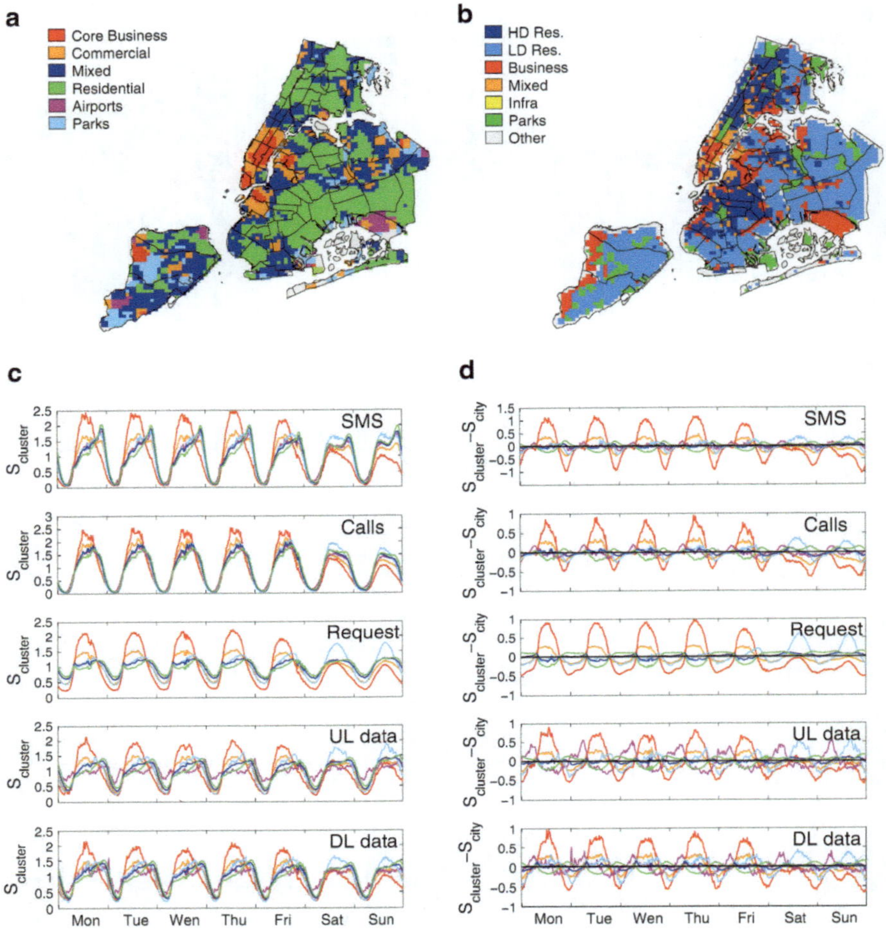

Fig. 15.7 New York clusters. (**a**) Spatial projection of $K = 6$ clusters, with their interpretation in the legend (see details in main text). (**b**) Actual land use maps as extracted from census data. (**c**) Signatures of the clusters in the different components of activity. (**d**) Deviations of the signatures compared to the whole city signatures displayed on Fig. 15.4. Colors on the signature plots match those on the cluster map (**a**), and *gray areas* correspond to zones with no recorded data

population and job densities within the clusters (when available), as well as confusion matrices highlighting the similarities between our clusters and the land use classes.

A few patterns are easy to interpret and compare with census data. For example:

- Cluster 1 (in red). In all cities, these clusters' signatures present high levels of activities during working hours and very low levels of activity in the evening

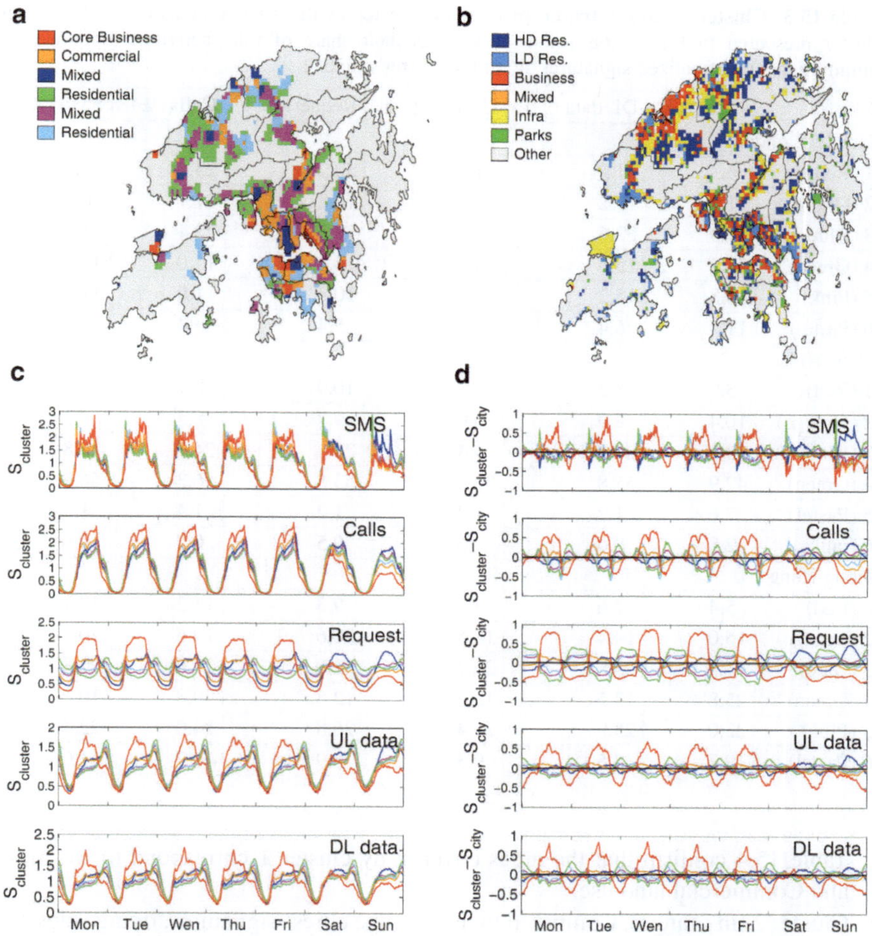

Fig. 15.8 Hong Kong clusters. (**a**) Spatial projection of $K = 6$ clusters, with their interpretation in the legend (see details in main text). (**b**) Actual land use maps as extracted from census data. (**c**) Signatures of the clusters in the different components of activity. (**d**) Deviations of the signatures compared to the whole city signatures displayed on Fig. 15.4. Colors on the signature plots match those on the cluster map (**a**), and *gray areas* correspond to zones with no recorded data

during the workdays and a huge weekday-to-weekend activity ratios. The combination of their small area shares (4–6 % of the cities' area as reported in Table 15.3) and their large activity share is consistent with a high concentration of activity and an identification as business cores. This is verified since the red clusters cover the 'City of London' district, the financial and decisional districts in New York (e.g., south Manhattan where Wall Street is located), and a large part of the Central district on Hong Kong Island. The confusion matrices of

Table 15.3 Clusters' mobile traffic properties. We report the relative areas covered by each cluster presented in Figs. 15.6a–15.8a, as well as their share of total activities (obtained by summing the unnormalized signatures A_i^λ of the cluster's pixels)

Cluster	Area (%)	DL data (%)	UL data (%)	Request (%)	Calls (%)	SMS (%)
Greater London						
1 (Red)	6.3	21.2	24.0	30.7	23.4	24.0
2 (Orange)	11.1	17.2	18.5	14.8	17.7	16.6
3 (Blue)	27.3	22.6	22.1	19.5	22.4	21.5
4 (Green)	23.3	15.6	13.6	14.4	15.0	14.7
5 (Pink)	18.8	12.2	11.6	10.3	11.0	11.4
6 (Pastel)	11.5	6.4	5.7	5.4	5.4	5.8
New York						
1 (Red)	3.9	8.7	11.0	10.9	8.7	8.1
2 (Orange)	10.4	8.9	10.0	9.7	8.5	8.3
3 (Blue)	32.1	24.7	24.4	24.1	24.3	24.8
4 (Green)	42.9	55.8	52.8	53.5	56.3	56.9
5 (Pastel)	7.6	1.2	1.2	1.2	1.4	1.3
6 (pink)	2.8	0.6	0.6	0.5	0.7	0.6
Hong Kong						
1 (Red)	5.4	7.6	7.4	9.3	11.2	10.7
2 (Orange)	15.0	19.4	18.5	20.6	19.6	20.7
3 (Blue)	7.3	23.8	24.7	22.0	27.0	22.8
4 (Green)	25.8	12.5	12.2	12.5	9.8	10.2
5 (Pink)	23.0	25.9	25.4	26.8	24.0	22.4
6 (Pastel)	19.7	8.5	9.3	6.9	6.7	11.0

Table 15.4 confirm that the areas covered by cluster 1 correspond to Business and Commercial land use.

- Cluster 2 (in orange). Compared to the average cities' signatures, orange clusters are characterized by higher levels of activity during weekday working hours and lower activity level in the weekend, similar (but to a lower extent) than what happens in cluster 1. On London and New York maps, these clusters mostly surround the red ones (notice also that in London the orange cluster covers Heathrow Airport on the western side of Greater London). Based on these characteristics orange clusters could be identified as commercial areas. This is confirmed by the confusion matrices, reporting low to average residential land use and higher than average Business land use in these clusters.
- Cluster 3 (in blue). These clusters have signatures similar to those of residential cluster 4, but with smaller deviations from the corresponding city signatures. These features suggest, as baked up by land use data, to interpret these clusters as a mixed area with strong residential component.
- Cluster 4 (in green). Compared to the average cities' signatures, these clusters display higher activity level at night and lower activity level on Monday to Friday business hours. The curves showing the deviations of their signatures compared

Table 15.4 Clusters' census properties. This table shows the average population and job densities on the area covered by each cluster presented in Figs. 15.6a–15.8a, interpolated from census values. Land use confusion matrices display, for each cluster, the percentage area that can be attributed to each of the seven land use classes: High-density Residential (HD Res), Low-density Residential (LD Res), Business and Commercial (Business), Mixed, Infrastructure and Government Facilities (Infra), Parks and Green Areas (Parks), and Other

Cluster	Pop/km^2	Job/km^2	HD Res	LD Res	Business	Mixed	Infra	Parks	Other
Greater London									
1 (Red)	8,310	48,390	0.00	0.00	73.64	–	1.82	5.45	19.09
2 (Orange)	9,450	9,492	3.59	1.03	37.44	–	18.46	5.64	33.85
3 (Blue)	9,977	4,084	18.66	5.66	14.26	–	14.88	11.53	35.01
4 (Green)	11,341	3,338	35.63	8.60	7.62	–	31.20	2.21	14.74
5 (Pink)	8,512	2,743	23.40	8.51	3.04	–	19.15	16.11	29.79
6 (Pastel)	10,257	2,430	55.94	16.34	1.49	–	17.82	1.98	6.44
New York									
1 (Red)	19,669	–	6.84	3.42	64.10	14.53	–	11.11	–
2 (Orange)	20,881	–	16.03	28.21	35.26	10.26	–	10.26	–
3 (Blue)	15,554	–	17.22	51.14	15.56	3.01	–	13.07	–
4 (Green)	17,310	–	26.65	52.21	10.49	2.41	–	8.24	–
5 (Pink)	3,002	–	1.20	24.10	65.06	0.00	–	9.64	–
6 (Pastel)	6,972	–	2.62	43.23	22.27	0.00	–	31.88	–
Hong Kong									
1 (Red)	24,722	–	6.74	16.85	32.58	–	22.47	4.49	16.85
2 (Orange)	24,336	–	17.41	8.10	22.67	–	19.03	7.29	25.51
3 (Blue)	27,570	–	14.05	3.31	27.27	–	23.97	13.22	18.18
4 (Green)	12,121	–	29.18	7.53	10.12	–	13.88	7.06	32.24
5 (Pink)	16,713	–	18.68	11.32	16.58	–	17.63	5.00	30.79
6 (Pastel)	10,828	–	26.77	6.15	8.31	–	20.62	4.00	34.15

to the cities' signatures remarkably mirror those corresponding to cluster 1 core business areas. These features suggest that these clusters correspond to purely residential areas, which is confirmed by the confusion matrices reporting high residential land use.

Other clusters have a more city-specific interpretation:

- Cluster 5 (in pink). These clusters have different interpretations in each city. In London, the pink cluster signatures are close to the green ones (residential), but the request and data signatures present specific peaks at commuting hours, consistent with the commuting hub nature of the area. In New York, the pink cluster mostly corresponds to the JFK airport, characterized by specific bumps of activity in the morning. In Hong Kong, these match a mixed area with a strong residential component.
- Cluster 6 (in pastel). In Greater London, the pastel cluster is another mainly residential cluster, in the southern part of Inner London. Its signatures' properties are

very similar to those of the London green cluster, with the exception of a specific peak of activity on Saturday just before lunch time. This peak could be explained by a recurring event (such as a market). In New York, a quick comparison with detailed maps of the city reveals that the pastel cluster corresponds to parks. The activity shares of this cluster are very low (in accordance with the fact that there are no people residing or working full time on the park premises), and its signature shows high activity levels during the weekends (when people may go for a walk within the nearer park). Finally in Hong Kong, the pink cluster appears to be a second residential one.

To summarize, each city can be characterized by a gradation of clusters. A feature common to all cities is the existence of core business areas and purely residential areas, an already known urban fact that we were able to check thanks to communication traffic data.

Compared to classical time-consuming and expensive field surveys, our approach makes it possible to build automatic, quick, and relatively cheap way of preparing land use maps of the cities. The maps we obtained are closely related to classic land use maps, especially for the distinction between business and residential areas. When it comes to other land use classes, some difference occurs. Indeed, while the classical land use classes are related to what the land looks like (is the neighborhood consisting of a retail store, bank, residential buildings, etc.), the clusters we found are based on communication data revealing human dynamic behaviors (like working in an office, shopping, eating, commuting, sleeping, etc.). Rather than emulating classical land use mapping, our approach thus produces a complementary point of view that enriches our understanding of the multiple dynamics at stake in the cities.

Although we used every type of activity in our clustering analyses, the similarities between the signatures of different types (calls, SMS, request, UL or DL data) displayed on Figs. 15.6c–15.8c suggest that our finding would not qualitatively change if we focused on only one activity type.

15.5.3 Revealing Universal Patterns

The results presented in the previous section suggest that mobile traffic patterns can reveal a concentric structure of cities into clusters that can be interpreted the same way. To what extent are these cities similar? In this section, we investigate this issue by making a transversal analysis of our three studied cities. We performed a K-means clustering analysis on all cities at once, grouping 500m by 500m grid pixels with similar signature patterns.

Figure 15.9 displays the results of this transversal clustering analysis. As before, we chose to display results obtained for $K = 6$.

- All previously identified core business centers are gathered into a single (here red) cluster, whose signature can be characterized as before.

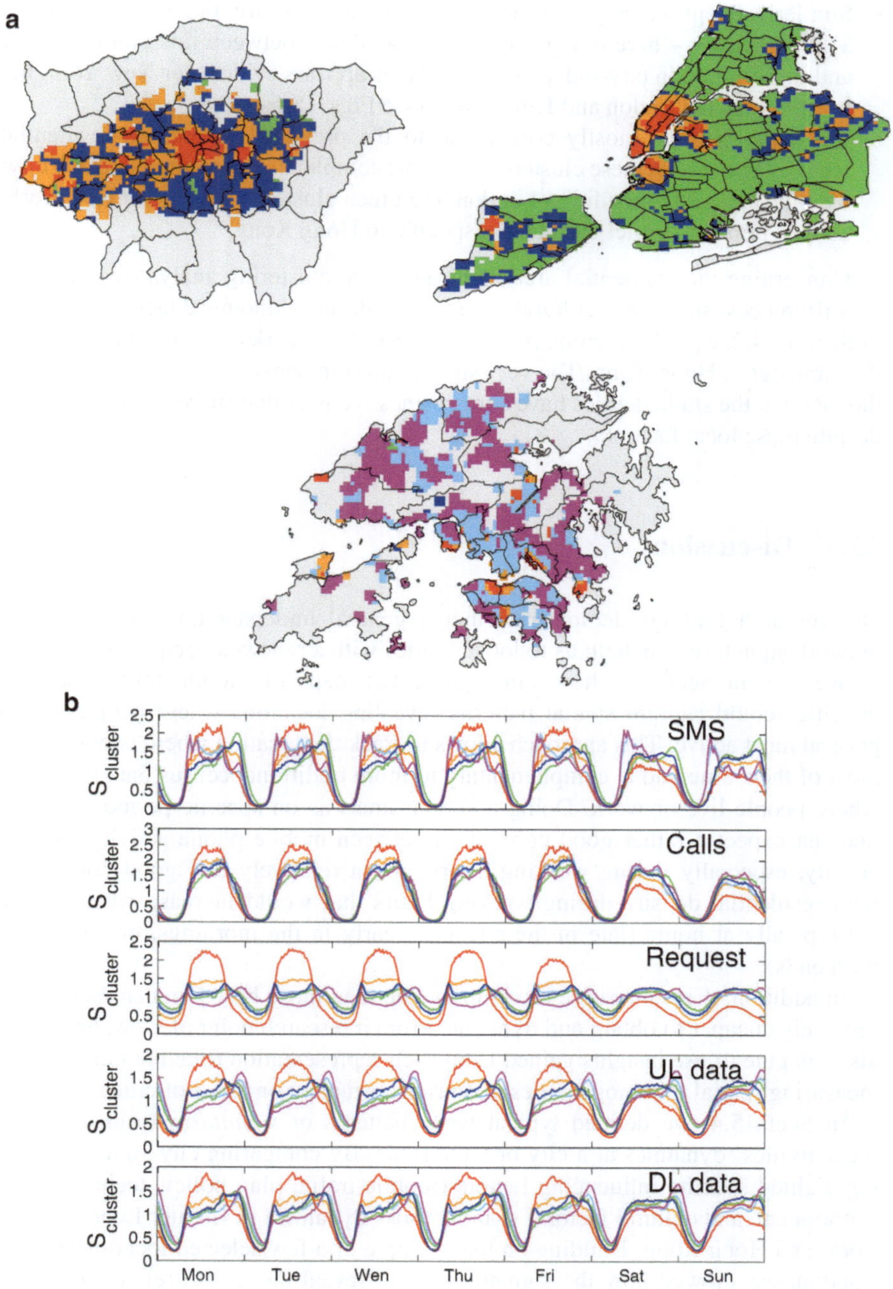

Fig. 15.9 Clustering the three cities at once. (**a**) Spatial projection of $K = 6$ clusters of all cities' counter data (*gray areas* correspond to zones with no recorded data). (**b**) Clusters' signatures. Colors on the plots match those on the map

- Similarly, the previously identified "commercial" areas are roughly gathered into a single cluster – here orange. The correspondence between this orange cluster and those found in city-independent analyses appears evident for New York, but less obvious in London and Hong Kong (see Figs. 15.6a–15.8a).
- The other clusters mostly correspond to the previously identified residential areas. Surprisingly, these clusters are almost completely concentrated in one city: the blue cluster is specific to London, the green cluster is specific to New York, and the pink and pastel clusters are specific to Hong Kong.

Concerning the residential area, the transversal clustering analysis emphasizes the differences due to local cultural, technological, and economical factors identified in Sect. 15.4.2, e.g., the evening peaks of SMS in New York or the evening peaks of data transfer in Hong Kong. The very strong and somewhat surprising result here is the fact that the studied cities have core business centers that share a similar pattern despite those local factors.

15.6 Discussion

Our research findings demonstrate that a general understanding of the mobile network signatures can help us to look at cities with a renewed perspective.

We saw in Sect. 15.3 how time-aggregated maps of mobile traffic inhomogeneities could capture spatial patterns revealing locations where people are in general most active. This approach allows to track the location where people spend most of their time and is complementary to more traditional census data recording where people live or work. Doing a similar analysis on specific periods of time, one can expect a rather good correlation between mobile phone activity and job density, especially during working hours, and a relatively less good correlation with residential density during working hours that would increase when people are typically at home (late in the evenings, early in the mornings, or during the weekends).

In addition, these maps are more up to date than those based on census polling, relatively cheaper to obtain and dynamic. From a research point of view, one could also imagine to use insights gained from such representation (like the Gini indices measuring spatial inhomogeneities) to enrich current taxonomies of cities.

In Sect. 15.4, we defined typical week patterns or *signatures* to characterize the activities' dynamics at a city or local scale. By comparing city signatures, we highlighted specific influencing factors (mobile traffic plan policy, technological, economical, and cultural factors) shaping those dynamics in Greater London, New York, and Hong Kong. Building on the example of a few selected locations within London, we showed how the signatures could reveal the nature (either financial, commercial, recreational, residential, or commuting hub) of the concerned areas. In general, the insights gained from the study of the typical week signatures could be used to optimize the overall network performance by informing the mobile operators of the actual typical usage for them to take proactive decisions upon. Diffusing the knowledge of the signature patterns could also generate new business ideas in the cities, by just allowing retailers to optimize scheduling based on the likelihood of people being in the vicinity of their stores.

We presented in Sect. 15.5 a clustering analysis process allowing automatic detection of locations with similar signatures within a city. Our findings showed that we could detect in each city a core business center, at least one pure residential area, one mixed area with a strong commercial component, and another mixed area with a strong residential component. In addition, we were able to detect other clusters of specific nature (commuting hub areas in London or the JFK airport in New York). The methods used could be easily pushed to generalize these findings. We stressed that our approach makes it possible to build an automatic, quick, and relatively cheap way of preparing maps that could complement and enrich classical land use maps based on surveys. Indeed, communication traffic data tell us about the actual dynamic behavior of people at each location, while land use maps rather tell us about the average type of behavior you can expect based on the urbanization levels and the type of buildings, shops, or infrastructures present at each location.

Our final finding was obtained by applying our clustering the procedure on the locations of all three cities at once. Quite surprisingly, we found that the core business centers of London, New York, and Hong Kong were gathered into a single cluster, which proves their high degree of similitude. On the other hand, the city has residential locations whose signatures are well distinct. To answer a question raised at the beginning of this chapter, it seems like globalization shapes the economical and political activity in the large cities' financial and decisional core centers, while individual activity patterns are still defined mostly by local factors.

On a broader note, as the digital databases are growing and our computational methods are improving, it may be tempting to multiply automatic procedures to generate lists of insights based on mobile traffic data. However, we argue that knowledge expertise is more than ever needed to understand, interpret, and critic these results.

Though this chapter presented new similarity measurements and links between three major cities, it also opens up the question of the universality of our findings. Is the common beat detected in the core areas a universal pattern common to all cities, or is it only peculiar to "occidental" developed cities? Is there any natural classification of the world's cities based on a similarity between peripheral residential signatures? It would be most interesting to study these questions by enlarging our datasets to include major cities from both developed and developing countries such as Paris, Mexico City, Shanghai, Rio de Janeiro, Sao Paulo, Lagos, or Mumbai. The challenge is now to gather a collection of mobile traffic data from all these cities before starting to build a yet-to-define comparative science of cities based on them.

Acknowledgements We thank Ericsson for providing datasets for this research and especially Dwight Witherspoon for the organizational support to the project. We also thank Christine Maynié-François for the stimulating discussions and thorough proofreading. We would further like to thank the National Science Foundation, the AT&T Foundation, the MIT SMART program, the Center for Complex Engineering Systems at KACST and MIT, Volkswagen ERL, BBVA, The Coca Cola Company, Expo 2015, Ferrovial, the Regional Municipality of Wood Buffalo, AIT, and all the members of the MIT Senseable City Lab Consortium for supporting the research.

References

Amini A, Kung K, Kang C, Sobolevsky S, Ratti C (2014) The impact of social segregation on human mobility in developing and industrialized regions. EPJ Data Sci 3(1):6

Andrienko G, Andrienko N, Fuchs G (2013) Multi-perspective analysis of d4d fine resolution data. Data for Development (D4D 2013)

Beaverstock JV, Smith RG, Taylor PJ (1999) A roster of world cities. Cities 16(6):445–458

Becker RA, Caceres R, Hanson K, Loh JM, Urbanek S, Varshavsky A, Volinsky C (2011) A tale of one city: Using cellular network data for urban planning. IEEE Pervasive Comput 10(4):18–26

Caceres R, Rowland J, Small C, Urbanek S (2012) Exploring the use of urban greenspace through cellular network activity. In: Second workshop on pervasive urban applications (PURBA), In conjunction with Pervasive, Newcastle, 2012

Calabrese F, Reades J, Ratti C (2010) Eigenplaces: segmenting space through digital signatures. IEEE Pervasive Comput. 9(1):78–84

Calabrese F, Colonna M, Lovisolo P, Parata D, Ratti C (2011) Real-time urban monitoring using cell phones: A case study in rome. IEEE Trans Intell Transp Syst 12(1):141–151

Candia J, González MC, Wang P, Schoenharl T, Madey G, Barabási A-L (2008) Uncovering individual and collective human dynamics from mobile phone records. J Phys A: Math Theor 41(22):224015

Eagle N, Pentland A (2006) Reality mining: sensing complex social systems. Pers Ubiquitous Comput 10(4):255–268

Frias-Martinez V, Soto V, Hohwald H, Frias-Martinez E (2012) Characterizing urban landscapes using geolocated tweets. In: 2012 international conference on privacy, security, risk and trust (PASSAT), and 2012 international confernece on social computing (SocialCom), Amsterdam. IEEE, pp 239–248

Gini CW (1912) Variability and mutability, contribution to the study of statistical distribution and relations. Studi Economico-Giuricici della R, Universita de Cagliari

Girardin F, Vaccari A, Gerber A, Biderman A, Ratti C (2009) Towards estimating the presence of visitors from the aggregate mobile phone network activity they generate. In International conference on computers in urban planning and urban management, Hong Kong

González M, Hidalgo C, Barabási A-L (2008) Understanding individual human mobility patterns. Nature 453:779–782

Jacobs-Crisioni C, Koomen E (2012) Linking urban structure and activity dynamics using cell phone usage data. In: Workshop on complexity modeling for urban structure and dynamics, 15th AGILE international conference on Geographic Information Science, Avignon

Kang C, Liu Y, Ma X, Wu L (2012) Towards estimating urban population distributions from mobile call data. J Urban Technol 19(4):3–21

Kang C, Sobolevsky S, Liu Y, Ratti C (2013) Exploring human movements in Singapore: a comparative analysis based on mobile phone and taxicab usages. In: Proceedings of the 2nd ACM SIGKDD international workshop on urban computing, Chicago. ACM, p 1

Kung KS, Sobolevsky S, Ratti C (2013) Exploring universal patterns in human home/work commuting from mobile phone data. PLoS ONE 9(6):e96180

Liu Y, Wang F, Xiao Y, Gao S (2012) Urban land uses and traffic 'source-sink areas': evidence from GPS-enabled taxi data in Shanghai. Landsc Urban Plan 106(1):73–87

Loibl W, Peters-Anders J (2012) Mobile phone data as source to discover spatial activity and motion patterns. In: Jekel T, Car A, Strobl J, Griesebner G (eds) G1_Forum 2012: geovizualisation, society and learning. Herbert Wichmann, VDE, Berlin/Offenbach, pp 524–533

Lorenz MO (1905) Methods of measuring the concentration of wealth. Publ Am Stat Assoc 9(70):209–219

Pei T, Sobolevsky S, Ratti C, Shaw S-L, Zhou C (2013) A new insight into land use classification based on aggregated mobile phone data. Int J Geogr Inf Sci 28(9):1–20

Ratti C, Williams S, Frenchman D, Pulselli R (2006) Mobile landscapes: using location data from cell phones for urban analysis. Environ Plan B: Plan Des 33(5):727

Ratti C, Sobolevsky S, Calabrese F, Andris C, Reades J, Martino M, Claxton R, Strogatz SH (2010) Redrawing the map of great britain from a network of human interactions. PloS One 5(12):e14248

Reades J, Calabrese F, Sevtsuk A, Ratti C (2007) Cellular census: explorations in urban data collection. IEEE Pervasive Comput 6(3):30–38

Reades J, Calabrese F, Ratti C (2009) Eigenplaces: analysing cities using the space-time structure of the mobile phone network. Environ Plan B: Plan Des 36(5):824–836

Rousseeuw P (1987) Silhouettes: a graphical aid to the interpretation and validation of cluster analysis. J Comput Appl Math 20:53–65

Rubio A, Sanchez A, Frias-Martinez E (2013) Adaptive non-parametric identification of dense areas using cell phone records for urban analysis. Eng Appl Artif Intell 26(1):551–563

Sobolevsky S, Szell M, Campari R, Couronné T, Smoreda Z, Ratti C (2013) Delineating geographical regions with networks of human interactions in an extensive set of countries. PloS One 8(12):e81707

Song C, Qu Z, Blumm N, Barabási A-L (2010) Limits of predictability in human mobility. Science 327(5968):1018–1021

Soto V, Frías-Martínez E (2011) Automated land use identification using cell-phone records. In: Proceedings of the 3rd ACM international workshop on MobiArch, Bethesda. ACM, pp 17–22

Sun J, Yuan J, Wang Y, Si H, Shan X (2011) Exploring space–time structure of human mobility in urban space. Phys A: Stat Mech Appl 390(5):929–942

Toole JL, Ulm M, González MC, Bauer D (2012) Inferring land use from mobile phone activity. In: Proceedings of the ACM SIGKDD international workshop on urban computing, Beijing. ACM, pp 1–8

Traag VA, Browet A, Calabrese F, Morlot F (2011) Social event detection in massive mobile phone data using probabilistic location inference. In: 2011 IEEE third international conference on privacy, security, risk and trust (PASSAT), and 2011 IEEE third international conference on social computing (SocialCom), Boston. IEEE, pp 625–628

Vieira MR, Frias-Martinez V, Oliver N, Frias-Martinez E (2010) Characterizing dense urban areas from mobile phone-call data: discovery and social dynamics. In: 2010 IEEE second international conference on social computing (SocialCom), Minneapolis. IEEE, pp 241–248

Chapter 16
Epilogue

Paul A. Longley

This is an important book that raises issues that will be of enduring importance in the coming years. There are three major reasons for this. First, the contributions provide further impetus to the current renaissance of quantitative computational modelling as a multidisciplinary research effort in a number of countries worldwide. This renaissance arises in large part from current preoccupations with Big Data and the realisation that we are creating data faster than we can realistically hope to analyse them. The second reason why this book is important is because, as various contributors make clear, our world is becoming increasingly urban, and understanding the locus of activities in urban environments is fundamental to rational organisation and planning if we are to harness the functioning of urban systems to the common good. The third reason is that the drive to develop successful computational approaches to urban analysis is inextricably bound up with the quest to develop and enhance successful urban environments.

Any urban system is of seemingly infinite complexity, and it follows that any representation of it is necessarily partial. The 'exa-flood' of data might seem to render such representations less partial, yet many of the contributions to this book suggest caveats to this, in a number of subtle and distinctive ways. Society's increasing dependence upon data is addressed in what is often termed *data science*. It has been argued that, in this 'Fourth Paradigm' for research, science is increasingly data-driven and is based on the vast quantities of data that are now becoming available; many of which are geographically referenced. Many of these Big Data sources receive attention in this book, including ground-based, airborne and space-based sensors of various kinds, mobile telephony, and social media data from a range of different sources. It is abundantly clear that new or improved computational approaches are required for storing, accessing, processing, and visualising these

P.A. Longley (✉)
Department of Geography, University College London, London, UK
e-mail: p.longley@ucl.ac.uk

M. Helbich et al. (eds.), *Computational Approaches for Urban Environments*,
Geotechnologies and the Environment 13, DOI 10.1007/978-3-319-11469-9_16

vast quantities of data and to accommodate the enormous data problems that are generated in data sharing, documentation, management, and archiving.

Nevertheless, understanding urban environments is about more than data reduction, as Behnisch et al.'s very useful multidimensional data analysis demonstrates through critical analysis of the role and remit of pattern detection and the identification of anomalies. Hagenauer is similarly reflective about the apparent problem of prespecifying the number of contextual neural gas clusters. In different ways, whether using real or synthetic data, both of these contributions beg the broader question, of what are Big data representative? – when data are harvested from populations that may not be clearly defined, when respondents may be self-selecting and distinct from nonrespondents, and apparent outliers may in practice be inliers that are drawing the distribution towards the true (unknown) population mean. Big Data rarely purport to be representative or inclusive, and an important objective of Big Data analytics should be to highlight, if not accommodate, the systematic biases that are present in them. Behnisch et al.'s contribution in particular begs the important question, how do emergent data structures relate to what we might understand as a data *infra*structure upon which we might represent an urban system?

It is also worth remembering that computational approaches are not restricted to secondary data and that 'Big' Data can be collected as part of a clearly specified research design. In this context, Yeboah et al.'s analysis of urban cycling demonstrates that computational research can fall within the prevailing scientific paradigm of hypothesis formulation, data collection and validation, analysis and report writing. From a substantive standpoint, it is also worth remembering that intra-urban travel usually manifests derived demand – and that the activity patterns of cyclists in relation to the broader morphology of the city presents an enticing research prospect.

New computational approaches are not, of course, universally synonymous with the data science paradigm, as contributions to this book make clear with respect to hedonic house price models (Razen et al.) and agent-based modelling (Sun and Manson). Razen et al. model housing demand using a sophisticated multilevel form that raises the issue of model parsimony in the trade-off between fidelity to the inherent complexity of the real world and the complexity, and hence intelligibility, of the models that we use to represent it. Sun and Manson's basic postulate is that straightforward (and hence parsimonious) search rules drive intra-urban residential mobility patterns and that modelling these using an agent-based approach can develop our understanding of city evolution. Rosner and Curtin's Urban Livability Index draws on half a century or so of thinking about indicators and links to Frankhauser's speculations about the importance of size, shape, scale, and dimension in thinking about urban forms. Waraich et al. demonstrate that microsimulation is now an established framework for trip distribution and travel demand modelling but also that it generates intense computational demands. All of these papers are convincing within their respective domains, but looking prospectively it would be interesting to identify whether the broader morphology of urban land use, and the patterns of human interactions that this generates, might be incorporated into these different approaches. The juxtapositioning of different functions within citywide morphology is an established tenet of many (e.g. cellular automata) models, and as new data sources become available, it may be interesting to investigate the relative importance of activities additional to night-time residence when investigating the dynamics of change in different cities.

In different ways, each of these contributions demonstrates that our representations of socioeconomic distributions are inevitably piecemeal and partial. By contrast, our ability to measure and monitor changes in the built forms of cities is approaching the point at which it can truly be described as comprehensive. Karantzalos makes clear that remotely sensed urban change detection is now a refined science. Berger et al. convincingly establish that this science provides the means by which data from multiple instruments can be integrated with one another. In both cases, there is a sense that we are creating and maintaining digital framework data – and that it is possible to build together a representation that is fit for clearly defined purposes. The ongoing quest remains to link improved two- and three-dimensional representations of the physical form of cities to their social, economic, and demographic functioning. Metadata often indicate that we are on much shakier ground when we seek to build socioeconomic data infrastructures. Tang's chapter broaches this issue, by supplementing satellite imagery with socioeconomic data. This provides an important waymarker, although as with some of the preceding chapters the representation of socioeconomic function remains limited to night-time residence or workplace statistics. A challenge for the future is to augment detailed, frequently updated, integrated morphological models to say more about urban function, and to reconceptualise cities as the locus of flows of individuals in order to work, play, and above all be creative.

Wang et al.'s use of telecoms data is an important step in this direction, though it also exemplifies many of the problems that arise when we seek to develop generalised time and motion studies of human activity patterns across the city. Privacy strictures will continue to restrict the remit of such studies until clear and robust guidelines can be developed concerning analysis and linkage of data sets through anonymisation – consistent with the imperatives of most data protection legislation and essential given the impossibility of obtaining universal consent. There is a Pandora's box of possible data sources that might further this objective and Schnebele et al.'s and Grauwin's contributions open it. Yet, looking to the future, both also raise issues of validity, reliability, and coverage that need to be addressed in the general sense.

These are speculations about the future that are stimulated by reading the individual papers, but in conclusion it is perhaps appropriate to consider the broader framework that links computational approaches with urban environmental analysis. It has been argued elsewhere that much recent computational research neither needs nor produces theory, but instead mines data for patterns that may be useful in solving humanity's problems. This data science perspective focuses upon the principles and techniques involved in managing the vast quantities of new data that are amenable to computationally intensive analysis. Yet it seems that familiarity with the generic principles of data science is not necessarily a sufficient qualification to better understand urban environments. In looking to the future, therefore, it is also important to look to the past – and best practices for systematic documentation of data (metadata), to accredited means of collecting primary data, and to best practices for archiving and sharing secondary sources.

The best data are a necessary but not sufficient prerequisite for the best decision-making and the best planning. In this context, Frankhauser and his colleagues have made a sustained and enduring contribution to our understanding of cities as organic entities, the morphologies of which grow from the 'bottom-up'. In this book, it is enlightening to see his proposals for planning regimes that emulate and harness this truism at the range of scales that can contribute to the sustainable city. His contribution demonstrates that our thinking need not be 'top-down' but that it nevertheless needs to be hypothesis driven.

Taken together, the chapters in this book provide an important milestone in our understanding of the contribution of data analytics to understanding not only the detailed form but also many of the detailed functions of cities. Why is the world increasingly urban? – fundamentally because cities are crucibles for innovation and creativity, in which human talent can flourish. There is no universal composite indicator for the urban forms and functions that bequeath and nurture human creativity, but the contributions to this book show how progress is being made on a number of fronts.

Index

M. Helbich et al. (eds.), *Computational Approaches for Urban Environments*, Geotechnologies and the Environment 13, DOI 10.1007/978-3-319-11469-9

MIX
Papier aus verantwortungsvollen Quellen
Paper from responsible sources
FSC® C105338

If you have any concerns about our products,
you can contact us on
ProductSafety@springernature.com

In case Publisher is established outside the EU,
the EU authorized representative is:
Springer Nature Customer Service Center GmbH
Europaplatz 3, 69115 Heidelberg, Germany

Printed by Libri Plureos GmbH
in Hamburg, Germany